# Chemistry of Semiconductors

# Chemistry of Semiconductors

By

## Sergio Pizzini ( 0000-0002-0542-3219)

*University of Milano-Bicocca, Italy*
*Email: sergiopizzini2011@gmail.com*

ROYAL SOCIETY
OF **CHEMISTRY**

Print ISBN: 978-1-83916-212-1
PDF ISBN: 978-1-83767-136-6
EPUB ISBN: 978-1-83767-137-3

A catalogue record for this book is available from the British Library

The Royal Society of Chemistry is a charity, registered in England and Wales, Number 207890, and a company incorporated in England by Royal Charter (Registered No. RC000524), registered office: Burlington House, Piccadilly, London W1J 0BA, UK, Telephone: +44 (0)20 7437 8656.

Visit our website at books.rsc.org

To my wife, Anna, with love.

# Preface

Silicon, germanium, and compound semiconductors, among which silicon carbide (SiC), gallium arsenide (GaAs), and gallium nitride (GaN) are the most representative examples, play an enduring role in the world economy, since they were, and still are, key to the development of modern microelectronics and optoelectronics, with a wealth of sister technologies.

The global value of the semiconductor market, at the time this book was written, amounts to around US$700 billion, twice the value of the global structural steel market (US$320 billion in 2020), as an impressive sign of the strategic role of this market segment, and it is expected to grow faster than other market segments.

As is well known, silicon is still the semiconductor material of choice for microelectronics, power electronics, and photovoltaic applications and, today, more than 8 million tons is annually available on the international market, as polycrystalline feedstocks, single-crystalline ingots, as standard-diameter size (300 mm) single crystal wafers, obtained by wire-saw ingot cutting, for microelectronic applications, and as standard M10 size (182 × 182 mm) for photovoltaic applications.

According to the Semiconductor Industry Association (SIA), the value of the silicon chip market (over 1 trillion semiconductor processors shipped in 2019) amounts to US$470 billion, and is expected to rise to US$730 billion in six years. Most of the polysilicon available is used today for the photovoltaic market, valued at US$54 billion in 2018, and projected to reach US$333 billion by 2026, with a global polysilicon capacity to reach 536 GW by the end of 2023.[1]

Chemistry of Semiconductors
By Sergio Pizzini
© Sergio Pizzini 2024
Published by the Royal Society of Chemistry, www.rsc.org

Compound semiconductors, strategic for optoelectronics (lasers and light emission diodes in the emission spectrum ranging from the visible to the deep UV), thin film solar cells, chemical sensors, and high frequency applications, represent today a growing fraction of the overall semiconductor value chain, valued at US$89.9 billion in 2019, and projected to reach US$212.9 billion by 2027.

The development of this impressive market is the result of more than half a century of research efforts carried out worldwide, devoted to the study of the chemical, physical and structural properties of semiconductors, their production feasibilities and to the processes for the fabrication of microelectronic and optoelectronic devices.

In this frame, chemistry has played a critical role, both in the experimental and theoretical aspects, thanks to centuries-long expertise in the synthesis of chemically pure materials, to the richness of spectroscopic techniques available or those *ad hoc* developed, and to the attitude to the resolution of structural or energetic problems with theoretical methodologies.

The aim of this book is to look to semiconductors with a mainly chemical approach, which does not neglect the intimate role played by physics, for a full understanding of semiconductor properties. Elemental and compound semiconductors will be considered in terms of their thermal stability and thermally induced defectivity, including the peculiar non-stoichiometry of the compound semiconductor phases. The equilibrium thermodynamics of point defects, and the structural and physical aspects of their equilibria in elemental and compound semiconductors is a case issue in this book, since defects dominate the properties of semiconductor phases, with a crucial influence on semiconductor doping processes.

Impurity contamination and impurity gettering processes will be considered in detail, given the detrimental effect of metallic impurities on the optoelectronic properties of semiconductors. Both will be deeply studied for their physicochemical aspects, and particular attention will be dedicated to the internal and external gettering of impurities in silicon for microelectronic and photovoltaic applications, where chemistry dominates.

The study of growth processes of both elemental and compound semiconductors is another fundamental issue of this book, which will be considered with particular attention in the case of silicon and several compound semiconductors, where important problems of practical and fundamental character are presented.

As expected, the physical properties of semiconductors depend on their interaction with a gaseous or vapour phase of their chemical constituents, at convenient high temperatures. The case of oxide semiconductors is an iconic example of the defect equilibria ruling their interaction with reducing and oxidizing atmospheres, which favours their application as chemical sensors.

Last, but not least, full attention will be devoted to nanostructured semiconductors, in view of their peculiar thermodynamic, structural, and physical properties, which depend on their size and surface.

As a personal choice, only silicon and ZnO will be considered in their nano-thermodynamical, nano-chemical and nanostructural aspects, with the hope of giving the reader a rational view of the complex features of these objects, and of the role of size in their physical properties.

This book is dedicated to my wife and colleague, Anna Cavallini, for the many years of research carried out, together, in the semiconductor field.

I also express my gratitude to a number of friends and colleagues worldwide, who helped me in various ways during my work in the preparation of this book. Among them, my gratitude goes to Stefan Estreicher, Gudrun Kissinger, Gabriella Borionetti, Emanuela Barbera, Marco Bernasconi, Jean-Luc Maurice, Bruno Ceccaroli, Jatindra Kumar Rath, Luciano Colombo, Jingyu Qin, Halvard Tveit, Angelo Costa, Maurizio Acciarri, Leonid Zhigiley, Rüdiger Paschotta, to my daughter-in-law Sara, and to Jenny Cossham.

Sergio Pizzini
University of Milano-Bicocca, Italy

# Reference

1. J. Scully, Global polysilicon capacities to reach 536GW by year-end 2023 – CEA, PV Tech, Oct 14, 2022, https://www.pv-tech.org.

# Abbreviations

| | |
|---|---|
| BiCMOS | Bipolar CMOS |
| CL | Cathodoluminescence (spectroscopy) |
| CMOS | Complementary metal–oxide–semiconductor |
| CVD | Chemical vapour deposition |
| CZ | Czochralski |
| DB | Dangling bonds |
| DE | Direct exchange |
| DFT | Density functional theory |
| DFT-LDA | Density functional theory, in the local density approximation |
| DLTS | Deep level transient spectroscopy |
| EC | Electrochemical (process) |
| $E_G$ | Energy gap |
| EG | Electronic grade |
| EMF | Electromotive force (measurements) |
| ESR | Electron spin resonance |
| EXG | External gettering (process) |
| FL | Fermi level |
| LID | Light induced degradation |
| FZ | Float zone |
| HREM | High resolution electron microscopy |
| HPSG | High pressure solution growth |
| HVPE | Hydride vapour phase epitaxy |
| IG | Internal gettering |
| IR | Infra-red (spectroscopy) |
| ITO | Indium tin oxide |

Chemistry of Semiconductors
By Sergio Pizzini
© Sergio Pizzini 2024
Published by the Royal Society of Chemistry, www.rsc.org

| | |
|---|---|
| LEC | Liquid encapsulation method |
| LED | Light emitting diode |
| L/S | Liquid to solid (growth) |
| LPE | Liquid phase epitaxy |
| MACE | Metal assisted chemical etching |
| MD | Molecular dynamics |
| MDZ | Magic denuded zone |
| MG | Metallurgical (silicon) |
| MOCVD | Metal–organic chemical vapour deposition |
| MOS | Metal/oxide/semiconductor |
| NC | Nanocrystal |
| NW | Nanowire |
| PAS | Positronic annihilation spectroscopy |
| PE | Pauling electronegativity |
| pBN | Pyrolytic boron nitride |
| PECVD | Plasma enhanced CVD |
| PL | Photoluminescence |
| QC | Quantum confinement |
| R&D | Research and development |
| RG | Relaxation gettering (process) |
| RX | X-ray |
| RTA | Rapid thermal annealing |
| PVT | Physical vapour transport |
| SC | Supercrystals |
| SIMS | Scanning ion mass spectroscopy |
| SL | Superlattice |
| SRH | Schottly–Read–Hall (recombination) |
| STC | Silicon tetrachloride |
| TCS | Trichlorosilane |
| TD | Thermal donors |
| TOF SIMS | Time of flight SIMS |
| TM | Transition metal (impurity) |
| TEM | Transmission electron microscope |
| TSSG | Top seeded solution growth |
| VB | Vertical Bridgman |
| VLS | Vapor–liquid–solid (growth process) |
| VSS | Vapor–solid–solid (growth process) |

# Contents

Chemistry of Semiconductors
By Sergio Pizzini
© Sergio Pizzini 2024
Published by the Royal Society of Chemistry, www.rsc.org

# 3   Physico-chemical Aspects of Growth Processes of Elemental and Compound Semiconductors    142

# 4   Chemistry of Semiconductor Impurity Processing    228

# 1 Thermochemistry of Semiconductors

## 1.1 Introduction

The feasibility of the growth of large, single crystalline ingots of elemental and compound semiconductors, and the success of further fabrication of microelectronic and optoelectronic components[1] on wafers wire-cut from them, entirely depend on the thermochemical properties of the semiconductor phases, which, in fact, determine their chemical and thermal stability in the wide range of temperatures and atmospheric conditions characteristic of the entire process.

Although thermal and chemical stability is relevant for elemental semiconductors, its impact is severe, or even critical, as we will see, in the case of the stoichiometric or non-stoichiometric phases[2,3] of compound semiconductors, with important consequences for the feasibility of their ingot growth and for microelectronic and optoelectronic component processing.

In this framework, thermodynamics is a precious research support, since it allows describing the equilibrium properties of semiconductor phases in their temperature, pressure and composition spaces, and the conditions of stability of a single-component or a multicomponent semiconductor phase as a function of the temperature, pressure and composition, using the Gibbs energies $G(T)^{\dagger}$ or the Gibbs formation energies $\Delta G_f(T, x)$ as critical parameters,

---

† Everywhere in this book we will omit the term "free" when dealing with the Gibbs energies.

---

Chemistry of Semiconductors
By Sergio Pizzini
© Sergio Pizzini 2024
Published by the Royal Society of Chemistry, www.rsc.org

experimentally determined by calorimetric and electrochemical measurements, or computationally calculated.[4]

We give as known by the readers of this book the fundamentals of equilibrium thermodynamics[5] of single and multicomponent phases, as well the properties of ideal or non-ideal solutions, although a short discussion on these topics is given in Appendix 1.1.

Thermodynamics have also to deal with the formation energies and concentration of native point defects in semiconductors, considered as equilibrium chemical species, which influence their structural, chemical and electrical properties, and play, as well, an essential role in the compensation of non-stoichiometry offsets, common for most compound semiconductors. For a full understanding of this last topic, a basic knowledge of semiconductor physics is also necessary, as well as of the principles on which density functional theory (DFT)[6] and *ab initio* calculation are based on, since they represent the essential instruments for the computation of semiconductor properties.

## 1.2 Thermochemical Properties of Elemental and Compound Semiconductors: A Key Issue for the Proper Design and Control of Their Synthetic Processes

It is well known that elemental and compound semiconductors used for microelectronic and optoelectronic applications need to be synthesized as ultrapure polycrystalline materials from ultrapure precursors, and grown as single crystal ingots, thin films, or nanocrystals.

Therefore, knowledge of their thermal and chemical properties and of their precursors[‡] (available in nature or synthetic) is essential not only to forecast the process conditions needed for their growth as ingots of single crystalline ultrapure materials, but to evaluate the features of the thermal and chemical processes which will be finally adopted for the final device fabrication.

The family of elemental and compound semiconductors includes a variety of materials presenting largely different chemical, physical and structural properties that predetermine the conditions of their preparation as single crystal ingots,[§] of their further processing and of

---

[‡] We call precursor a natural or synthetic compound which can be used to synthetize a semiconductor.

[§] These are mandatory for most microelectronic or optoelectronic applications.

their final technological application, making a careful physico-chemical approach the origin of success.

As an example, the high melting points of most common semi-conductors, which are 1211.25 K for Ge, 1687 K for silicon, 3823 K for diamond,[¶] 1511 K for GaAs and 3110 K for SiC, and the high sub-limation and/or dissociation pressures of most common compound semiconductors (arsenides, phosphides and nitrides) makes the de-sign of their growth processes as single crystal ingots very demanding.

In the case of melt growth, liquid to solid (LS) processes, critical parameters are the thermodynamic stability of the material and its sublimation pressures at the melting temperature, on which its mass losses would depend, and its chemical reactivity with respect to po-tential environmental degradation.

It is a fortunate circumstance that the vapour pressure of elem-ental and of several compound semiconductors of industrial inter-est is sufficiently low at their melting temperatures (see Table 1.1) to allow their growth from a stable melt under normal pressure conditions.

As we will see in Chapters 2 and 3 for the case of compound semiconductors spontaneously decomposing at or below the melting temperature, melt stabilization could be, however, obtained by applying a suitable hydrostatic overpressure.

## 1.2.1 Thermochemical Properties of Elemental Semiconductors Ge, Si and Diamond

Diamond, silicon and germanium present a fully covalent bonding character,[‖] as well as a cubic diamond structure at ambient pressure and temperature, but a variety of their polymorphs could be gener-ated by varying the temperature and applied pressure, as can be seen from their phase diagrams[13,17] displayed in Figures 1.1 and 1.2.

Diamond is the cubic, high-pressure polymorph of carbon, whose temperature/pressure relationship is defined by two heterogeneous equilibria[13] with solid graphite and liquid carbon, as can be seen in Figure 1.1.

The temperatures and pressures involved in the phase transforma-tions of diamond make it almost technically unfeasible to directly use these transformations for the industrial synthesis of diamond crystals.

---

[¶] The most recent experimental value of the melting temperature of diamond, determined with a multi-anvil apparatus, is (5968 ± 457 K) at a pressure of 15 GPa.[198]

[‖] We will see that this is not the case for compound semiconductors.

**Table 1.1** Properties of selected semiconductors of technological interest. Data from ref. 7-12.

| Material | Melting point (K) | Heat of fusion (kJ mol$^{-1}$) | Vapor pressure at MP (atm) | Density at 298 K (g cm$^{-3}$) | Energy gap (eV) |
|---|---|---|---|---|---|
| Ge | 1210 | 34.7 | $3.45 \times 10^{-6}$ at 1608 K | 5.323 | 0.66 |
| Si | 1690 | 50.21 | 0.644 Pa at 1850 K | 2.3296 | 1.12 |
| C (diamond) | 3823 | 120 | | 3.52 | 5.47 |
| 3C-SiC | 3100 | n.a. | $\cong 10^{-5}$ | 3.166 | 2.3 |
| InSb | 800 | 25.5 | $4 \times 10^{-8}$ | 5.775 | 0.16 |
| GaSb | 985 | 25.1 | $10^{-6}$ | 5.61 | 0.73 |
| GaAs | 1510 | 87 | 1 | 5.32 | 1.43 |
| InP | 1333 | 62.7 | 27.5 | 4.81 | 1.34 |
| GaP | 1730 | 117.6 | 32 | 4.14 | 2.26 |
| GaN | 2790 | | 45 kbar | 6.15 | 3.23 (cubic) 3.45 (wurtzite) |
| HgSe | 1063 | n.a. | | 8.245 | |
| HgTe | 943 | n.a. | 12.5 | 8.63 | |
| CdSe | 1623 | 44.8 | 0.3 | 5.81 | 1.71 |
| CdTe | 1314 | 43.5; 64.852 | 0.65 | 5.86 (zinc blende) | 1.47 |
| ZnSe | 1373 | 56 | 0.5 | 5.26 | 2.82 |
| ZnTe | 1513 | 51.00 | 0.6 | 5.65 | 2.39 |

Extended polymorphism is, instead, the case of silicon and germanium, both of which are stable with the cubic diamond structure at ambient pressure. At least four stable polymorphs are known for germanium,[14] and 11 high pressure polymorphs for silicon, of which seven are well defined[15] and are still the object of research interest.[16]

For silicon and germanium, the heterogeneous equilibria of technological interest, see Figure 1.2, concern the cubic phases stable at ambient pressure up to a critical pressure, and the Si II and Ge II phases, crystallizing with the β-SN structure, stable at applied pressure larger than 12 GPa for silicon and 10 GPa for germanium, both of which stable in equilibrium with a liquid phase.[17,18]

Different from diamond, their temperature/pressure conditions relative to the heterogeneous equilibria among the cubic diamond phase and the liquid phase are entirely compatible for the synthesis of Si and Ge crystals with liquid to solid (LS) crystallization processes.

The chemical stability of diamond, silicon and germanium in oxidizing environments, which are systematically involved in all the steps of their synthesis and of further device fabrication** processes, has been the subject of notable interest in the literature, with

---

** The use of argon gas protective atmospheres cannot avoid the presence of traces of oxygen.

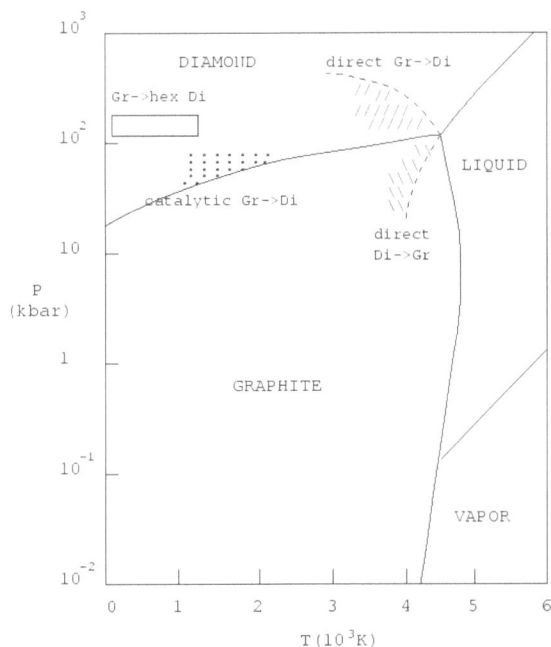

**Figure 1.1** Phase diagram of carbon. Reproduced from ref. 13 with permission from Elsevier, Copyright 1989.

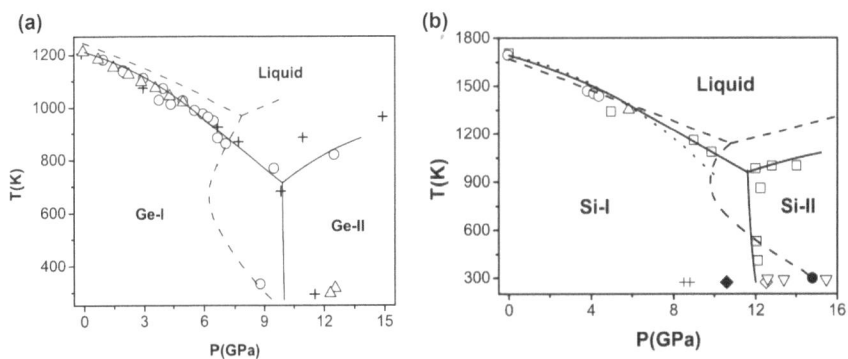

**Figure 1.2** (a) Phase diagram of Ge. Reproduced from ref. 17 with permission of Elsevier, Copyright 2004. (b) Phase diagram of silicon. Reproduced from ref. 18 with permission from Elsevier, Copyright 2004.

particular attention paid to the case of diamond, in connection with its optoelectronic properties, and for the case of silicon, since its oxidation features depend strongly on temperature, with critical changes in the structural, morphological and electronic properties of the silicon oxide films, which are used to mask limited regions of the device.

Concerning the case of diamond, Howe *et al.*[19] demonstrated that the oxidation route strongly depends on the oxygen coverage. In fact, at high temperatures (973 K and higher) and at low oxygen coverage (0–50%), diamond converts first to $sp^2$-bonded amorphous carbon and then the surface oxidation occurs with the formation of CO and $CO_2$. At zero oxygen coverage, and at higher temperatures (1600 °C) amorphous carbon undergoes a graphitization process, with the formation of crystalline graphite.

Joshi *et al.*,[20] who carried out oxidation experiments of natural diamond and of chemical vapor deposited (CVD) diamond films in pure oxygen at 600–700 and 800 °C, showed that no intermediate steps of graphitization are involved in the oxidation process carried-out with pure oxygen, and that oxidation proceeds preferentially on (100) planes in the case of natural diamond, and on surface defects (grain boundaries) in the case of polycrystalline diamond films. In both cases, the surface oxidation proceeds with the formation of local morphological changes and surface faceting. A faster oxidation of (111) surfaces is, instead, favoured by high $CO_2$ pressures according to Fedortchouk *et al.*[21]

The stable oxide phase of silicon is $SiO_2$, with a Gibbs energy of formation of $-853.355$ kJ mol$^{-1}$ at 298 K, and $-612.544$ kJ mol$^{-1}$ at the melting temperature of silicon.

Also for Ge the stable oxide phase is the dioxide $GeO_2$, with a Gibbs energy of formation of $-516.671$ kJ mol$^{-1}$ at 298 K and $-351.802$ kJ mol$^{-1}$ at the Ge melting point.

It is therefore understandable that at any temperature between room temperature and their melting, spontaneous surface oxidation would occur for both Si and Ge, also in neutral environments contaminated by traces of oxygen, with increasing oxidation rates depending on temperature and oxygen pressure.

Taking into consideration the case of silicon, its oxidation occurs rapidly at room temperature when a silicon sample is exposed to oxygen or air, with the formation of a native oxide film 0.5–1 nm thick, which slowly increases up to 1–2 nm, with a rate that depends on the oxygen diffusivity in the oxide.

According to Kim and Carpenter,[22] who carried out microdiffraction and high resolution electron microscopy (HREM) measurements, the native oxide is completely amorphous with the composition of the sub-oxide SiO, although Al-Bayati[23] demonstrated that the oxide composition varies with the depth, being that of SiO at the silicon/oxide interface.

Thick oxide layers could be grown by high temperature oxidation, with growth rates that are limited by the diffusion of interstitial oxygen (as molecular oxygen) in the growing layer, according to Deal and Grove,[24] and confirmed by recent literature reports.[25] Also in the high temperature-grown $SiO_2$ layers the presence of suboxides at the $Si/SiO_2$ interface is evidenced by photoemission spectroscopy.[26] High temperature (800–1000 °C) silicon oxidation, carried-out in wet and dry conditions,[24–27] was used, at the beginning of the transistor era, to grow thick (up to 100 nm), amorphous $SiO_2$ layers, with excellent dielectric quality, to mask limited regions of the device.

Today, ultrathin (1–2 nm) silicon oxide layers are used to mask the gate of silicon MOSFET transistors with a size of 4 nm.

Different from silicon, the high temperature oxidation of germanium is rate-limited by a different mechanism, see Figure 1.3, which involves the molecular oxygen dissociation to atomic oxygen at the $O_2/GeO_2$ interface, the diffusion of atomic oxygen *via* an oxygen vacancies mechanism in the $GeO_2$ layer, and the oxidation of Ge at the $Ge/GeO_2$ interface with formation of oxygen vacancies $(V_O)$[28]

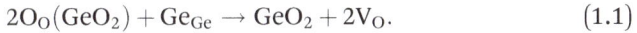

$$2O_O(GeO_2) + Ge_{Ge} \rightarrow GeO_2 + 2V_O. \tag{1.1}$$

The use of $GeO_2$ layers in germanium MOSFET transistors is, however, presented by the lack of thermodynamic stability of the $Ge/GeO_2$ interface, which leads to the formation and sublimation of GeO

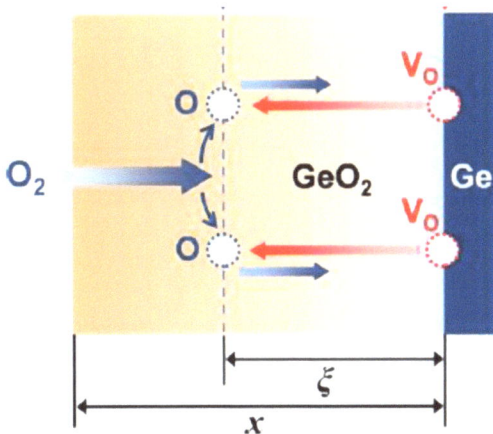

$$GeO_2 \rightleftharpoons GeO + \frac{1}{2}O_2 \tag{1.2}$$

**Figure 1.3** Schematic features of germanium oxidation. Reproduced from ref. 28 with permission from AIP Publishing, Copyright 2017.

(with a value of Gibbs free energy of formation of GeO of $-33.502$ kJ mol$^{-1}$ at 298 K, which negligibly varies with the temperature[29]), and to a considerable degradation of the Ge/GeO$_2$ interface.[30]

This process has been discussed in terms of GeO disproportionation by Wang *et al.*,[31] with the same conclusions regarding the thermodynamic instability of the Ge/GeO$_2$ interface.

## 1.2.2 Thermodynamic Stability of Compound Semiconductors

Compound semiconductors of the II–VI and III–V groups present, in general, a lower thermodynamic stability than elemental semiconductors. This condition might lead, at temperatures close or even lower than the melting temperatures, to extreme equilibrium partial pressures of their components, leading to their decomposition, especially in the case of the nitride members of the group.

Thermodynamic stability conditions of a compound with a stoichiometric MX composition,[††] whose equilibrium formation reaction is given by the reaction[‡‡]

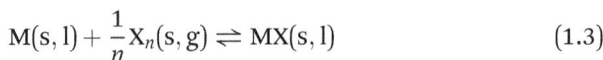

$$M(s, l) + \frac{1}{n}X_n(s, g) \rightleftharpoons MX(s, l) \tag{1.3}$$

are defined by the following equilibrium constant, when the equilibrium pressure of the metal is negligible with respect to the equilibrium pressure of the non-metal, which is often the case,

$$K_{eq} = \frac{a_{MX}}{a_M p_{X_n}^{1/n}} = \exp{-\frac{\Delta G_f(T)}{RT}} = p_{X_n}^{-1/n} \tag{1.4}$$

where $\Delta G_f(T)$ is the Gibbs energy of formation of the solid phase at the temperature $T$, $p_{X_n}$ is the equilibrium partial pressure of the non-metallic species $X_n$, and $a_M$ and $a_{MX}$ are the activities[§§] of the metal and of the stoichiometric compound MX, taken equal to 1. When the

---

[††] However, the formal analysis will be the same for a semiconductor of the II–VI family.
[‡‡] Where $n$ gives the molecular composition of the species X (N, P, As, Sb), and g, l, s mean gas, liquid and solid).
[§§] As is known, the thermodynamic activity of a chemical species i in solution is connected to its chemical potential by the relationship $\mu_i = RT \ln a_i$ and $a_i = \gamma_i x_i$, where $\gamma_i$ is an activity coefficient that measures the deviations from the ideal. The activities of a pure substance is taken equal to 1.

equilibrium pressure of the metal is not negligible, we need to write the equilibrium constant with the following equation

$$K_{eq} = \frac{a_{MX}}{p_M p_{X_n}^{1/n}} = \exp - \frac{\Delta G_f(T)}{RT} = p_M p_{X_n}^{-1/n} \qquad (1.5)$$

where $p_M$ is the equilibrium pressure of the metal.

Different from oxides, with their Gibbs energies of formation $\Delta G_f(T)$ of the order of hundreds of kJ mol$^{-1}$, and with their equilibrium oxygen pressures ranging from 1 bar to $10^{-100}$ bar, the Gibbs energies of the formation of III–V compound semiconductors are much less, see Table 1.2[32,33] (with the exception of AlN), and tend to zero[¶¶] at temperatures lower than their estimated melting temperatures, leading to their decomposition

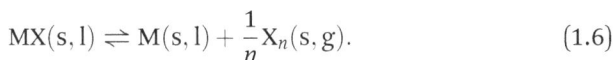

$$MX(s,l) \rightleftharpoons M(s,l) + \frac{1}{n} X_n(s,g). \qquad (1.6)$$

The decomposition of a solid MX phase, at any temperature up to the melting temperature, might be suppressed by applying an hydrostatic counter-pressure, using as a working fluid the gaseous, non-metallic component $X_n$ of the MX phase (nitrogen in the case of nitrides) to balance the decomposition pressure. Apparently, this is only possible if reaction (1.6) is reversible or behaves as reversible,[46] and if a reservoir of the metal and a convenient supply of the component $X_n$ at various pressures are available.

**Table 1.2** Properties of III–V semiconductors.

| | $T_m$ (K) | $\Delta G_f$ (kJ mol$^{-1}$) at 298.5 K | Structure | $E_G$ (eV) | Ref. |
|---|---|---|---|---|---|
| AlN | 2473 | $-316.2 + 0.1157T$ | Wurtzite | 6.02–6.28 | 35 and 36 |
| GaAs | 1511 | $-70.374$ | Zinc-blende | 1.43 | 38, 39, 41, 43 and 44 |
| GaP | 1730 | $-91.317$ | Zinc-blende | 2.26 | 44 |
| GaN | 2573 | $-77.741$ | Wurtzite/ zinc-blende | $3.452 \pm 0.001$ | 37, 42 and 44 |
| InAs | 1215 | $-53.286$ | Zinc-blende | 0.354 | 38 and 44 |
| InN | 1373 | $+15.676$ | Wurtzite | 1.89–2.05 | 35 and 44 |
| InP | 1333 | $-77.046$ | Zinc blende | 1.344 | 35, 40, 41, 44 and 47 |
| InSb | 798 | $-25.389$ | Zinc blende | 0.17 | 35 and 44 |

[¶¶] From basic thermodynamics we know that the stability of a compound occurs when its Gibbs formation energy is less than zero.

A different, and more convenient, way to overcome the problem is to hermetically encapsulate the MX phase with a suitable liquid and dynamically balancing the dissociation pressure with an equivalent pressure of inert gas.[34]

This will enable the control of the equilibrium pressure of the gaseous component $X_n$ at the interface between the MX phase and the liquid encapsulant. We will see in Chapter 3 that this technique is in widespread use for the growth of compound semiconductors under the name of the liquid encapsulation method, called LEC when used with the Czochralski growth process, see Chapter 3.

It should be mentioned that since the melting point $T_m$ of a generic MX phase depends on the pressure, according to the Clausius–Clapeyron equation

$$\frac{dT_m}{dP} = \frac{T\Delta V_m}{\Delta H_m} = \frac{\Delta V_m}{\Delta S_m} \tag{1.7}$$

(where $\Delta V_m$ is the volume variation on melting and $\Delta H_m$ and $\Delta S_m$ are the enthalpy and entropy of fusion), the equilibrium melting temperature of a MX phase that decomposes before melting depends on the mechanical pressure applied to stabilize the phase.

It is, therefore, apparent that the establishment of thermodynamic equilibrium of these systems at their melting temperatures, under liquid encapsulation conditions, implies that both hydrostatic and mechanical equilibria should be simultaneously satisfied.

We will consider here the case of Al, Ga and In nitrides,[48] see Figure 1.4, that shows that the decomposition pressures are

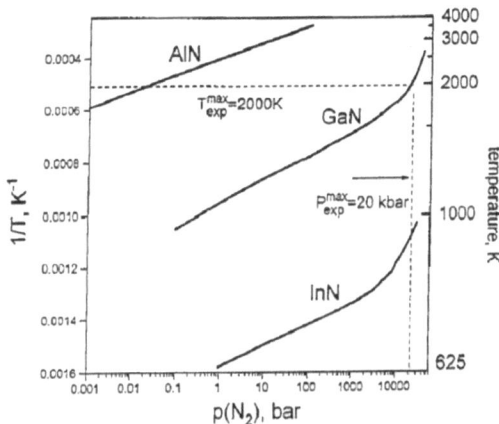

**Figure 1.4**  Equilibrium pressure of $N_2$ over III–V nitrides. Reproduced from ref. 48 with permission from Elsevier, Copyright 1997.

higher than 20 kbar for both InN and GaN at temperatures well below their estimated melting point (1370 and 2573 K), while AlN begins to decompose only above its melting point (2473 K). The low thermal stability of InN is well understood if we look at its Gibbs energy of formation, which is $+15.676$ kJ mol$^{-1}$ at room temperature.

The dotted lines in the figure define the experimental range of temperatures (2000 K) and pressures (20 kbar) allowable by the high-pressure system using nitrogen gas, which represent the best solution for operating under full hydrostatic and mechanical equilibrium conditions. It is apparent that for neither of the semiconductors of Figure 1.4 does the nitrogen balancing system allow reaching their melting temperatures.

To this end, well designed high-pressure mechanical systems should be used, which must provide conditions of gas-tightness of their sample cells to fulfil the required equilibrium conditions. We will see that equilibrium conditions with mechanical systems are not systematically obtained at the extreme pressure and temperature of the experiments carried out on high melting nitrides.

Phosphides and arsenides behave like nitrides, although in a range of milder temperatures and pressures, as can be seen for the case of InP[49] in Figure 1.5, which shows that the decomposition starts at 850 °C, below its melting point, which lies at 1062 °C, and in Figure 1.6, which displays the equilibrium pressures of Ga and of As and P polymeric vapour species over GaAs and GaP along their liquidus curves.[50]

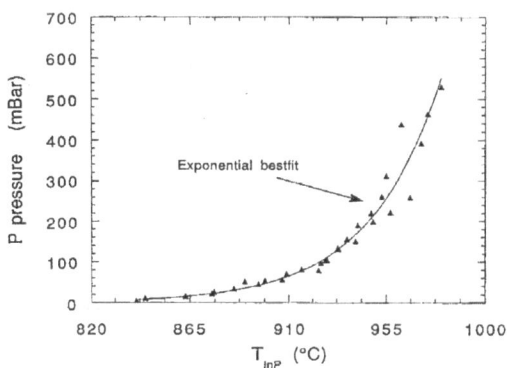

Figure 1.5 Temperature dependence of the phosphorus pressure of solid InP (the melting temperature of InP is 1335 K). Reproduced from ref. 49 with permission from AIP Publishing, Copyright 1994.

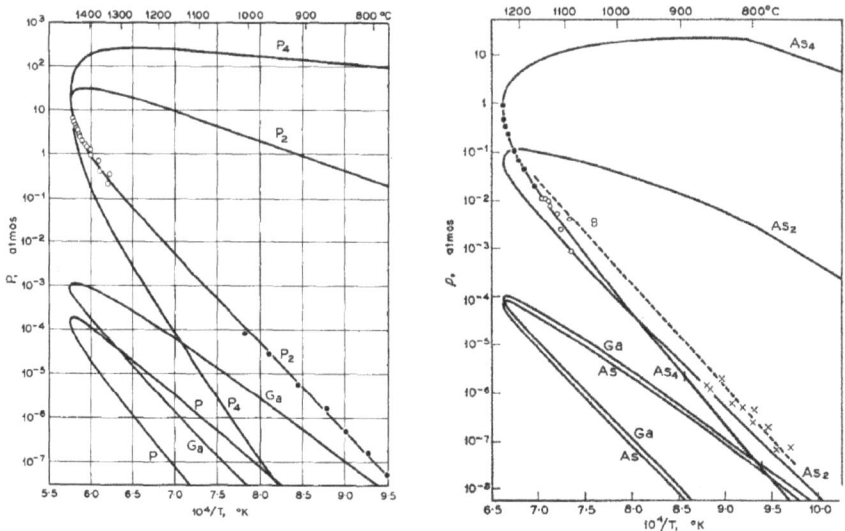

**Figure 1.6**   Temperature dependence of the equilibrium partial pressures of Ga, $P_n$ and $As_n$ over GaP and GaAs along their liquidus curves. Reproduced from ref. 50 with permission from Elsevier, Copyright 1965.

From Figure 1.6 one can recognize that at the melting point of GaAs (1238 °C) the total As pressure is around 1 bar, at the melting point of GaP (1457 °C), the total $P$ pressure is more than 100 bar, while the Ga pressure amounts $10^{-4}$ bar at the melting temperatures of GaAs and GaP.

It is, eventually, apparent that the contribution of the congruent sublimation

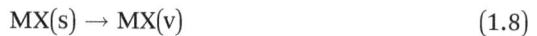

$$MX(s) \rightarrow MX(v) \tag{1.8}$$

on the mass loss of most of the compound semiconductors of the III–V group becomes appreciable with the increase of the temperature,[51] as is shown in Figure 1.7.

Due to the poor thermal stability of III–V compounds, their growth is carried out under moderate or high pressures (see Table 1.2) with a pressure balancing method,[34] as will be discussed in Chapter 3.

### 1.2.3   A Case Study, GaN

Among covalent semiconductors, the case of GaN, extensively used for blue and green light emitting diodes (LEDs) and lasers, which granted the 2014 Nobel Prize to Isamu Akasaki, Hiroshi Amano and

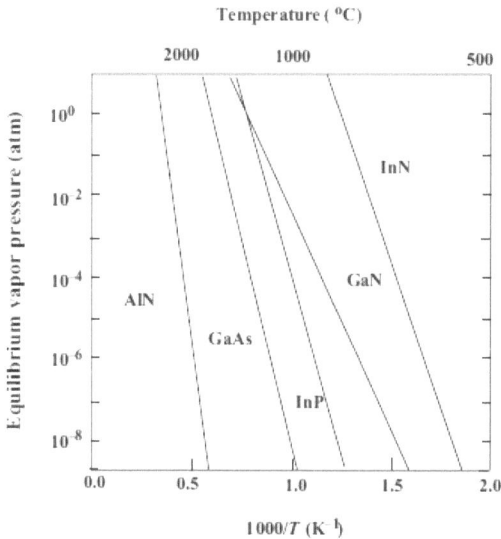

**Figure 1.7** Temperature dependence of the congruent equilibrium vapour pressures of GaAs and of nitride semiconductors. Reproduced from ref. 51 with permission from Elsevier, Copyright 1992.

Shuji Nakamura, deserves major attention since its thermal stability and melting features have for years motivated great theoretical and experimental interest, which favoured the set-up of detailed knowledge of its thermodynamic properties.

As an example, the Gibbs energy of formation $\Delta G_f$ of GaN has been determined both by electromotive force (EMF) measurements on a solid galvanic cell up to 1500,[52,53] and with equilibrium nitrogen pressures measurements up to 1900 K,[42] using a nitrogen gas high pressure system up to 25 kbar and a well designed mechanical hemispherical anvil apparatus potentially working up to 60 kbar, which allows the establishment of full thermodynamic equilibrium conditions.

On that experimental basis,[52] see Figure 1.8, the temperature dependence of the Gibbs energy of formation $\Delta G_f$ of GaN is given by the equation

$$\Delta G_f \left( \frac{kJ}{mol} \right) = -128.749 + 115.029T \tag{1.9}$$

which leads to a thermodynamic decomposition temperature of GaN at 1119 K.[52] This temperature is much lower than its melting point, estimated to be 2791 K by Van Vechten,[54] who used a semi-empirical method, or $T_m = 2825 + 210P - 5P^2$ (K) at pressures higher

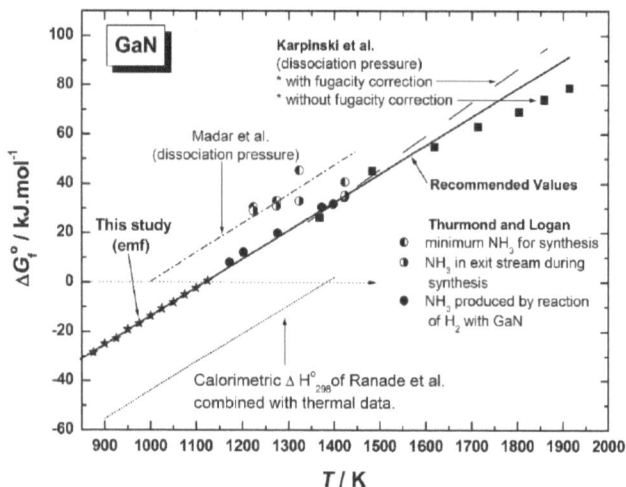

**Figure 1.8** Temperature dependence of the Gibbs energy of formation ($\Delta G_f$) of GaN. Reproduced from ref. 52 with permission from Springer Nature, Copyright 2007.

than 9 GPa, calculated by Harafuji,[55] who used a molecular dynamic method (MD), which accounts for the presence of the two-phase, liquid–solid equilibrium at the melting temperature.[‖ ‖]

Since one, obviously, expects that with the increase in the temperature the equilibrium $N_2$ pressures would gradually increase, the application of a pressure balance would be a rational criterium for its thermodynamical stabilization, potentially up to temperatures corresponding to its melting point.

On that basis, Karpinski *et al.*[42] and Krukowski *et al.*[56] were able to show, using a high-pressure nitrogen balancing system and GaN powder samples, that thermodynamic stabilization of the solid GaN phase does, in fact, occur when the equilibrium $N_2$ pressure was balanced by a suitable nitrogen overpressure. They eventually argued that the melting temperature of GaN should be higher than 2500 K, because no sign of melting was observed in samples brought up to this temperature, and that the equilibrium pressure of nitrogen at the melting point of GaN should be, therefore, higher than 20 kbar.

The *P–T* diagrams of GaN in the pressure limit of 20 kbar (2 GPa), deduced from these results,[56] are reported in Figures 1.9 and 1.10, which show that the solid GaN phase is experimentally (and thermodynamically) stable at 2000 K under a pressure of 20 kbar.

---

[‖ ‖] Very often melting is theoretically evaluated by neglecting the influence of the liquid phase in equilibrium with the solid at the melting temperature.

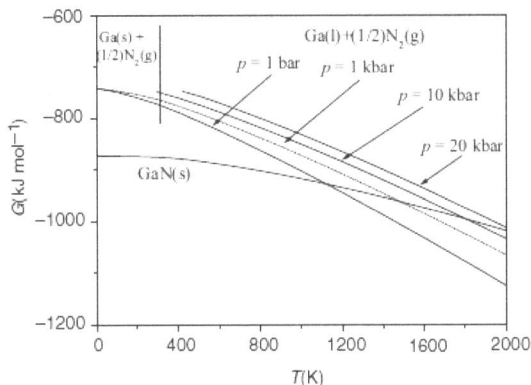

**Figure 1.9** Temperature dependence of the equilibrium condition of solid GaN and its precursors. Reproduced from ref. 56, https://www.academia.edu/25208911/Blue_and_UV_Semiconductor_Lasers, under the terms of the CC BY 4.0 license, https://creativecommons.org/licenses/by/4.0/.

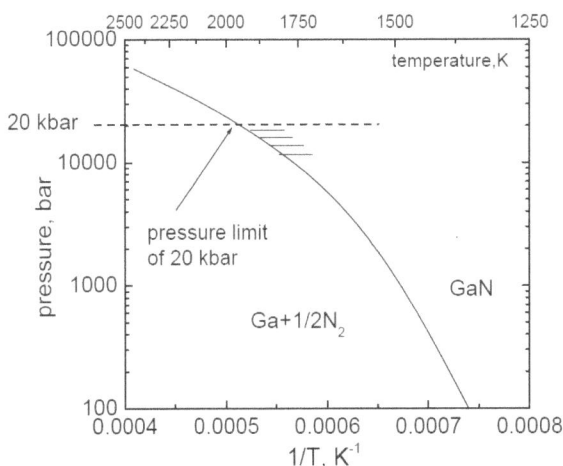

**Figure 1.10** Pressure–temperature diagram of GaN. Reproduced from ref. 56, https://www.academia.edu/25208911/Blue_and_UV_Semiconductor_Lasers, under the terms of the CC BY 4.0 license, https://creativecommons.org/licenses/by/4.0/.

Further studies on the decomposition process of GaN were carried out by Porowski[48] and Grzegory *et al.*,[57] by Peshek *et al.*[58] and by Davydov *et al.*,[59] who carried-out a preliminary thermodynamic assessment of the Ga–N system.

As noted by Jakob *et al.*,[53] a critical problem concerning studies addressed at the measurement of the decomposition and melting equilibria of GaN is the slow kinetics of the process, which could make it difficult to obtain reliable equilibrium values.

The decomposition process

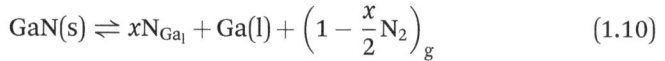

$$GaN(s) \rightleftharpoons xN_{Ga_l} + Ga(l) + \left(1 - \frac{x}{2}N_2\right)_g \qquad (1.10)$$

does, in fact, occur with the formation of a diluted solution of nitrogen in liquid gallium (or of GaN in Ga), whose concentration $xN_{Ga_l}$ increases with temperature (and the applied $N_2$ pressure) being 0.5 at% at 1900 K (see Figures 1.11 and 1.12), 3.7 at% at $P = 6$ GPa and $T = 2800$ K, 11 at% at 8 GPa and $T = 3150$ K, and 17 at% at 9 GPa and 3400 K.[48]

Not only the stability of the GaN phase depends on the nitrogen pressure, but also its melting temperature, since melting occurs under a three-phase equilibrium condition

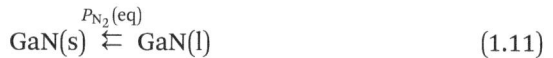

$$GaN(s) \overset{P_{N_2}(eq)}{\leftarrow} GaN(l) \qquad (1.11)$$

where $P_{N_2}(eq)$ is the nitrogen equilibrium pressure at the melting temperature, different from the standard pressure (1 bar) at which melting is measured for most materials.

We do expect, therefore, that the melting temperature of the thermodynamically stable GaN phase would depend on the pressure applied to stabilize the phase.

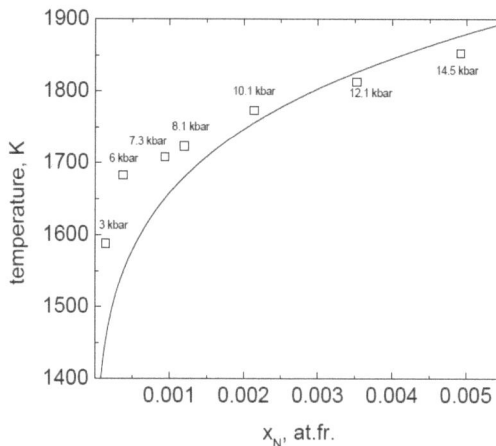

**Figure 1.11**  Temperature and pressure dependence of the nitrogen solubility in liquid gallium. Reproduced from ref. 48 with permission from Elsevier, Copyright 1997.

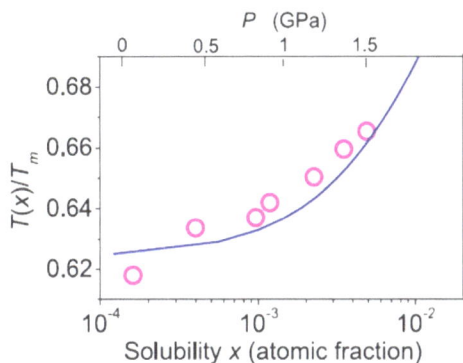

**Figure 1.12** Temperature and pressure dependence of the nitrogen solubility in liquid gallium. The melting temperature is assumed to be 2971 K. Reproduced from ref. 45 with permission from Elsevier, Copyright 2015.

The effect of pressure on the melting temperature of a solid phase that is thermodynamically stable at its standard melting temperature can be quantitatively evaluated with the Clausius–Clapeyron eqn (1.7), using known or calculated values of the volume variation $\Delta V_m$ on melting and of experimental values of the enthalpy $\Delta H_m$ or entropy $\Delta S_m$ of fusion, although for decades the Simon–Glatzel equation[60,61]

$$T_m(P) = T_m^{\circ}\left(1 + \frac{\Delta P}{\pi}\right)^{1/b} \tag{1.12}$$

has been used, where $T_m^{\circ}$ is the melting temperature at the standard pressure $P_m^{\circ}$, $\Delta P = P - P_m^{\circ}$, and $\pi$ and $b$ are adjustable parameters.

Considering that the parameter $\pi$ in eqn (1.12) could assume positive and negative values, this equation could fit not only the behaviour of materials whose melting temperature increases with pressure, but also that of materials like silicon and germanium, whose melting temperatures decrease with increase in pressure due to the negative variation of the volume on melting $\Delta V_m$, associated with the formation of a liquid metallic phase.

Among other improved variants available,[62] the Simon–Glatzel equation has been recently implemented by Drozd-Rzoska et al.[63] by adding a term $\exp\left(-\dfrac{\Delta P}{c}\right)$

$$T_m(P) = T_m^{\circ}\left(1 + \frac{\Delta P}{\pi + P_m^{\circ}}\right)^{1/b} \exp\left(-\frac{\Delta P}{c}\right) \tag{1.13}$$

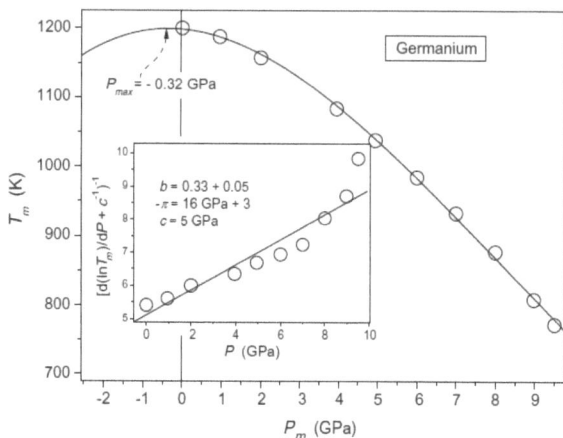

**Figure 1.13**   Temperature dependence of the melting temperature of germanium. In the inset the parametrization parameters are reported,
obtained using a differential analysis of the experimental data.
Reproduced from ref. 63 with permission from Elsevier,
Copyright 2007.

(where $P_m^\circ$ is the equilibrium pressure at the melting temperature
and $\pi$, $c$ and $b$ are again adjustable parameters), which allows displaying a monotonical increase in the melting temperature with
pressure up to a maximum, followed by a temperature decrease,
which better accounts for the actual behaviour of a series of inorganic
and organic materials.

For materials like germanium (see Figure 1.13) and silicon, whose
melting temperature decreases with the increase in the applied
pressure just above the standard pressure, the pressure dependence
of the melting temperature is formally described starting from a
maximum sitting at a negative pressure.

The pressure dependence of the melting temperature of covalently
bonded materials has been also theoretically studied by Van Vechten,[54]
Harafuji *et al.*[55] and by Nord *et al.*[64]

Van Vechten[54] applied the Phillips theory of electronegativity[65] to
calculate the pressure dependence of the melting enthalpy $\Delta H_m$, of
the melting entropy $\Delta S_m$, and of the melting temperature
$T_m(P) = \dfrac{\Delta H_m(P)}{\Delta S_m(P)}$,*** which allows building the phase diagram of the
particular system studied, accounting for the volume change on
melting and for the physical nature of the liquid phase.

---

*** At the melting temperature the Gibbs free energy of melting $\Delta G_m = 0$.

The reliability of the method is demonstrated by the good or excellent fit of the experimental and theoretical behaviour of gallium antimonide (GaSb), germanium and silicon[54] displayed in Figure 1.14, for which the metallic properties of the liquid phases are well known, together with the properties of the solid phase in equilibrium with the liquid.

Analogous behaviour is predicted by the calculated phase diagram of GaN[54] displayed in Figure 1.15, which suggests that the melting

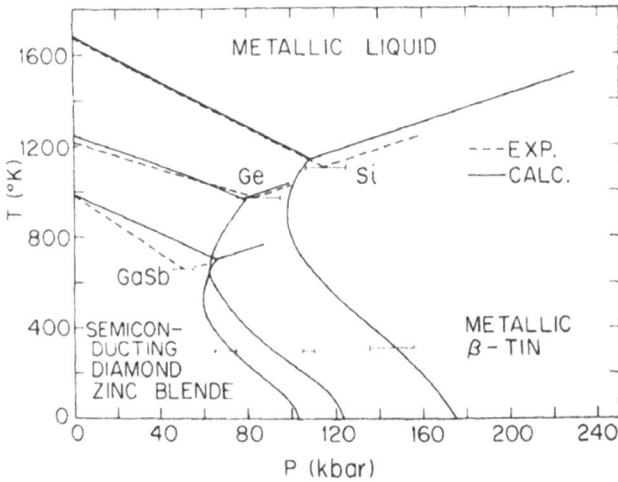

Figure 1.14 Phase diagrams of GaSb, germanium and silicon. Reproduced from ref. 54 with permission from American Physical Society, Copyright 1973.

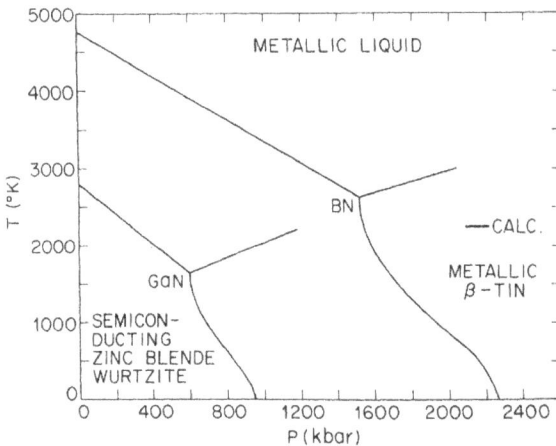

Figure 1.15 Phase diagrams of GaN and BN. Reproduced from ref. 54 with permission from American Physical Society, Copyright 1973.

temperature of GaN is around 2800 K at 1.6 Gbar, and increases with the increase of the pressure, in the hypothesis that the GaN liquid phase is metallic (*i.e.* $\Delta V_{l,s} < 0$), in the absence, however, of experimental data.

Harafuji *et al.*[55] carried out the same study on GaN, using a molecular dynamics (MD) approach, assuming conditions of equilibrium of a three-phase system, consisting of a solid phase with wurtzite structure, of a liquid with a structure obtained by disordering the wurtzite structure of the solid, and of a nitrogen gas atmosphere.

The results of the simulation are reported in Figure 1.16, which shows that a discontinuity associated with melting does occur around 3500 K, under a nitrogen pressure of 9 GPa.

Only few experimental works devoted to the determination of the melting temperature of GaN are available in the literature, of which the most recent one by Sokol *et al.*,[66] carried out using a multi-anvil apparatus and graphite or boron nitride (BN) cells, validates the results of Karpinski *et al.*,[42] with the conclusion that congruent melting of GaN should occur above 2500 K under an applied pressure higher than 7.5 GPa.

Utsumi *et al.*,[37,67] were, instead, able to melt congruently GaN using a leak-free, multi anvil system, an electrically heated BN cell and powder samples that facilitate obtaining full equilibrium conditions. They demonstrated that GaN decomposition does systematically occur by heating the sample under an applied pressure lower than

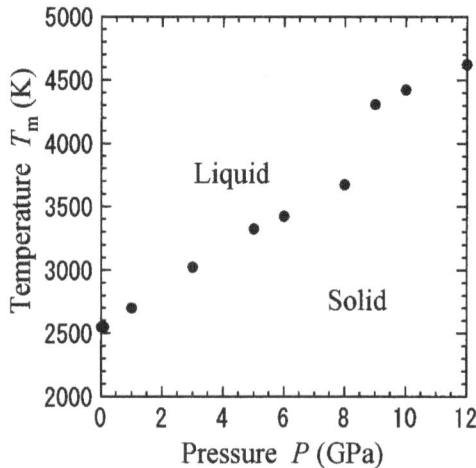

**Figure 1.16**  Calculated pressure dependence of the melting temperature of GaN. Reproduced from ref. 55 with permission from AIP Publishing, Copyright 2004.

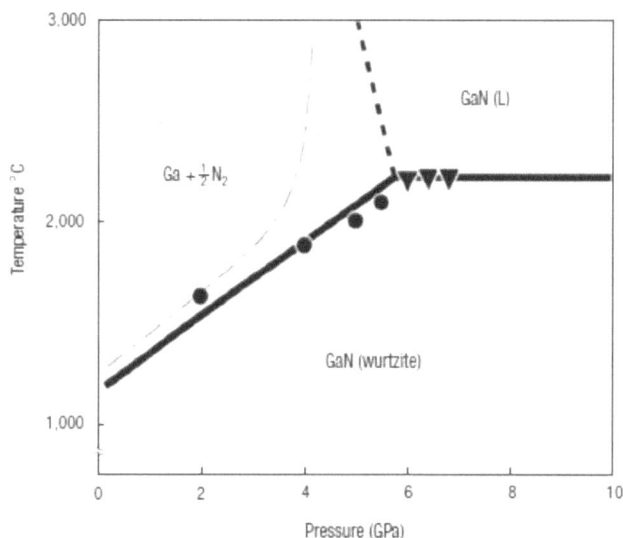

**Figure 1.17** Experimental results on the pressure dependence of the decomposition and melting temperatures of GaN. Reproduced from ref. 37 with permission from Springer Nature, Copyright 2003.

6 GPa, while congruent melting occurs at ~2473 K (Figure 1.17) at a pressure of 6 GPa. Apparently, the melting temperature remains almost constant with a further increase in the pressure.

While these last results do agree well with the Van Vechten suggestion that the melting point of GaN should be lower than 2700 K, those of Porowski *et al.*[45] displayed in Figure 1.18 are in disagreement.

Their attempt was to carry out direct measurement of the melting temperature of GaN by mechanically pressurizing up to 9 GPa and heating up to 3400 K small solid GaN samples (0.76 cm³), consisting of a mixture of single crystals platelets of GaN and of GaN powder, placed in a tantalum-graphite capsule.

The experimental results of this study (stars ★ and triangles △ in Figure 1.18) show that GaN does not show congruent melting, but only decomposition up to pressures of 9 GPa and to temperatures close to 3500 K. The linear extrapolation of experimental data (dotted curve of Figure 1.18) to cross the melting curve modelled with the DR equation and fitted by the Harafui results,[55] leads to a congruent melting temperature of ~4500 K under an applied pressure of 12 GPa, well above the congruent melting temperature value of 2473 K experimentally determined by Utsumi *et al.*[37]

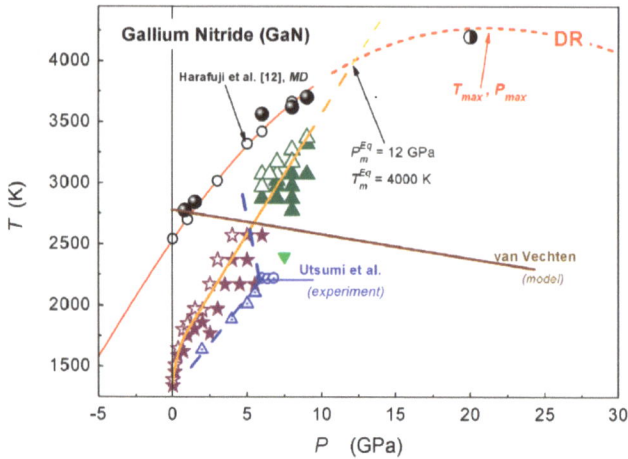

**Figure 1.18** Experimental temperature decomposition range (stars ★ and triangles △) of GaN and theoretical dependence (DR curve) of the melting temperature on pressure (solid curve with circles ○, ●). Reproduced from ref. 45 with permission from Elsevier, Copyright 2015.

These results show that slow equilibrium kinetics, poor equilibrium conditions realized with mechanical pressurizing systems, and extreme temperature and pressure conditions, make the experimental study of the decomposition and melting of GaN very cumbersome.

## 1.3   Thermally Induced Disorder in Elemental and Compound Semiconductors

Topological defects are standard components of condensed matter systems, including the black holes in the universe (see Figure 1.19), which is the largest condensed mass system we know.[68]

Topological defects are all kinds of point defects (vacancies, interstitials, antisites[†††]) and extended defects (dislocations, stacking faults, precipitates) that violate the order of a crystalline phase and take origin either from a thermally induced site-exchange process (the point defects), or from a strain-induced lattice disruption (dislocations, stacking faults) or by a secondary phase segregation process in the presence of impurities with a concentration exceeding their solubility.

---

[†††] Consisting, in the case of an AB compound, of atoms of A in the sublattice of B($A_B$) or atoms of B in the A sublattice $B_A$.

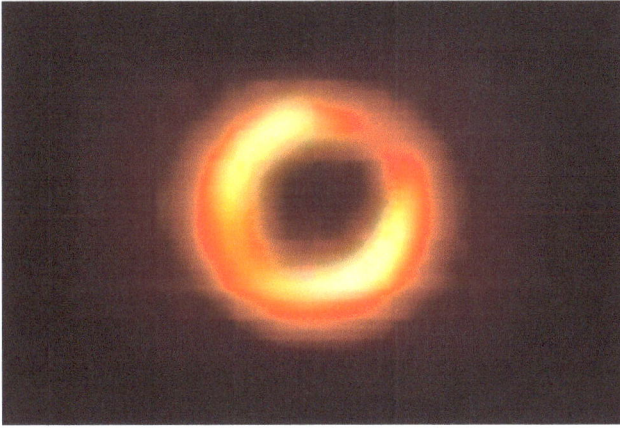

**Figure 1.19** Image of the giant black hole at the centre of the M87 galaxy. Reproduced from https://solarsystem.nasa.gov/resources/2319/first-image-of-a-black-hole with permission from NASA, Copyright 2019 Event Horizon Telescope Collaboration.

Among all kinds of defects, only point defects are equilibrium properties, since their concentration is ruled by a thermodynamic equilibrium process, granted by a well-defined Gibbs energy of defect formation, which foresees an increase of their concentration with increases in temperature.

We can thoroughly discuss this process for vacancies,[‡‡‡] whose formation in an elemental semiconductor, silicon as an example, could be formally written with the following reaction

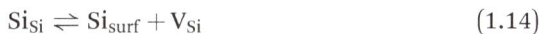

$$\text{Si}_{\text{Si}} \rightleftharpoons \text{Si}_{\text{surf}} + \text{V}_{\text{Si}} \tag{1.14}$$

using the Kröger–Vink nomenclature introduced for ionic solids,[69] but applied as well to semiconductor phases and systematically used in this book.

In close analogy with defects in ionic solids, if the concentration of vacancies $[\text{V}_{\text{Si}}]$ is very low with respect to the density of bulk silicon sites in crystalline silicon $(4.9995 \times 10^{22}$ cm$^{-3})$ and at the silicon surface[§§§] $[(100)6.78 \times 10^{14}$ cm$^{-2}$; $(110)9.59 \times 10^{14}$ cm$^{-2}$; $(111)7.83 \times 10^{14}$ cm$^{-2}]$, one could formally rewrite eqn (1.14), assuming the bulk silicon density undisturbed as the silicon surface density

$$\text{nil} \rightleftharpoons \text{V}_{\text{Si}} \tag{1.15}$$

---

[‡‡‡] However, the issue would be the same for all kinds of point defects.
[§§§] This depends on the specific surface structure.

since the measured silicon vacancy concentration is $4.1 \times 10^{13}$ cm$^{-3}$ at 800 °C.[70]

Under these boundary conditions, the temperature dependence of the concentration of vacancies [$V_{Si}$] should be given by the following equation

$$[V_{Si}] = \exp - \frac{\Delta G_f}{k_B T} \qquad (1.16)$$

where $\Delta G_f$ is the Gibbs energy of formation of silicon vacancies (in eV) at the temperature $T$, and $k_B$ is the Boltzmann constant.

Several physico-chemical properties of crystalline solids, such as the self-diffusion,[¶¶¶] or the diffusion of impurities in semiconductors, and the deviations from the stoichiometry of compound semi-conductors, might be only understood by assuming the presence and the active role of thermal-disorder-induced point defects, vacancies and interstitials,[‖‖] which not only favour the dynamics of the mass transfer processes in a crystalline lattice, occurring with jumps be-tween neighbouring lattice sites or along interstitial paths, but also behave, in fact, as true equilibrium pseudo-chemical species.

An extended discussion on this topic in this book would be in-appropriate, since several thousands of papers and hundreds of books have been published on the argument, of which we give only a limited list.[71–81]

Nevertheless, a short digression on this subject is necessary to see whether general relationships occur between the concentration, the configuration and the physico-chemical properties of point defects, and the crystallographic structure and the chemical composition of the semiconductor phase in which they are embedded.

In view of the difficulty to be experimentally detected as individual species in equilibrium conditions, given their extremely low concen-trations, defects remain, however, elusive particles, whose presence, structure and properties have been theoretically predicted in great detail and implemented by the knowledge of their spectroscopic features, starting from the work of Watkins,[82] who assigned to the isolated silicon vacancy the electron spin resonance (ESR) spectrum of silicon irradiated at 40 K with 1.5 MeV electrons,[****] and whose

---

[¶¶¶] Self-diffusion is called the process of diffusion of a natural isotope (also radioactive) of the elemental component(s) of the semiconductor matrix.

[‖‖] We will call self-interstitials the interstitial atoms of a semiconductor matrix.

[****] Irradiation with energetic electrons is a way to create point defects, out of equilibrium conditions, in semiconductors.

activation energy of motion was estimated to be $0.33 \pm 0.03$ eV. Irradiation with energetic particles and photons is, in fact, necessary to increase the concentration of point defects above their equilibrium values up to the detection levels of the spectroscopic techniques used, while cryogenic temperatures are essential to maintain their integrity.

Since the work of Watkins, few additional spectroscopic techniques are now available for the direct point defect detection, among which positronic annihilation spectroscopy (PAS) is specific for vacancies.

An indirect, but immediate proof of their existence and of their properties is given by results of self-diffusion experiments carried-out at different temperatures on elemental and compound semi-conductors, using as diffusing species the natural isotopes (also radioactive) of the elemental component(s)[††††] of the semiconductor matrix, which allow following with dedicated spectroscopic tools (scanning ion mass spectroscopy (SIMS), deep level transient spec-troscopy (DLTS), IR spectrometry) the diffusion paths at different temperatures and then to measure the self-diffusion coefficient $D^{SD}$ (cm$^2$ s$^{-1}$). From these results it becomes systematically apparent that the self-diffusion as well as the impurity diffusion in semi-conductors is thermally activated with activation energies $E^*$, which obviously depend on the nature of the diffusing species and of the semiconducting phase

$$D_i^{SD} = A \, \exp - \frac{E^*}{kT}. \tag{1.17}$$

As an example, Figure 1.20 displays the temperature dependence of the self-diffusion coefficient of Si in silicon[83] and Figure 1.21 that of carbon in natural diamond,[84] while Figure 1.22 displays the tem-perature dependence of the diffusion coefficients of several impur-ities in silicon,[85] showing that the activation energies of impurity diffusion in a silicon crystal strongly depends on the identity of the diffusing species, and that an exponential temperature dependence is systematically observed in a full range of temperatures.

These features might be explained by assuming the occurrence of a thermally activated mass-transfer process mediated by equilibrium point defects, vacancies and/or self-interstitials, supposed to be in-trinsic, equilibrium constituents of the material,[86] and whose role is fundamental in the engineering of electronic materials,[87] because it

---

[††††] In a compound semiconductor A$_x$B$_y$ the self-diffusion coefficient of both A and B species can be measured.

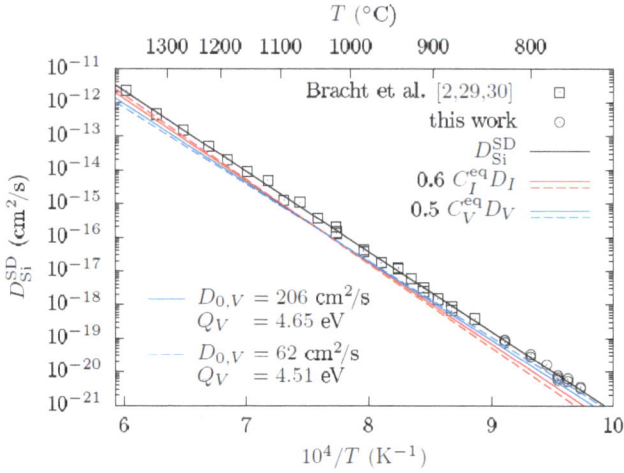

**Figure 1.20** Arrhenius plot of the Si self-diffusion in undoped silicon. Reproduced from ref. 83 with permission from American Physical Society, Copyright 2016.

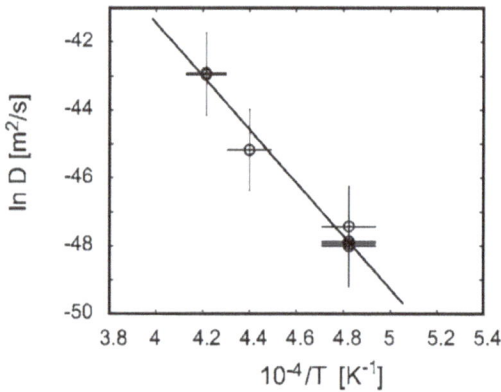

**Figure 1.21** Arrhenius plot of the carbon self-diffusion coefficient in diamond at 10 GPa. Reproduced from ref. 84 with permission from American Physical Society, Copyright 2005.

provides a method, as an example, of impurity doping semiconductor materials.

The physical arguments beyond this model are essentially based on the assumption, confirmed by theoretical calculations,[88–96] that the transfer of an atomic species X by a direct exchange (DE) process of neighbour atoms in a perfect solid lattice network is energetically less favourable than that occurring *via* a jump to a nearby available empty position, which we call a vacancy, or *via* a jump between two interstitial positions.

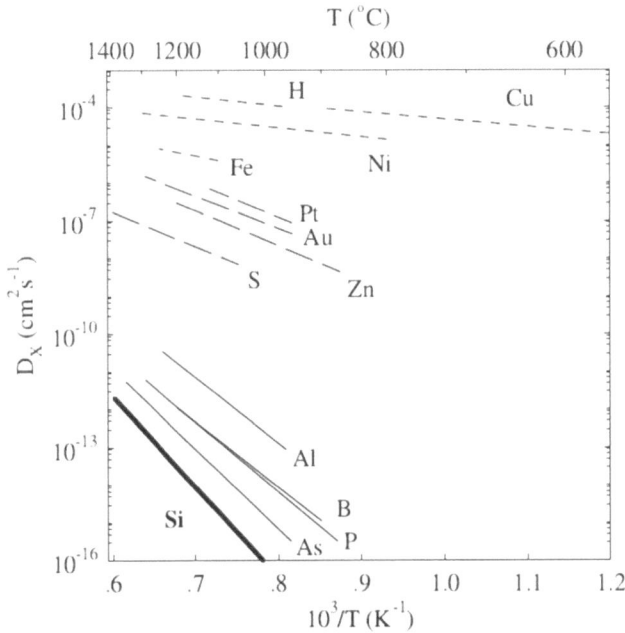

**Figure 1.22** Temperature dependence of diffusion coefficients of foreign atoms in silicon. Reproduced from ref. 85 with permission from Taylor & Francis, Copyright 2015.

It is further assumed:

- That vacancies and self-interstitials are equilibrium defects populating the semiconductor lattice.
- That vacancies originate from a thermally induced shifts of an atom $X_X$ sitting in a lattice position to a surface site leaving an empty lattice position, with a Schottky type of process

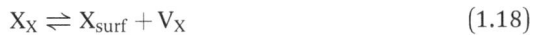

$$X_X \rightleftharpoons X_{surf} + V_X \qquad (1.18)$$

where any physical surface or interface of a crystalline phase is supposed to behave as an infinite source or sink for defects.

- That interstitials originate from a thermally induced shifts of an atom $X_X$ sitting in a lattice position on an (empty) interstitial site leaving an empty lattice position, with a Frenkel type of process

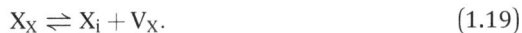

$$X_X \rightleftharpoons X_i + V_X. \qquad (1.19)$$

- That these defects behave as equilibrium properties of the system, *i.e.* that their concentration $x_D^{eq}$ depends on the temperature with a law of the type

$$x_D^{eq} = \exp - \frac{\Delta G_D^f}{kT} = A \exp - \frac{\Delta H_D^f}{kT} \qquad (1.20)$$

where the term $\Delta G_D^f$ represents the Gibbs formation energy of the defect, the constant $A$ is a formation entropy term, and the term $\Delta H_D^f$ represents the enthalpy of formation of the defect.

- That a perfect solid is thermodynamically less stable than a defective solid, since defects (and impurities) add a configurational entropy term of the type $\Delta S_{conf} = \sum_j x_j \ln x_j$ to the free energy $\Delta G°$ of the perfect solid.

If all these assumptions are satisfied, we could expect that the self-diffusion $D_i^{SD}$ of a species i mediated by vacancies and by self-interstitials would be given by the following relationship

$$D_i^{SD} = f_V x_V^{eq} D_V + f_I x_I^{eq} D_I \qquad (1.21)$$

where $f_V$ and $f_V$ are correlation coefficients that account for the individual weight of each defect on the diffusion process.

In turn, each individual defect contribution to the diffusion process should be given by a relationship of the type

$$f_D x_D^{eq} D_D = f_D \exp \frac{\Delta S_D^m + \Delta S_D^f}{k} \exp - \frac{\Delta H_D^m + \Delta H_D^f}{kT} = A \exp - \frac{\Delta H_D^m + \Delta H_D^f}{kT} \qquad (1.22)$$

where the terms $\Delta S_D^m$ and $\Delta S_D^f$ are diffusion and defect formation entropy terms, $\Delta H_D^m$ is a diffusion enthalpy term and $\Delta H_D^f$ is the enthalpy of formation of the defect $D$, which foresees an Arrhenius-type dependence of the self-diffusion on temperature, in good agreement, see again Figures 1.20 and 1.21, with the experimental temperature dependence of the self-diffusion coefficient of Si in silicon[83] and of carbon natural diamond,[84] and Figure 1.22 for the diffusion coefficients of impurities in silicon.[85]

However, since the measured activation energy of a diffusion process mediated by a single defect is the sum of a diffusion enthalpy term and of a defect formation enthalpy, and the situation is more intriguing if both vacancies and interstitials are involved in the diffusion process, as is the case of silicon,[83] germanium, and of most compound semiconductors, self-diffusion measurements alone cannot deliver information on the formation energy and concentration of defects. Therefore, an independent evaluation of the diffusion enthalpy and/or a careful theoretical modelling of the process is necessary[83] for the quantitative determination of the defect formation energy and of the defect concentration as a function of the temperature.

As a result of the studies carried out in the last six decades, a vast amount of experimental evidence and of theoretical calculations, carried out with *ab initio* methods, is today available, which allows us to confirm that intrinsic point defects do behave, in fact, as stable, quasi-chemical equilibrium species in elemental and compound semiconductors, with almost consolidated values of their formation energies and then of their temperature-dependent concentrations.

Since point defects in semiconductors may behave as charged (donor or acceptor) species, and compound semiconductors phases may deviate from the stoichiometry, as we will see in Section 1.5, their formation enthalpy $\Delta H_{D,q}$ does depend both on their charge $z = (0, \pm 1, \pm 2, \ldots)$, and thus on the Fermi energy $E_F$, which is the energy of the electron reservoir with which electrons are exchanged, and on the composition of the semiconductor, and, therefore, on the chemical potential $\mu_i$ of the lattice species involved in the defect formation[97]

$$\Delta H_{D,q}(E_F, \mu, z) = \Delta H_D^\circ + \sum_i n_i \mu_i + z q E_F \qquad (1.23)$$

where $\Delta H_D^\circ$ is the formation enthalpy of the neutral defects and the term $\sum_i n_i \mu_i$ represents the energy gain associated with the stoichiometry of the phase in the case of compound semiconductors. It is a common practice to use DFT and *ab initio* techniques for the calculation of the formation energy of the defects as a function of the Fermi energy and of the composition of the phase, although only in few cases is the full dependence of $\Delta H_{D,q}$ on the phase composition available, since, generally, the formation energies of defects in non-stoichiometric compound semiconductors are calculated for metal-rich or metal poor conditions.

As an example, the calculated formation energies (or the formation enthalpies[‡‡‡‡]) of vacancies, self-interstitials and split-vacancies and interstitials[§§§§] in silicon[95] and of common defects in GaAs[98] at 0 K are displayed in Figures 1.23 and 1.24, respectively.

From the calculated values of formation enthalpies of defects at 0 K, their Gibbs formation energy at ambient temperature, or at every process temperature, can be evaluated using the standard equation

$$\Delta G_f = \Delta H_f - T \sum \Delta S \qquad (1.24)$$

where the enthalpy term is considered independent of temperature and the entropy term $\sum \Delta S$ includes the thermal and the configurational contributions to the entropy; these last are due to the contribution of a statistical distribution of defects in a semiconductor matrix.

It is apparent from Figure 1.23 that, in the 0 K limit, neutral and charged interstitials with different configurations are the defect

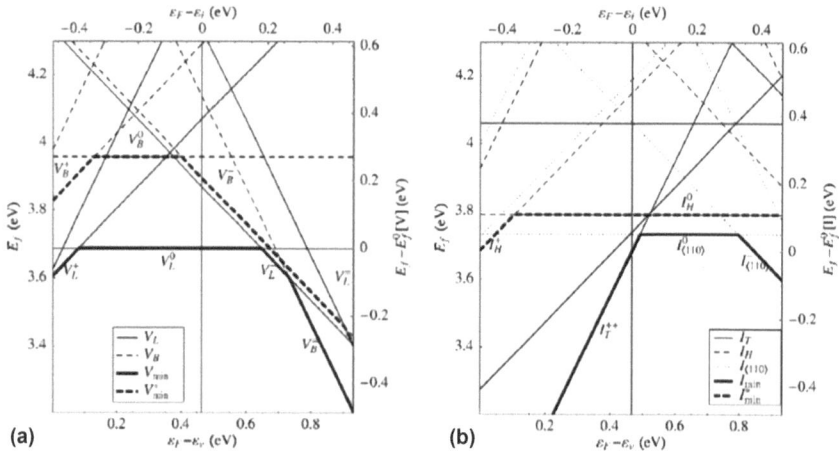

**Figure 1.23**   Formation energies of neutral and charged vacancies (a) and interstitials (b) in silicon [$V_L^0$ neutral lattice vacancy, $V_B^0$ neutral split vacancy, $I_H$ hexagonal, $I_T$ tetragonal and split $I_{100}$ interstitials]. Reproduced from ref. 95 with permission from American Physical Society, Copyright 2005.

species present in silicon,[95] with formation energies ranging between of 3.4 and 4.2 eV, depending on the Fermi level.

The case of GaAs,[95] see Figure 1.24, is complicated by the presence, in addition to Ga vacancies $V_{Ga}$ and As vacancies $V_{As}$, of As and Ga antisites [$As_{Ga}$ (conventionally called EL2), $Ga_{As}$], consisting in the localization of Ga in a lattice position of As, or of As in the lattice position of Ga, as the dominant defects in As-rich and Ga-rich conditions, with the lowest formation energies both in Ga-rich and As-rich conditions. We will discuss this case in full in Section 1.5, dedicated to the role of defects on the compensation of non-stoichiometry in compound semiconductors.

As expected, defects might also, additionally, arise as the result of the interaction of a solid phase with a gaseous atmosphere, as happens with oxides in equilibrium with an oxygen atmosphere.[99] In this case the equilibrium conditions might be satisfied by the formation of oxygen interstitials or oxygen vacancies, generated by the following equilibrium reactions

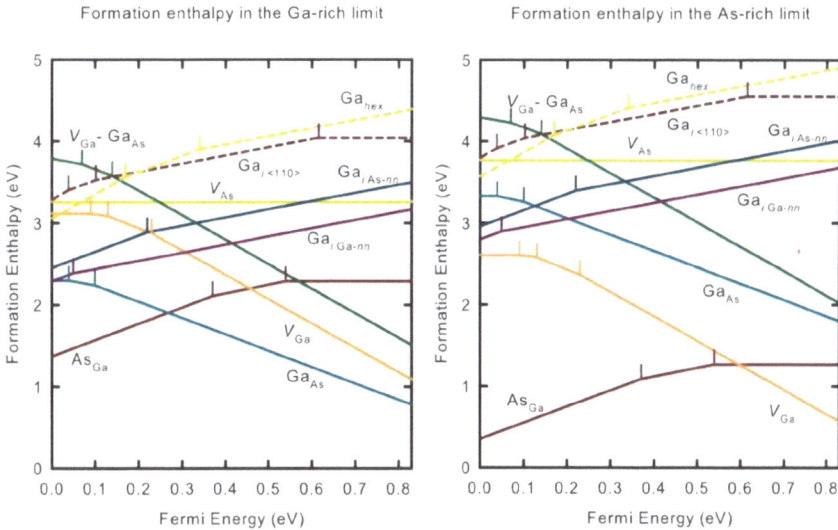

$$\frac{1}{2}O_2 \rightleftharpoons O_i \tag{1.25}$$

$$O_O \rightleftharpoons V_O + \frac{1}{2}O_2 \tag{1.26}$$

Figure 1.24    Enthalpy of formation of neutral defects in GaAs as a function of the Fermi Level. $Ga_{hex}$ is a Ga interstitial in a hexagonal position, $Ga_{i/100}$ is a Ga–Ga split interstitial and $Ga_{iGa-nn}$ is a Ga interstitial in the Ga lattice. Reproduced from ref. 98 with permission from AIP Publishing, Copyright 2011.

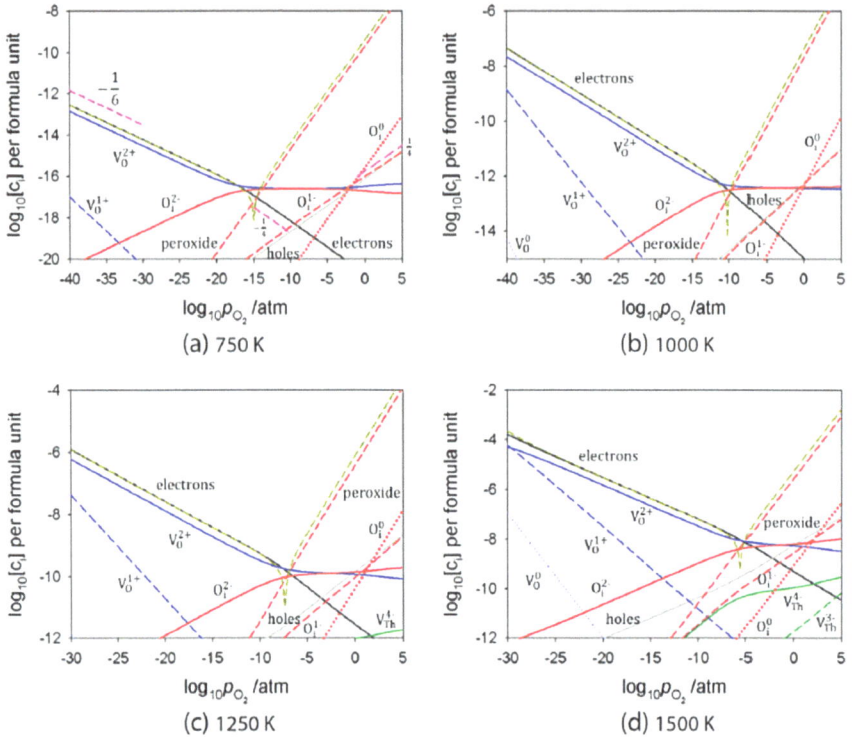

**Figure 1.25**   DFT calculated concentration of charged defects and electrons in non-stoichiometric thoria, as a function of partial pressure of oxygen (a) 750 K (b) 1000 K (c) 1250 K (d) 1500 K. Reproduced from ref. 99 with permission from Elsevier, Copyright 2014.

whose concentration should be influenced, at constant temperature, by the partial pressure of oxygen. Figure 1.25[99] illustrates the effect of partial pressure of oxygen at four different temperatures on the calculated concentration of defects in thorium oxide.[¶¶¶¶] Here, in addition to the effect of the partial pressure of oxygen on the concentration of oxygen vacancies and oxygen interstitials, which are the predominant defects, its effect on the concentration of electrons and holes is also evident, which arises from the ionization of oxygen defects.

A list of calculated formation energies of neutral vacancies in undoped homopolar and compound semiconductors is reported in Table 1.3, but more details on their dependence on doping and on non-stoichiometric deviations are reported in Section 1.5.

---

¶¶¶¶Thorium oxide is a ionic oxide with the fluorite structure, taken here as an example of the effect of the interaction with an external atmosphere on the formation of point defects.

**Table 1.3** Formation energies of neutral vacancies in elemental semiconductors, calculated with *ab initio* methods[95,101,102] and in III–V and II–VI semiconductors calculated with the extended Hückel method.[100,103–105] Data from ref. 95 and 100–105.

| | $E_{V_A}$ (eV) | $E_{V_B}$ (eV) | Ref. |
|---|---|---|---|
| Si | 3.69 | | 95 |
| Ge | 2.33 | | 101 |
| Diamond | 7.31 | | 102 |
| GaSb | 5.07 | 4.11 | 100 |
| InSb | 4.72 | 4.15 | 100 |
| GaAs | 5.29 | 4.59 | 100 |
| GaN | 8.40 | 3.16 | 103 |
| GaP | 5.73 | 5.23 | 100 |
| AlN | 10.20 | 4.60 | 103 |
| InP | 5.89 | 5.70 | 100 |
| InAs | 5.14 | 4.85 | 100 |
| ZnTe | 4.08 | 2.81 | 100 |
| ZnO | 3.54 | 0.73 | 104 and 105 |
| CdTe | 3.65 | 2.60 | 100 |
| ZnS | 5.65 | 4.33 | 100 |
| CdS | 5.22 | 4.0 | 100 |
| PbTe | 4.20 | 4.13 | 100 |
| PbS | 5.80 | 5.41 | 100 |

## 1.4 Ionicity of Compound Semiconductors and Its Effect on Their Thermodynamic, Spectroscopic, and Structural Properties

Different from elemental semiconductors, the chemical bonds in compound semiconductors are partially ionic,[106–111] and could be parametrized with an ionicity $f_i$ value, defined as the ratio

$$f_i = \frac{\left(\Delta E_{sp^3}\right)^2}{E_G^2} \tag{1.27}$$

where $\Delta E_{sp^3}$ is the offset of the anions and cations hybrid energy levels due to the ionicity of the bonds and $E_G^*$ is the energy separation between sp$^3$ hybrid bonds and antibonding states, which in turn is the sum of two terms

$$E_G^* = \left[\left(\Delta E_{sp^3}\right)^2 + \left(E_h\right)^2\right]^{1/2} \tag{1.28}$$

of which the second term $E_h$ is a bond hybridization contribution.‖‖‖

---

‖‖‖ Full details of this issue are reported in the Christensen work.[112]

**Table 1.4**  Calculated ionicities of II–VI and III–V semiconductors.
Data from ref. 112.

| Compound | $d$ (Å) | $\Delta E_{sp^3}$ (eV) | $E_h$ (eV) | $f_i$ |
|---|---|---|---|---|
| CdSe | 2.63 | 8.35 | 3.62 | 0.841 |
| BeO | 1.65 | 13.9 | 9.21 | 0.798 |
| CdS | 2.53 | 5.90 | 4.67 | 0.794 |
| CdTe | 2.81 | 6.35 | 3.77 | 0.739 |
| InAs | 2.62 | 5.06 | 0.357 | 0.553 |
| AlP | 2.36 | 5.16 | 6.04 | 0.421 |
| GaP | 2.36 | 5.00 | 6.66 | 0.361 |
| GaAs | 2.45 | 4.29 | 6.40 | 0.310 |
| SiC | 1.88 | 7.47 | 9.27 | 0.177 |
| C | 1.54 | 0.00 | 13.31 | 0.000 |
| Si | 2.35 | 0.00 | 6.82 | 0.000 |
| Ge | 2.45 | 0.00 | 6.38 | 0.000 |

While in elemental semiconductors the $\Delta E_{sp^3}$ term vanishes, and the ionicity turns to zero, in compound semiconductors both terms contribute to $E_G^*$ and go to the ionicity values calculated by Christensen *et al.*,[112] displayed in Table 1.4 for some common compound semiconductors.

It is apparent from Table 1.5, which displays the actual values of bond lengths, ionicities, melting temperatures and energy gaps of II–VI and III–V semiconductors, that ionicity provides, as already noted by Catlow and Stoneham,[113] an excellent parametrization criterium for structural, thermodynamic and spectroscopic properties of all the families of II–VI and III–V semiconductors, including the oxides, despite some uncertainty of their ionicity values.

One can, in fact, see that with a decrease in the ionicity, both the melting temperatures and the energy gaps decrease, as do the energy gaps.

Furthermore, as Table 1.6 shows, the formation energies of vacancies in both sublattices of compound semiconductors are also closely related to the ionicity, decreasing with the decrease in the ionicity factor, with a marked deviation in the case of lead compounds.

It can be eventually demonstrated, that the thermodynamic stability of compound semiconductors can also be quantitatively discussed using the ionicity as a criterium, and that the different stabilities of crystal structures with coordination number (NC) = 4

**Table 1.5** Bond lengths ($d$), ionicities ($f_i$), melting temperatures ($T_m$) and energy gaps ($E_G$) of II–VI and III–V semiconductors. Data from ref. 95, 110, 112 and 114–116.

| Compound | $d$ (Å) | $f_i$ | $T_m$ (°C) | $E_G$ (eV) | Ref. |
|---|---|---|---|---|---|
| AlP | 2.36 | 0.421 | 2530 | 2.45 | 112 |
| AlAs | 2.43 | 0.274 | 1740 | 2.14 | 112 |
| AlSb | 2.66 | 0.250 | 1060 | 1.6 | 112 |
| GaN | 1.94 | 0.5 | 2500 | 3.4 | 114 and 115 |
| GaP | 2.36 | 0.361 | 1457 | 2.26 | 112 |
| GaAs | 2.45 | 0.310 | 1240 | 1.43 | 112 |
| BeO | 1.65 | 0.798 | 2530 | 10.6 | 112 |
| MgO | 1.819 | 0.841–0.911 | 2852 | 7.3–7.8 | 112 |
| ZnO | 1.98–2.0 | 0.616 | 1975 | 3.3 | 116 |
| CdO | 2.35 | 0.785 | 1427 | 2.31 | 116 |
| SnO | 2.231 | n.d. | 1080 | 0.7 | 112 |
| CdS | 2.53 | 0.794 | 1750 | 2.42 | 112 |
| CdSe | 2.63 | 0.699 | 1240 | 1.70 | 112 |
| CdTe | 2.81 | 0.60 | 1092 | 1.51 | 110 |
| ZnS | 2.34 | 0.764 | 1830 | 3.54 | 112 |
| ZnSe | 2.45 | 0.630 | 1525 | 2.70 | 112 |
| ZnTe | 2.64 | 0.609 | 1238 | 2.24 | 112 |
| InP | 2.54 | 0.534 | 1062 | 1.29 | 112 |
| InAs | 2.62 | 0.357 | 942 | 0.354 | 112 |
| InSb | 2.81 | 0.303 | 527 | 0.17 | 112 |

**Table 1.6** Ionicity-dependent formation energies of vacancies in III–V and II–VI semiconductors.

| | $E_{V_A}$ (eV) | $E_{V_B}$ (eV) | $f_i$ | Ref. |
|---|---|---|---|---|
| GaP | 5.73 | 5.23 | 0.361 | 112 |
| GaAs | 5.29 | 4.59 | 0.310 | 112 |
| GaSb | 5.07 | 4.11 | 0.108 | 112 |
| InP | 5.89 | 5.70 | 0.534 | 112 |
| InAs | 5.14 | 4.85 | 0.553 | 112 |
| InSb | 4.72 | 4.15 | 0.303 | 112 |
| PbS | 1.78 | 2.08 | 0.39 | 112 and 117 |
| PbSe | 1.83 | 2.15 | 0.23 | 112 |
| PbTe | 1.90 | 2.14 | n.d. | 112 |
| ZnS | 4.96 | 5.71 | 0.685 | 112 |
| ZnSe | 4.48 | 3.23 | 0.630 | 112 and 118 |
| ZnTe | 4.08 | 2.81 | 0.560 | 112 and 118 |
| CdS | 4.28 | 5 | 0.794 | 112 |
| CdSe | 3.66 | 3.75 | 0.699 | 112 and 119 |
| CdTe | 3.65 | 2.60 | 0.60 | 100 and 112 |

(zinc blende and wurtzite) and those with NC = 6 (rock salt) originate from the competition between covalent sp$^3$ bonding and ionic bonding, which increases with the increase in the ionicity.

On that basis, Yeh *et al.*[120] were able to show that a clear correlation occurs between Pauling's electronegativity (PE),***** the Christensen ionicity, and the thermodynamical stability of the structures of II–VI and III–V semiconductor compounds, which crystallize either with the wurtzite or the zincblende, and the rock-salt structure, and that these predictions are theoretically supported.

It has been found,[120] as an example, that the calculated energy difference $\Delta E_{W-ZB}$ between the wurtzite and zincblende structures of compound semiconductors is linearly correlated with the difference in Pauling's electronegativity $\Delta_{PE}$,[115] see Figure 1.26a and with the difference in their tetrahedral radii, see Figure 1.26b.

It has also been found that above a critical ionicity value $f_i = 0.786$ ($\Delta_{PE} = 1.4$) the thermodynamically stable structure is the rock-salt one, with a coordination number NC = 6. Below this critical value compound semiconductors crystallize with tetrahedrally co-ordinated bonds (NC = 4), but take the cubic zincblende structure for

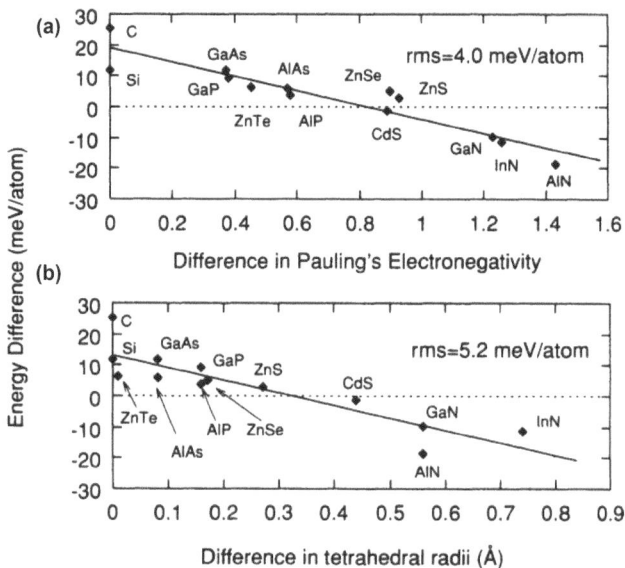

**Figure 1.26**   Calculated total energy differences $\Delta E_{W-ZB}$ (meV per atom) between the zincblende (diamond) and wurtzite structures of some common semiconductors, displayed as a function of the Pauling electronegativity (a) and of the difference of the tetra-hedral radii of A and B atoms (b). Reproduced from ref. 120 with permission from American Physical Society, Copyright 1992.

-----

***** For the relationships between Pauling's electronegativity and ionicity see Hidaka,[110] and Catlow and Stoneham.[113]

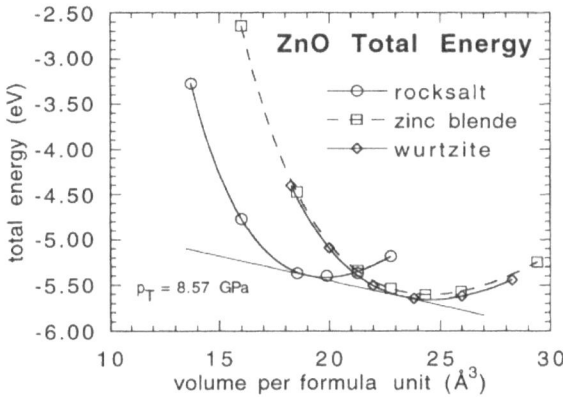

**Figure 1.27** Calculated energy of the wurtzite, zinc blende, and rock salt structures of ZnO. Reproduced from ref. 122 with permission from American Physical Society, Copyright 1993.

$0.8 \leq \Delta_{PE} > 1.4$ and the hexagonal wurtzite structure above an ionicity value $f_i = 0.786$ $(\Delta_{PE} \geq 0.8)$.

A key case is represented by the semiconductor oxides BeO, ZnO and CdO, with their very high ionicity factors (see Table 1.5). BeO and CdO, in fact, crystallize with a rock-salt structure, while ZnO crystallizes with a wurtzite structure, is metastable with the zincblende structure and crystallizes with the rock salt structure under applied kbar pressures.[121]

The calculated stability of the different ZnO phases under an applied pressure of 8.57 GPa[117] is reported in Figure 1.27, which shows that the three structures compete in an energy range of 200 meV.[122]

# 1.5 Deviation from the Stoichiometry of Compound Semiconductors

For most compound semiconductors, the equilibrium phase is not a line-phase of stoichiometric composition, but a homogeneous, non-stoichiometric phase, stable in a finite range of compositions and temperatures, whose width might also depend on the pressure applied to balance their decomposition.[†††††]

It is also known, see Table 1.7, that the maximum width of the homogeneity region of compound semiconductors qualitatively

---

[†††††] We saw in Section 1.1 that several compound semiconductors decompose before melting, but that their decomposition might be suppressed by the application of a pressure that allows thermodynamic stabilization of the phase.

**Table 1.7** Homogeneity range $\Delta\delta_{max}$ of several non-stoichiometric compound semiconductors. Data from ref. 123.

| Material | GaAs | InP | CdS | CdSe | CdTe | PbTe |
|---|---|---|---|---|---|---|
| $\Delta\delta_{max}$ | $\sim 4 \times 10^{-5}$ | $\sim 5 \times 10^{-5}$ | $\sim 1 \times 10^{-4}$ | $\sim 5 \times 10^{-4}$ | $3 \times 10^{-4}$ | $\sim 1 \times 10^{-3}$ |
| Region of max excess | $\sim$As-rich | In-rich | Cd-rich | Cd-rich | Te-rich | $\sim$Te-rich |
| Ionicity | 0.310 | 0.421 | 0.60 | 0.73 | 0.84 | $>0.8$ |

depends on the ionicity,[123] and systematically increases with increases in the ionicity. It is, as well, known that deviations from the stoichiometric composition affect not only the type of conductivity, the current carrier concentration and the efficiency of dopant incorporation, but also the generation of dislocations and their mobility,[‡‡‡‡‡] with remarkable degradation of the material properties and of the devices prepared with these materials.[41,124]

It was eventually agreed that stoichiometry offsets are accommodated by point defects, their type and concentration depending on their formation energies and on the chemical excess of a component in the homogeneous phase. The stoichiometric offsets, therefore, could be discussed with a comprehensive equilibrium thermodynamics approach.

Deviations from the stoichiometry in compound semiconductors have been the subject of dedicated experimental and theoretical studies in the last decades, addressed at a critical understanding of their features, which should provide also means for their technological management.

To discuss the non-stoichiometry issues of compound semiconductors we can assume that the physico-chemical properties of a homogeneous, non-stoichiometric $A_{1+x}B$ or $AB_{1+y}$ binary phase, characterized by the presence of an excess or deficiency of A (metallic) or B (non-metallic) species in the lattice as a consequence of the interaction with the ambient during their synthesis or across subsequent thermal treatments, can be formally accounted for by assuming the occurrence of the following set of reactions, which involve the A and B species in the corresponding sublattices

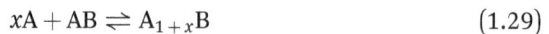

$$xA + AB \rightleftharpoons A_{1+x}B \tag{1.29}$$

---

[‡‡‡‡‡] This is as well as the incorporation of the excess component under form of inclusions and the nucleation of second phase precipitates during the cooling process of the as-grown crystals.

$$AB \rightleftharpoons A_{1-x}B + xA \qquad (1.30)$$

$$xB + AB \rightleftharpoons AB_{1+x} \qquad (1.31)$$

$$AB \rightleftharpoons AB_{1-x} + xB \qquad (1.32)$$

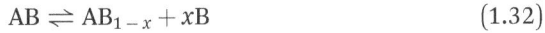

with a gain or loss of matter.

In most practical situations, a single chemical species (the non-metallic B one) is exchanged during the synthetic process or after further thermal annealing, as in the case of GaAs, where only $As_4$ vapours are exchanged, see Section 1.5.1, allowing consideration of only reactions (1.31) and (1.32).

As we will see in the case of ZnO in Section 1.5.3, both Zn and O might be instead exchanged with the occurrence of the full set of reactions (1.29) and (1.31).

Since the formation of any mass-defect/excess of A or B in the homogeneous AB phase is compensated by the formation of point defects, vacancies, interstitials and antisites in the two sublattices, we could write the following set of equations to account for all the possible thermodynamic equilibria involving neutral defects, on the assumption that the formation of B-interstitials $B_i$ is the thermo-dynamically favourite process

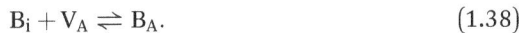

$$B(g) \rightleftharpoons B_i \qquad (1.33)$$

$$A_A \rightleftharpoons A_i + V_A \qquad (1.34)$$

$$B_B \rightleftharpoons B_i + V_B \qquad (1.35)$$

$$nil \rightleftharpoons V_A + V_B \qquad (1.36)$$

$$A_i + V_B \rightleftharpoons A_B \qquad (1.37)$$

$$B_i + V_A \rightleftharpoons B_A. \qquad (1.38)$$

Here, eqn (1.33) involves the equilibrium of interstitial B species with an external gaseous phase consisting of B molecules, eqn (1.34) and (1.35) represent the process of formation of a Frenkel pair, eqn (1.36) represents the formation of a Schottky pair, and eqn (1.37) and (1.38) the formation of antisites.

Since the activity of B in the external phase could be expressed in terms of the partial pressure $p_B$ of the B species, when A is a pure solid

$(a_A = 1)$, we have for the activity (and concentration, since we deal with dilute solutions of defects) of interstitials of B

$$a_{B_i} \approx x_{B_i} = \exp - \frac{\Delta G_f^{B_i}}{RT} p_B^{-1} \tag{1.39}$$

where $\Delta G_f^{B_i}$ is the Gibbs energy of formation of interstitials of B.

If we substitute eqn (1.35) into eqn (1.33), a relationship is obtained for the equilibrium constant relative to the process of formation of vacancies of B

$$K_{V_B} = p_B V_B \tag{1.40}$$

and eqn (1.40) into eqn (1.41)

$$V_A V_B = K_s \tag{1.41}$$

we eventually obtain the corresponding relationship for the formation of vacancies of A $(V_A)$

$$V_A = \frac{K_s}{K_{V_B}} p_B \tag{1.42}$$

where $K_{V_B} = \exp - \dfrac{\Delta G_f^{V_B}}{RT}$ and $\Delta G_f^{V_B}$ is the Gibbs energy of formation of vacancies of B, $K_s = \exp - \dfrac{\Delta G_f^S}{RT}$ and $\Delta G_f^S$ is the Gibbs energy for the formation of a Schottky pair of A and B vacancies.

Similar conclusions can be obtained on the assumption that the equilibrium

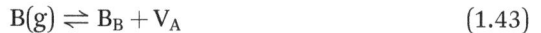

$$B(g) \rightleftharpoons B_B + V_A \tag{1.43}$$

will predominantly occur with the formation of vacancies of A, favoured with respect to the formation of B interstitials $(B_i)$, as well as for the case that both the species A and B are exchanged with the gas phase.

The defect equilibria considered above account for defects in their neutral configuration and for a formally stoichiometric phase, while in practical cases one should rely on non-stoichiometric and doped phases, and with the formation of charged defects, as we saw in Section 1.3, eqn (1.23) and Figures 1.23–1.25.

To gain better insight into the question, we will consider three semiconductor materials of large technological interest, GaAs, CdTe and ZnO, all of them being a very convenient case study since they are thermodynamically stable in a wide range of temperatures and applied pressures as homogeneous, non-stoichiometric phases, and their defect equilibria have been studied in great detail, also in the presence of dopants.

## 1.5.1 Defects in Non-stoichiometric GaAs

Before Nakamura's development of nitride-based blue, green and white light emitting diodes in 1991,[125,126] GaAs was already the material of choice for red LEDs and laser applications, and still represents an important fraction of the compound semiconductors value chain, being also used as large diameter, dislocation-free substrates for homo- or heteroepitaxial applications, and for its hybrid or monolithic integration on silicon chips for 4G phones.

For this reason, GaAs has been the subject of systematic and comprehensive studies that succeeded in delivering a detailed knowledge of its thermochemical properties.

The homogeneity range of GaAs is shown to depend on temperature,[127,128] see Figure 1.28(a) and on the As pressure applied to balance the decomposition pressure,[129] as seen in Figure 1.28(b).

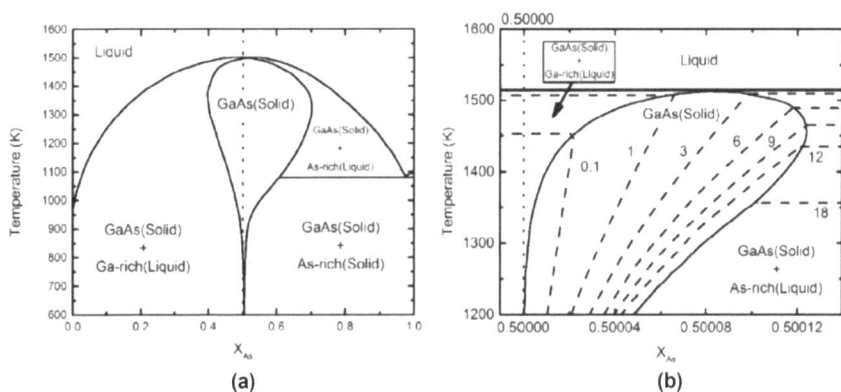

**Figure 1.28**    (a) Phase diagram of the Ga–As system. The real extension of the non-stoichiometric region is displayed on top, in terms of molar fraction of As. (b) Section of the phase diagram of the GaAs system in the As-rich region. The dashed lines display the extent of deviation from the stoichiometry as function of the temperature and at constant values of the total pressure of the equilibrium As species (As, $As_2$ and $As_4$) (in bar). Reproduced from ref. 129 with permission from Elsevier, Copyright 2016.

It presents a maximum deviation from the stoichiometry at 1380 K, with a Ga-rich extreme at 50.002% Ga, and an As-rich extreme at 50.012% As.

It is apparent from Figure 1.28(b) that a GaAs phase with an almost stoichiometric composition is stable only at temperatures lower than 1100 °C, that a shift towards As-rich conditions occurs by increasing the total As pressure from 0.1 bar to 18 bar, and that the congruent melting temperature occurs in correspondence with the isobar at 3 bar with an As excess of $x_{As} \approx 8 \times 10^{-4}$.

The thermodynamic stability of the GaAs phase[§§§§§] as a function of the composition could be discussed starting with the assumption, taken by most authors up to today,[98,130–135] that in the hypothetical stoichiometric GaAs phase, as well as in the non-stoichiometric phases, the defects present are Ga vacancies ($V_{Ga}$), As vacancies ($V_{As}$), Ga and As interstitials ($Ga_i$ and $As_i$), Ga and As antisites ($Ga_{As}$ and $As_{Ga}$), and $[Ga_{As}-V_{Ga}]$ complexes, as already shown in Section 1.3.

All these defects could be present as neutral and charged species, and therefore their formation energies and their concentration should depend on the temperature, on the doping level, *i.e.* on the Fermi level, and on the stoichiometry offset, this last expressed in terms of the chemical potential $\mu_{Ga}$ and $\mu_{As}$ of Ga and As in the GaAs non-stoichiometric phase, as shown by eqn (1.23). If we neglect the contribution of non-stoichiometry, the formation enthalpy $\Delta H_f$ of a defect takes the form[137]

$$\Delta H_f(E_F, z) = \Delta H_f^\circ + zqE_F + \sum E_{D,i} \tag{1.44}$$

where $\Delta H_f^\circ$ is the formation enthalpy of the neutral defect, $\pm zq$ is the charge of the defect and $E_{D,i}$ is the ionization energy of the $i$th level of the defect.

Using density functional theory (DFT) within the local density approximation, Schick *et al.*[98] calculated the formation energies of these defects in the Ga-rich and As-rich regions of GaAs as a function of the Fermi energy, and their results are displayed in Figure 1.24.

Apparently, in the 0 K limit, As vacancies, $V_{As}$, behave as neutral defects, since their formation energy is independent of the Fermi energy and are also the defects presenting almost the largest values of their formation energy in both Ga-rich and As-rich GaAs.

Given that the energy gap of GaAs is 1.519 at 0 K (and 1.522 eV at 300 K),[138,139] one can see that for undoped and p-type GaAs, in the

---

[§§§§§] For the case of nitrides see ref. 136.

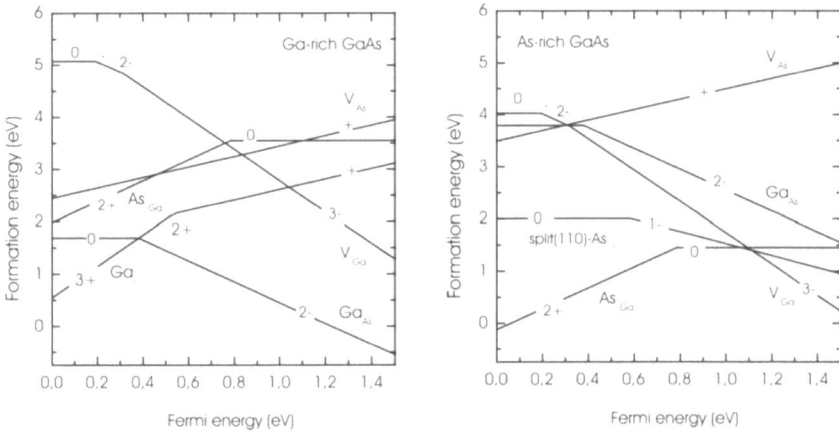

**Figure 1.29** Calculated dependence of the enthalpy of formation of charged defects in Ga-rich and As-rich GaAs. Reproduced from ref. 140 with permission from Vladimir Bondarenko.

Ga-rich limit, Ga antisites $Ga_{As}$ and Ga vacancies $V_{Ga}$ are the energetically favoured defects.

On the As-rich limit, the formation of Ga vacancies $V_{Ga}$ and $As_{Ga}$ antisites is instead energetically favoured.

Complementary results were reported by Bodnarenko,[140] see Figure 1.29, which displays the formation energies of defects in their charge states in the full range of Fermi energy levels, in Ga-rich and As-rich configurations. It appears that charged Ga interstitials $Ga_i^{3+}$ and Ga antisites $Ga_{As}^{2-}$ exhibit the lowest formation energies in Ga-rich GaAs, while charged As antisites $As_{Ga}^{2+}$ and Ga vacancies $V_{Ga}^{3-}$ present the lowest formation energies in As-rich GaAs. The only significant difference between Schick's and Bondarenko's results is the predominance of Ga antisites $Ga_{As}^{2-}$ vs. Ga interstitials $Ga_i^{3+}$ in the Ga-rich limit, understandable if the formation energy of Ga antisites from Ga interstitial, see eqn (1.37), is very low.

These results are compatible with those of Schultz and Lilienfeld[132] displayed in Table 1.8, which shows that the calculated values of the energy of formation of $Ga_{As}$ and $As_{Ga}$ in all charged states are lower than those of vacancies and interstitials, suggesting their predominant presence.

They show, also, that neutral and charged Ga vacancies in n-type GaAs complement $Ga_{As}$ and $As_{Ga}$ as predominant defects, while $V_{Ga}$ are unstable in p-type conditions, in good agreement, again, with the results of Figure 1.28.

Eventually, Schick *et al.*[98] were also able to calculate the concentration of defects at 1100 K, and in the full range of deviation from the

**Table 1.8**  Calculated formation energies (in eV) of common defects in the As-rich limit, using DFT under the local density approximation (LDA) or the Perdew–Burke–Ernzerhof (PBE) formulation. Data from ref. 132.

| Charge state | $As_{Ga}$ | $Ga_{As}$ | $V_{Ga}$ | $V_{As}$ | $Ga_i$ | $As_i$ |
|---|---|---|---|---|---|---|
| **LDA-3d** | | | | | | |
| n-type | 1.48 | 1.10 | −0.11 | 1.57 | 3.86 | 2.88 |
| Neutral | 1.48 | 2.80 | 2.50 | 3.41 | (4.10) | 3.51 |
| p-type | 0.24 | 2.03 | 1.90 | 2.54 | 2.14 | 2.44 |
| | | | | | | |
| **LDA** | | | | | | |
| n-type | 1.50 | 1.45 | 0.01 | 1.88 | 4.07 | 2.88 |
| Neutral | 1.50 | 3.19 | 2.69 | 3.55 | (4.53) | 3.58 |
| p-type | 0.29 | 2.53 | 2.02 | 2.57 | 2.23 | 2.54 |
| | | | | | | |
| **PBE** | | | | | | |
| n-type | 1.27 | 1.50 | −0.02 | 1.80 | 3.90 | 3.85 |
| Neutral | 1.27 | 3.20 | 2.65 | 3.44 | (4.20) | 3.46 |
| p-type | 0.04 | 2.47 | 1.31 | 2.38 | 2.07 | 2.24 |

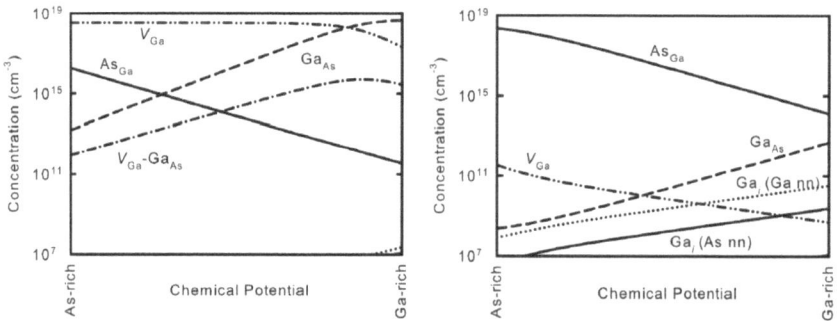

**Figure 1.30**  Calculated concentration of defects in GaAs at 1100 K for a n-doped sample with a concentration of donors $N = 10^{19}$ (left) and for a p-doped sample with a concentration of acceptors $N = 10^{19}$ (right). Reproduced from ref. 98 with permission from AIP Publishing, Copyright 2011.

stoichiometry for p-type and n-type GaAs, see Figure 1.30, showing that Ga vacancies $V_{Ga}^{3-}$ present the largest concentration in n-type GaAs, while As antisites $As_{Ga}^{2+}$ are the prevalent defects in p-type GaAs.

It is possible, therefore, to conclude that within the error bars associated with calculations, non-stoichiometry of GaAs is dominated by $Ga_{As}$ and $As_{Ga}$ antisites, with vacancies playing a complementary role.

Among the few experimental results available in the literature concerned with the actual concentration of defects in GaAs,[137,140–146] the concentration of Ga vacancies was experimentally detected by

Gebauer *et al.*[137] and Bodnarenko *et al.*[140,146] at the Positron Laboratory of the University of Halle (Germany), using positron annihilation spectroscopy (PAS), a powerful technique that uses positrons as tracers, usually obtained by the radioactive decay of the $^{22}$Na isotope, and delivers information about the size of vacancies, their charge state, their concentration and chemical surrounding, but is unable to distinguish the sublattice to which the vacancy pertains.

For this reason, the investigation on vacancies in GaAs should be carried out in As-rich or Ga-rich GaAs, to identify the presence of Ga vacancies or As vacancies, respectively.

Within the results obtained, it was found, as an example, that at 1100 °C in n-type GaAs samples doped with Te, the concentration of Ga vacancies (see Figure 1.31) increases with the increase of the doping level and with the increase of As vapor pressure, and could be fitted by a $p_{As}^{1/4}$ power law

$$x_{V_{Ga}} = K p_{As}^{1/4} = \exp -\frac{\Delta G_f}{kT} p_{As}^{1/4} \tag{1.45}$$

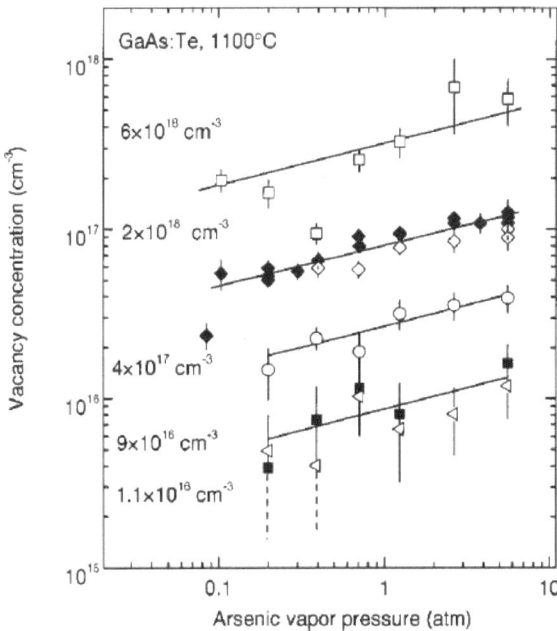

**Figure 1.31** Concentration of Ga-vacancies in Te-doped GaAs at 1100 °C as a function of the As-vapour pressure. Reproduced from ref. 137 with permission from American Physical Society, Copyright 2003.

following the reaction

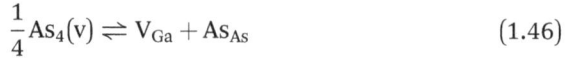

$$\frac{1}{4}As_4(v) \rightleftharpoons V_{Ga} + As_{As} \qquad (1.46)$$

considering that the As vapours are dominated by the $As_4$ species.

Extrapolating the concentration of vacancies given in Figure 1.31 to undoped conditions, the authors obtained a value of the Gibbs energy of formation of the neutral Ga vacancy, $\Delta G_f$ of 3.2 eV $\pm$ 0.5 eV, reasonably close to the calculated values reported in Table 1.8 (2.50–2.69 eV) and in Figure 1.29,[134] where a value of 4 eV has been evaluated for the formation of a neutral Ga vacancy in As-rich GaAs.

## 1.5.2  Defects in CdTe

The case of CdTe merits consideration in this book for its applications in radiation detectors and solar cells, favoured by its energy gap larger than silicon (1.6 eV), which allows, as an example, the operation of radiation detectors of improved spatial and energy resolution in clinical studies[147] and the fabrication of solar cells with 21.5% conversion efficiency.[148,149]

It is however believed that a better control of the role of impurities, non-stoichiometry, and crystal disorder in CdTe crystals, which are the real limit of the potentialities of CdTe solar cells, would allow reaching higher efficiencies. This is one of the reasons why the thermochemical properties and the defect chemistry of CdTe were carefully investigated.

The equilibrium conditions of the vapour phase with CdTe along the liquidus curves are displayed in Figure 1.32.[150] It appears that, different from GaAs where the equilibrium partial pressure of $As_n$ vapour species dominate the composition of the vapour phase, see Figure 1.6, for CdTe the Cd vapour pressure is higher than that of the dimeric $Te_2$ species at every temperature. Furthermore, though several polymeric $Te_n$ species coexists at equilibrium, the $Te_2$ species is the Te-dominant species in equilibrium over CdTe.

The phase diagram of Cd–Te system, see Figure 1.33(a),[148] shows the presence of a well centred CdTe phase, whose homogeneity range and equilibrium conditions with the vapour phase were experimentally studied by Greenberg.[151–153]

The results are reported in Figure 1.33(b), which displays with open squares the composition of the gas phase in equilibrium with samples of different composition,[152] and with the solid curve the contour

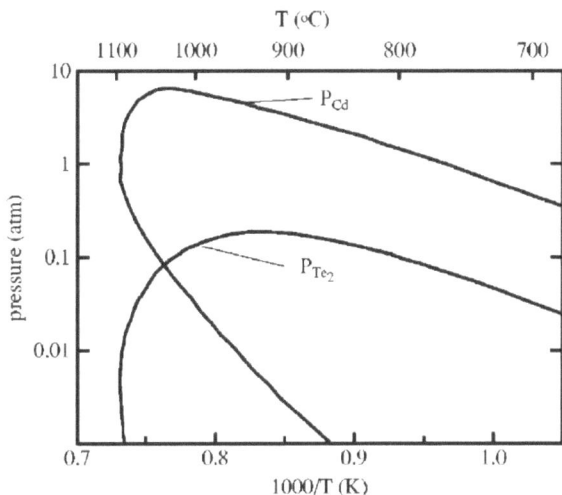

Figure 1.32 Temperature dependence of the equilibrium partial pressures of Cd and Te$_2$ over CdTe, along the liquidus curve of CdTe. Reproduced from ref. 150 with permission from American Physical Society, Copyright 1999.

of the homogeneity region, whose maximum width lies at about 1150 °C. In addition, the composition of CdTe samples exhibiting congruent sublimation is given by the S–V line.[¶¶¶¶¶] It is, therefore, apparent that the congruent melting composition is very close to the stoichiometric composition.

In a more recent work, Avetissov et al.[154] studied the homogeneity limits and the non-stoichiometry of CdTe polycrystalline samples, prepared by sublimation under vacuum of a CdTe powder in the temperature range 970–1365 K.

The results of this work are displayed in Figure 1.34, together with a full range of literature results which include those of the Greenberg group. It is possible to note the very good fit of the Aventissov and Greenberg results concerning the homogeneity range of the Cd-rich region and the composition of the phases exhibiting congruent sublimation (dotted S–V curve).

The nature and concentration of defects involved in the compensation of non-stoichiometry of CdTe, a highly ionic semiconductor ($f_i = 0.739$), have also been the subject of several theoretical and of (a few) experimental studies.[155–161]

It appears, see Figure 1.35,[162] that in Te-rich and Cd-rich CdTe, Cd interstitials Cd$_i$ and Cd vacancies V$_{Cd}$ are the most stable defects,

---

[¶¶¶¶¶] Details about the measurements are reported in the Greenberg paper.

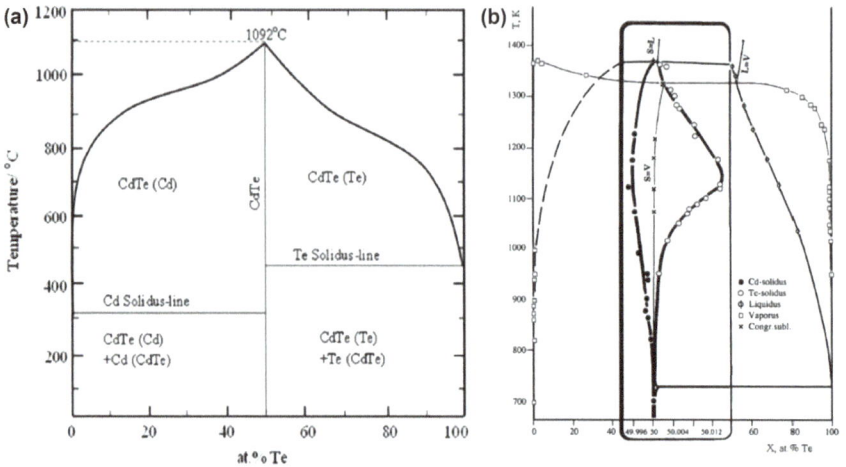

**Figure 1.33**    (a) Phase diagram of CdTe. (b) Phase diagram of CdTe with the non-stoichiometry phase given in an enlarged scale in the centre of the image. The dotted *S–V* curve displays the composition of the phases exhibiting congruent sublimation. (a) Reproduced from ref. 148, https://doi.org/10.3390/en8065440, under the terms of the CC BY 4.0 license, https://creativecommons.org/licenses/by/4.0/. (b) Reproduced from ref. 152 with permission from Elsevier, Copyright 1996.

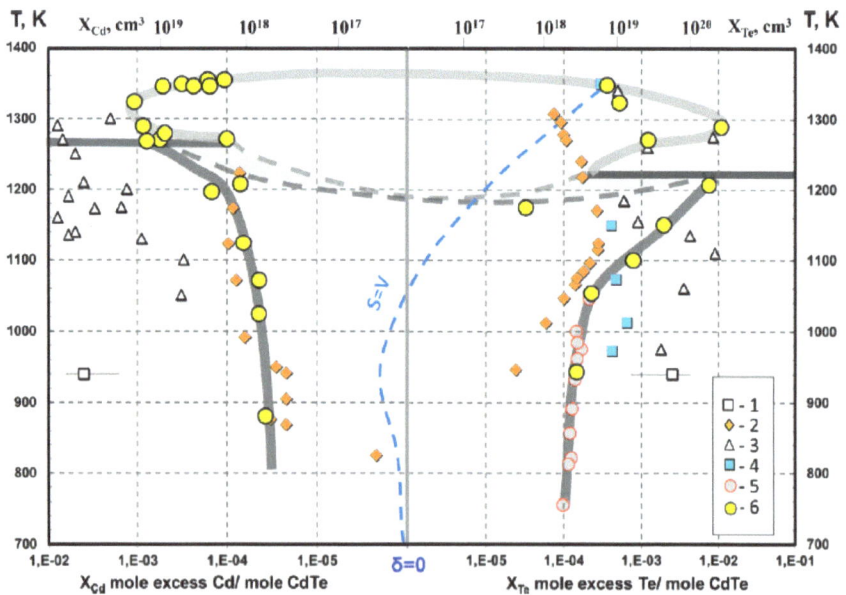

**Figure 1.34**    Cumulative data concerning the homogeneity region of CdTe. Reproduced from ref. 154 with permission from the Royal Society of Chemistry.

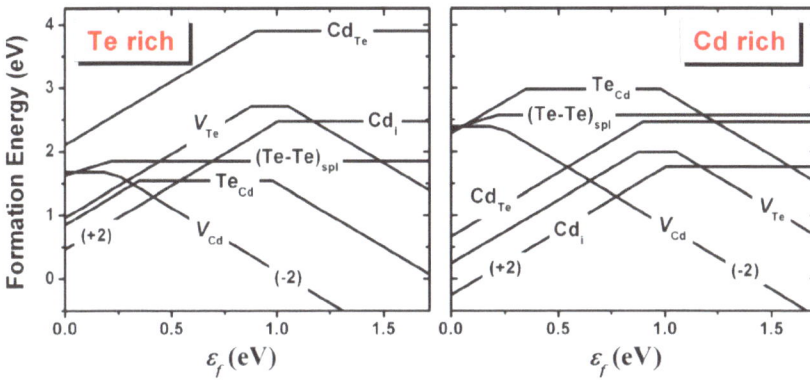

**Figure 1.35** Density functional calculated formation energy (in eV) of defects in Te-rich and Cd-rich CdTe. Reproduced from ref. 162 with permission from AIP Publishing, Copyright 2008.

together with $Cd_{Te}$ and $Te_{Cd}$ antisites and tellurium vacancies $V_{Te}$, which exhibit slightly higher formation energies.

The calculated dominance of Cd vacancies $V_{Cd}$ in CdTe is confirmed by positron annealing spectroscopy, which clearly detects the presence of the Cd vacancy in undoped CdTe,[156] in substantial agreement with the results of a more recent work of Linstrom,[158] which reports for the formation energy of neutral Cd vacancies a value of 1.8 eV.

## 1.5.3 Defects in Metal Oxides, the Case of ZnO

The entire metal oxide semiconductor family merits our attention because of their high ionicity, their high thermodynamic stability and their large band gaps, as well as for the strong role of point defects in their interaction with a reacting atmosphere.[163–166]

Since it would be impossible to take into detailed consideration all oxide semiconductors, only the case of ZnO will be discussed, which is sufficiently representative of the entire family.

Being optically transparent in the visible and presenting high electron mobility, ZnO is, in fact, the prototype of the group of transparent conducting oxides (TCO) used as conducting windows in solar cells.[116]

In addition, its high energy gap (3.34 eV), comparable with that of GaN ($E_G = 3.452$ eV) and its high exciton binding energy (60 meV), larger than both the room temperature thermal energy (25 meV) and the exciton binding energy in wurtzite GaN, which is $23.5 \pm 0.5$ meV,[116,167] would allow technological applications of its stimulated excitonic emission. We will see in Chapters 2 and 5 that the fabrication of stable p/n junctions and thus of ZnO LED and lasers, is still unfortunately frustrated by the difficulties encountered in the p-type doping of ZnO.

ZnO is also a very stable, ionic ($f_i = 0.616$) oxide semiconductor, with large values of Gibbs formation energy $\Delta G_f$ (ZnO) $= -354.560 + 107.8T$ (K) kJ mol$^{-1}$,[168] or $-364.34 + 136.39$ kJ mol$^{-1}$ (1073–1273 K),[169] and the electrical conductivity changes induced by its interaction with oxidizing or reducing atmospheres, in various microstructural or nanostructural configurations, make ZnO an active substrate for chemical sensors.

It is almost obvious, on the basis of previous considerations on the role of defects in compound semiconductors, that defects should be involved in the conductivity changes of ZnO in reducing and oxidizing environments.

The role played by defects in the properties of ZnO has been, in fact, discussed in hundreds of papers, starting from the 1950s, but we will rely here on few of them, dealing with the role of defects on the thermodynamic stability of ZnO in vacuum, and in oxygenated and reducing atmospheres.

As a first example, the results of *ab initio* computations carried out by Janotti and Van de Walle[170] displayed in Figure 1.36 for ZnO in

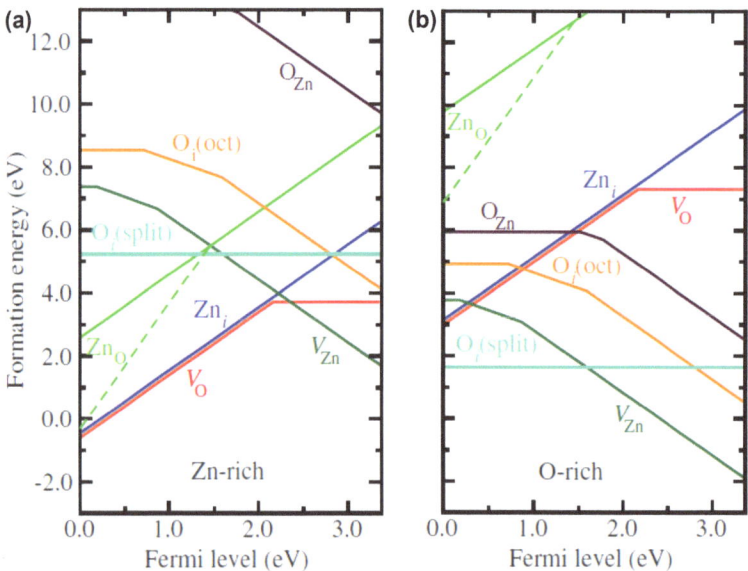

**Figure 1.36** Formation energy of defects in Zn-rich (a) and O-rich (b) ZnO. Reproduced from ref. 170 with permission from IOP Publishing, Copyright 2009.

Zn-rich or oxygen-rich conditions, suggest that oxygen vacancies $V_o$ and zinc interstitials $Zn_i$ are the most stable defects in Zn-rich conditions, while zinc vacancies $V_{Zn}$ are the most stable defects in oxygen-rich conditions, together with oxygen interstitials $O_i$, in the split configuration. It is also obvious that these defects are charged, and that their concentration depend on doping.

In qualitative agreement with these theoretical results, Parmar *et al.*[171] were able to show that high concentrations (up to $10^{20}$ cm$^{-3}$) of Zn vacancies $V_{Zn}$ are present at the surface of (oxygen rich) ZnO single crystals, thermally annealed in an atmosphere of pure oxygen at 1370 and 1470 K, using positron annealing spectroscopy (PAS) to measure their concentration, see Figure 1.37. They also demonstrated, using Auger spectroscopy to measure the concentration of oxygen and Zn along the volume of the material, that ZnO remains virtually stoichiometric after these oxidizing treatments.

Another result of these PAS measurements was that the surface $V_{Zn}$ concentration increases with the temperature of the thermal treatment, and that a gradient of $V_{Zn}$ concentration sets up in the material, leading to a minimum of $V_{Zn}$ concentration around $10^{15}$ cm$^{-3}$ deep ($>5000$ nm) into the bulk of the sample, a value almost independent of the temperature of the treatment. It was also observed that the Zn-vacancy concentration of oxygen-annealed ZnO crystals was converted to this concentration level after one hour annealing at 1000 K in air, which should correspond to the equilibrium concentration of Zn vacancies in ZnO at 1000 K.

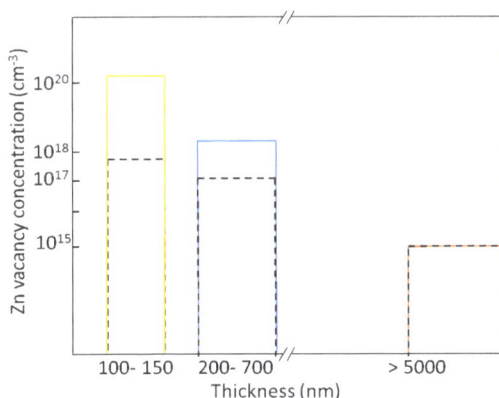

**Figure 1.37** Zinc-vacancies concentration profiles at different depths of ZnO crystal samples after annealing at 1273 K (broken lines) and 1473 K (solid lines). Reproduced from ref. 171, https://doi.org/10.1038/s41598-018-31771-1, under the terms of the CC BY 4.0 license, https://creativecommons.org/licenses/by/4.0/.

These experimental results could be accomplished by assuming that by annealing ZnO in oxygenated atmospheres the formation process of Zn-vacancies at the sample surface occurs in thermodynamic equilibrium conditions following the reaction

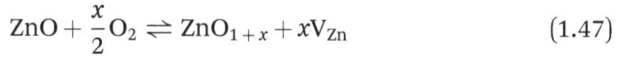

$$ZnO + \frac{x}{2}O_2 \rightleftharpoons ZnO_{1+x} + xV_{Zn} \qquad (1.47)$$

with an experimental value of the enthalpy of formation of Zn vacancies $\Delta H_{V_{Zn}}^f = 61.7506 \text{ kJ mol}^{-1}$ [171] (0.64 eV), corresponding to the formation of charged vacancies calculated by Janotti and Van de Walle[170] for n-doped conditions.

Also the thermal decomposition of ZnO, which spontaneously occurs at temperatures well below its melting temperature (MP 2248 K), with the formation of a slightly non-stoichiometric phase, is a source of point defects.

Allsopp and Roberts,[172] as an example, studied the thermal decomposition of sintered ZnO samples under the hypothesis that decomposition occurs with the formation of metallic Zn and oxygen gas

$$ZnO \xrightarrow{T,O_2(p)} Zn + \frac{1}{2}O_2 \qquad (1.48)$$

and determined the excess Zn present in these samples as a function of the temperature and of the covering atmosphere by measuring the hydrogen evolved when the thermally annealed samples were etched with a hydrochloric solution.

It was shown that the amount of Zn took the highest values (14.3–18 ppmw) when the samples are heated in an Ar atmosphere at 1563 K and the lowest when the samples were heated in oxygen (0.8–5.6 ppmw)-Ar atmosphere at the same temperature. It was also observed that the excess Zn was concentrated at the surface of the samples, suggesting a diffusion of Zn from bulk regions, where the thermal ZnO decomposition also occurs. The qualitative conclusion of this work is that the thermal decomposition of ZnO occurs with the formation of an oxygen-rich $ZnO_{1+x}$ oxide and of zinc interstitials $Zn_i$

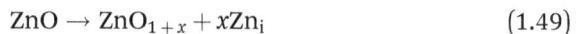

$$ZnO \rightarrow ZnO_{1+x} + xZn_i \qquad (1.49)$$

which diffuses to the surface of the sample, where it segregates as metallic Zn, and where it might be dissolved by HCl.

The hypothesis of Zn segregation was confirmed by Secco,[173] also under vacuum conditions, who demonstrated that ZnO powder decomposes above 1073 K under vacuum, with net weight losses, and the formation of Zn deposits at the surface of the sample and at the surface of the vacuum vessel where the experiments are carried out.

The thermal decomposition of ZnO was, therefore, supposed to occur following the reaction

$$ZnO(s) \rightleftharpoons Zn(v) + \frac{1}{2}O_2 \tag{1.50}$$

and the Arrhenius plot of the experimental values of the decomposition constant $K_p$ of reaction (1.50), at constant initial $p_{Zn}$, in the temperature range 1073–1273 K, *i.e.* above the melting point of Zn (692.5 K), yields a reaction enthalpy of $-100.416$ kJ mol$^{-1}$, which does not correspond to the enthalpy of thermal dissociation of ZnO, nor to the enthalpy of sublimation of ZnO, that is $-188.28$ kJ mol$^{-1}$.

Instead, this value is close to the enthalpy of sublimation of Zn

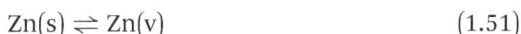

$$Zn(s) \rightleftharpoons Zn(v) \tag{1.51}$$

which is $-130.4$ kJ mol$^{-1}$, suggesting that the Zn sublimation behaves as the rate determining step of the process under eqn (1.50).

A similar behaviour of ZnO was described by Imoto *et al.*,[174] who found, also, that the decomposition rate was proportional to the surface area of the samples and thermally activated, with an activation energy of 40.58 kJ mol$^{-1}$, measured in the temperature range 933–1093 K.

Both authors suggest that the thermal dissociation of ZnO occurs with the cooperation of a point defect equilibrium involving the formation of metallic zinc and oxygen vacancies at the surface of the sample

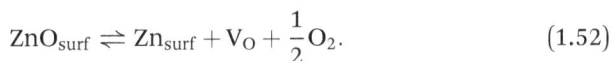

$$ZnO_{surf} \rightleftharpoons Zn_{surf} + V_O + \frac{1}{2}O_2. \tag{1.52}$$

A later work of Lott *et al.*[175] supports the hypothesis of Zn volatilization under thermal annealing of non-stoichiometric ZnO samples. They, in fact, show that Zn-saturated ZnO samples, heated in the temperature range 500–800 °C, are in equilibrium with a zinc partial pressure corresponding to the equilibrium partial pressure of Zn over liquid zinc, as can be seen in Figure 1.38, considering that the partial pressure of Zn above liquid Zn is $2.55 \times 10^{-4}$ bar at 690 K.[176]

A different hypothesis about the thermodynamics of ZnO decomposition in oxygenated atmospheres was suggested by Shi *et al.*,[177]

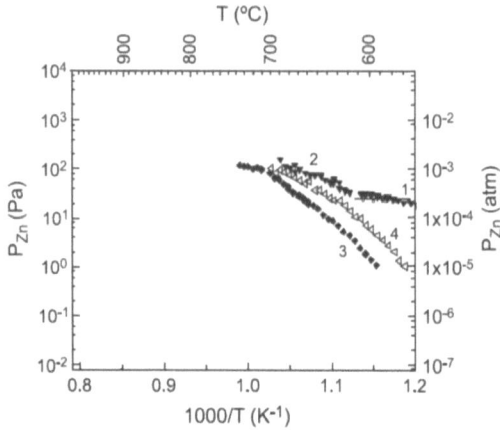

**Figure 1.38** Partial pressure of Zn in equilibrium with: (1) Zn saturated ZnO, (2) ceramic ZnO, (3) and (4) non-stoichiometric ZnO powder. Reproduced from ref. 175 with permission from Elsevier, Copyright 2004.

who used a microgravimetric method to determine the weight losses caused by the thermal decomposition of ZnO in a range of applied oxygen pressure between 1 and $10^{-6}$ bar, by varying the temperature between 673 and 1473 K. Measurable weight losses of the ZnO samples were observed even under an oxygen pressure of 1 atm and interpreted as due to a loss of oxygen and to the formation of an oxygen-poor $ZnO_{1-x}$ phase and of oxygen vacancies $V_O$

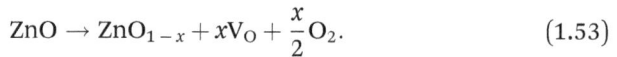

$$ZnO \rightarrow ZnO_{1-x} + xV_O + \frac{x}{2}O_2. \tag{1.53}$$

The experimental dependence of the parameter $x$, measured by the deviation from the stoichiometry of the ZnO phase, on the temperature and on the partial pressure of oxygen is displayed in Figure 1.39.

Figure 1.39 shows the increase of the $x$ values with the increase of temperature and with the decrease of the partial pressure of oxygen, and the Arrhenius-type of dependence of $x$ on the temperature, with a slope corresponding to the enthalpy of formation of oxygen vacancies $\Delta H_{V_O}$

$$x \approx \exp - \frac{\Delta H_{V_O}}{kT}. \tag{1.54}$$

A change of slope is observed at a temperature $T_P$ that occurs between 934 and 1234 K, depending on the oxygen pressure in equilibrium.

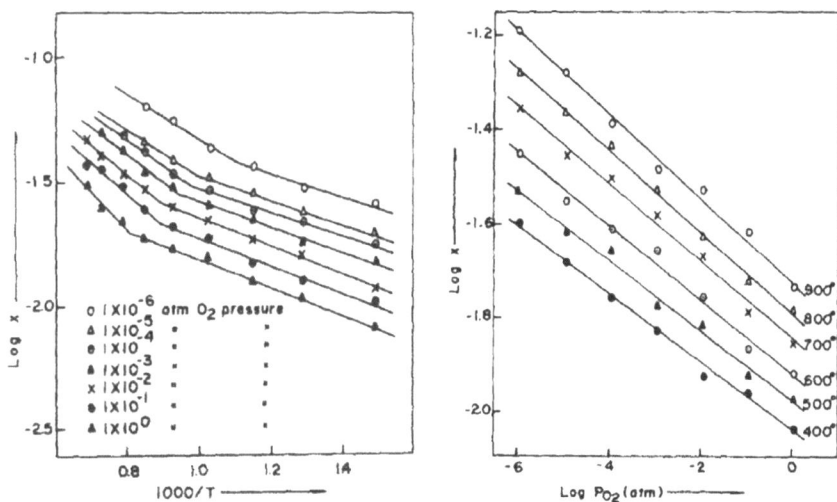

Figure 1.39  Dependence of the deviations from the stoichiometry of ZnO samples on the temperature (left) and on the oxygen pressure (right). Reproduced from ref. 177 with permission from Elsevier, Copyright 1976.

The formation enthalpies values $\Delta H_{V_O}$ deduced by these measurements in the low temperature regime $(T < T_P)$, substantially independent of the partial pressure of oxygen, could be estimated to be around $-30.46$ kJ mol$^{-1}$ (0.32 eV), close to the value of 0.5 eV calculated by Ehrard[178] for the ionized oxygen vacancy $V_O^{2+}$.

The hypothesis of oxygen vacancy $V_O$ formation in a range of oxygen poor-atmospheres is supported only by electron spin resonance (ESR) measurements,[179,180] since PAS measurements cannot detect the presence of oxygen vacancies in ZnO, because their lifetime (160 ps) is too close to the bulk lifetime of ZnO (158 ps).

ESR results demonstrate, in fact, the systematic presence of ionized oxygen vacancy $V_O^+$ in various samples of heat-treated ZnO, and also allows the evaluation of the value of its ionization energy[179] (0.44 eV), that well compares with results of Hall measurements on the $V_O^+$ centre carried out by the same authors.

It seems, therefore, established that Zn interstitials $Zn_i$ are involved in the thermal decomposition of ZnO, which leaves a substantially stoichiometric phase, whereas oxygen vacancies $V_O$ are, involved in the process of formation of a non-stoichiometric phase when ZnO is heat treated in oxygen poor atmospheres or under vacuum. Apparently, Zn vacancies $V_{Zn}$ arise instead by treatment of ZnO in oxygen rich atmospheres. The same conclusions could also be inferred from the measurements of oxygen diffusion in ZnO carried out by

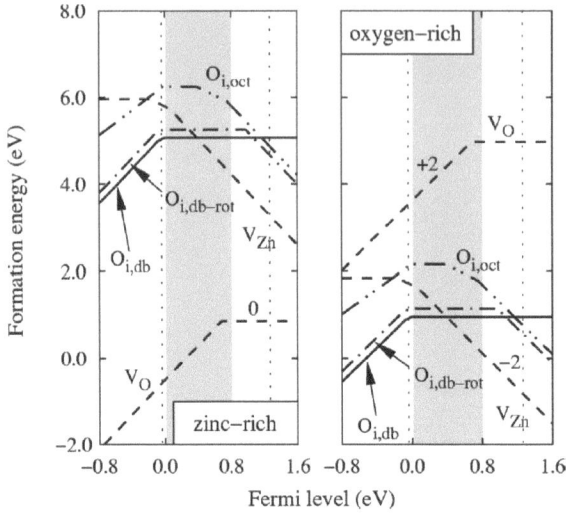

Figure 1.40   First principles calculated formation energies of defects in zinc-rich and oxygen-rich ZnO. Reproduced from ref. 178 with permission from American Physical Society, Copyright 2005.

Moore and We[181] and from the results of the first principles studies of Ehrard *et al.*[178] displayed in Figure 1.40, that show that $V_o$ cause the dominant defects in zinc-rich (oxygen poor) ZnO, while $V_{Zn}$ are the dominant defects in oxygen-rich ZnO.

The nature of defects arising by the interaction of hydrogen with ZnO has also been investigated for years,[179–181] as well the electrical conductivity changes induced by their ionization, measured by ZnO-based chemical sensors in equilibrium with hydrogen-inert gas mixtures.

We will account here for the results of electrical conductivity measurements carried out by Pizzini *et al.*[82] with simple devices based on thin, Al-doped,|||||| microcrystalline ZnO films, in the temperature range 500–600 K, typical of operational applications of chemical sensors.

It was found that the electrical conductivity changes measured after equilibration of ZnO with hydrogen–argon mixtures show an exponential dependence with the temperature, with a slope of the Arrhenius plot that indicates the presence of a defect species with a donor level at $E_C - 0.06$ eV.

This defect, however, does not correspond to either oxygen vacancies $V_o$ or to Zn interstitials $Zn_i$, which are the ESR-active donor

---

|||||| Al is an n-type dopant for ZnO.

species found in $ZnO^{179,180}$ after reaction with hydrogen, generated according to the following equations

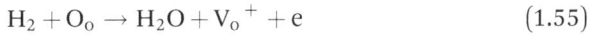

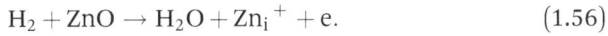

$$H_2 + O_o \rightarrow H_2O + V_o{}^+ + e \tag{1.55}$$

$$H_2 + ZnO \rightarrow H_2O + Zn_i{}^+ + e. \tag{1.56}$$

In fact, the oxygen vacancy $V_o$ behaves as a shallow donor with a donor level at $E_C - 0.9$ (eV),****** like the Zn interstitial $Zn_i$, with a donor level at $(E_C - 0.46 \text{ eV})$,[183] see Figure 1.41.

The problem finds a solution if we consider that hydrogen is soluble in ZnO, demonstrated among others by Gaspar *et al.*,[184] who showed, using scanning ion mass spectrometry (SIMS), that pure hydrogen or hydrogen diluted in argon dissolves in thin ZnO films (180 nm thick) across the entire thickness of the film during their deposition at room temperature from a sputtering source, with a maximum concentration at the surface, see Figure 1.42.

We could, therefore, suggest that ZnO not only reacts with hydrogen with the formation of oxygen vacancies and zinc interstitials, but dissolves hydrogen as interstitial hydrogen $H_i$

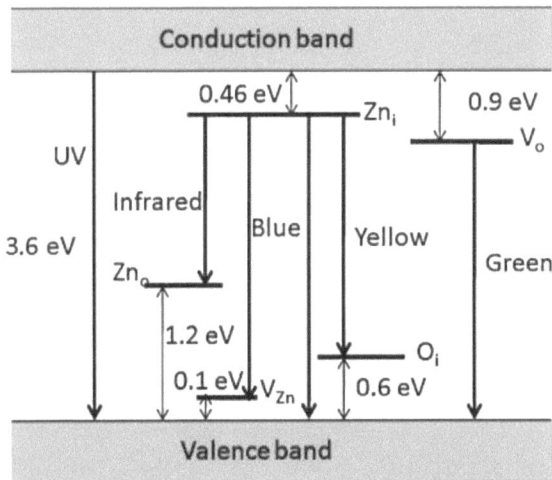

$$\frac{1}{2} H_2 (ZnO) \leftrightarrow H_i \tag{1.57}$$

**Figure 1.41** Room temperature energy levels of defects in zinc oxide measured with cathodoluminescence spectroscopy. Reproduced from ref. 183 with permission from IOP Publishing, Copyright 2008.

****** Or $E_c - 0.44$ eV according to ref. 179.

**(a)**

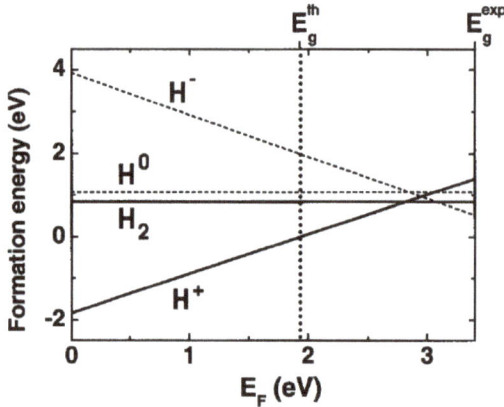

**(b)**

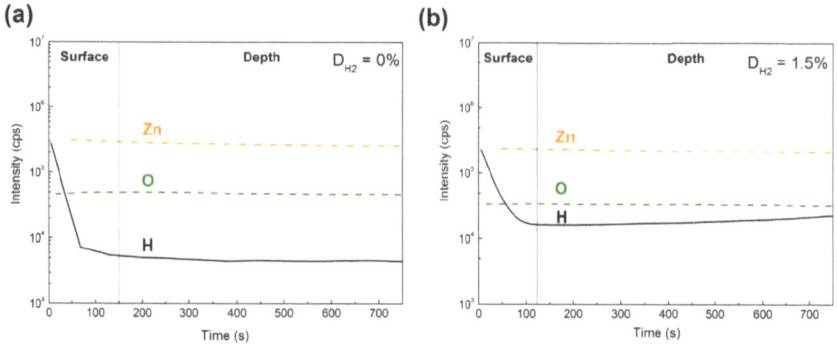

**Figure 1.42**   SIMS-measured hydrogen concentration in ZnO under (a) pure
hydrogen, (b) 1.5% in Ar. Dotted lines display the XPS values for
Zn and O, D means dilution of hydrogen in argon. Reproduced
from ref. 184, https://doi.org/10.1016/j.solmat.2017.01.030, under
the terms of the CC BY 4.0 license, https://creativecommons.org/
licenses/by/4.0/.

**Figure 1.43**   Formation energy of interstitial hydrogen in zinc oxide. Reproduced from ref. 185 with permission from American Society of
Physics, Copyright 2000.

which behaves as a shallow donor

$$H_i \rightarrow H_i{}^+ + e \tag{1.58}$$

leading to the conductivity transients experimentally observed, in
good agreement with the results of van de Walle,[185] see Figure 1.43,
which suggest hydrogen as the cause of intrinsic n-type doping of
ZnO.

Furthermore, the experimental values[186] of the ionization energies
of the two hydrogen interstitial species formed by the interaction of

hydrogen with ZnO, the hydrogen bond centred $H_{BC}$ interstitial, with an ionization energy of 53 meV, and the hydrogen–oxygen vacancy centre $H_o$ with an ionization energy of 47 meV, are very close to the ionization energy of 66 meV of the defect centre inferred from the electrical conductivity measurements.[182]

It seems, therefore, established that oxygen vacancies, zinc interstitials and hydrogen interstitials are the defect species generated by the equilibration of ZnO with hydrogen, and that hydrogen interstitials are the donor species giving rise to the electrical conductivity transients.

# Appendix 1.1   Equilibrium Thermodynamics of Multicomponent Phases

Let us first consider the simplest case of a phase behaving as a solution of two elemental components A and B, with very close physical, chemical and structural properties, *i.e.* very close values of the lattice constants, atomic radii and electronic properties, such to grant conditions of mutual solubility in all proportions and virtual absence of pairwise chemical interactions, *i.e.* conditions of ideal behaviour.

In this case, the thermodynamic activities of the components would coincide with their atomic fractions, $a_i = x_i$, and the molar free energy of the solution is given by the equation

$$G_{sol} = x_A \mu_A + x_B \mu_B = x_A \mu_A^\circ + x_B \mu_B^\circ + RT(x_A \ln x_A + x_B \ln x_B) \qquad (1.59)$$

where $\mu_i = \mu_i^\circ + RT \ln x_i$ and the excess energy of mixing $\Delta G_{mix}$ is given by

$$\Delta G_{mix} = G_{sol} - \left( x_A \mu_A^\circ + x_B \mu_B^\circ \right) = RT \left[ x_A \ln x_A + x_B \ln x_B \right] \qquad (1.60)$$

showing that the Gibbs energy of mixing has a purely configurational entropy character

$$\Delta G_{mix} = -T \Delta S_{conf} \qquad (1.61)$$

as could be deduced using simple mathematics starting from the Boltzmann equation which would easily lead to the equation holding for the configurational entropy

$$\Delta S_{conf} = R(x_A \ln x_A - x_B \ln x_B). \qquad (1.62)$$

For an ideal solution, therefore, the enthalpy of mixing $\Delta H_{mix} = 0$.

For solutions displaying significant deviations from the ideal, the activities $a_i = \gamma_i x_i$ should be used instead of the concentrations in eqn (1.57), and the excess Gibbs energy of mixing is given by

$$\Delta G_{\text{mix}} = -RT(x_A \ln \gamma_A x_A - x_B \ln \gamma_B x_B). \tag{1.63}$$

In this case, one might apply the regular solution model, originally proposed by van Laar,[188,189] Hildebrand,[190] Fowler and Guggenheim,[191] and Guggenheim,[192] for whom the excess Gibbs energy of solution is ruled by pair-like interactions between neighbour atoms in solution and is given by the following equation

$$\Delta G_{\text{mix}}(x, T) = x_A x_B \Omega - T\Delta S_{\text{conf}} - x_A x_B \eta T \tag{1.64}$$

where $\Omega = ZN\left[H_{AB} - \left(\dfrac{H_{AA} + H_{BB}}{2}\right)\right]$ is an interaction coefficient, taken as the difference between the binding enthalpy of unlike species in solution $H_{AB}$ and the average of the binding enthalpies of like pairs, and $ZN$ is the product of the coordination number $Z$ and the total number $N$ of atoms in solution.

Here the term $x_A x_B \Omega$ is the enthalpy of mixing $\Delta H_{\text{mix}}$ now different from zero and $x_A x_B \eta$ is a non-configurational entropy term arising from vibrational and electronic contributions.

The model assumes the absence of elastic contribution terms to the enthalpy of mixing, due to lattice mismatch of the components, as well as the random distribution of components atoms in a rigid lattice.

On that basis, the $\Omega$ values for pair-like interactions between elements of groups III, IV and V of the periodic table calculated by Stringfellow[193] are available.

The Gibbs energy of mixing is, then, given by the following equation

$$\Delta G_{\text{mix}}(x, T) = x_A x_B \Omega - x_A x_B \eta = x_A x_B (\Omega - \eta T). \tag{1.65}$$

To account for the non-configurational entropy term $x_A x_B \eta$, the vibrational and electronic contributions should be known or calculated with a quantum mechanical approach,[194] but it is common practice to neglect the non-configurational entropy term and consider for the Gibbs energy of the solution and for the excess Gibbs energy of mixing the sole enthalpic terms

$$G_{\text{sol}}(x, T) = x_A x_B \Omega - TS_{\text{conf}} \tag{1.66}$$

$$\Delta G_{\text{mix}} = \Delta H_{\text{mix}} = x_A x_B \Omega. \tag{1.67}$$

It should be noted that the interaction coefficient $\Omega$ term is assumed here to be temperature-independent, but it is known that the correct modelling of compound semiconductor alloys often requires the use of temperature dependent interaction coefficients.

Furthermore, as only the solid, or only the liquid or both solutions might be regular, and for both positive and negative $\Omega$ values, the modelling of phase equilibria systems with $\Omega$ values different from zero requests a solution for all the eight domains,[195] as shown in Table 1.9.

If the excess energy of mixing of a regular solution is given only by the enthalpic term $\Delta H_{mix} = x_A x_B \Omega$ (and the non-configurational entropy terms are negligible), it is easy to get the values of the activity coefficients of the components in solution and, thus, to evaluate the chemical potentials from the experimental (coming from calorimetric measurements, or from the calculated values of $\Delta H_{mix}$.

In fact, from eqn (1.64)

$$\frac{\delta \Delta G_{mix}}{\delta x_B} = RT \ln \gamma_B \tag{1.68}$$

and eqn (1.67)

$$\frac{\partial \Delta G_{mix}}{\partial x_B} = \frac{\partial \Delta H_{mix}}{\partial x_B} = x_A x_B \Omega + \left( (1 - x_B) \left( \frac{\partial}{\partial x_B} \right) x_A x_B \Omega \right) = (1 - x_B)^2 \Omega \tag{1.69}$$

one obtains the equations holding for the activity coefficients for the B species

$$\ln \gamma_B = \frac{(1 - x_B)^2 \Omega}{RT} \tag{1.70}$$

for the A species

$$\ln \gamma_A = \frac{x_B^2 \Omega}{RT} \tag{1.71}$$

and for the chemical potentials

$$\mu_A(x, T) = \mu_A^\circ + x_B^2 \Omega + RT(x_A \ln x_A) \tag{1.72}$$

$$\mu_B(x, T) = \mu_B^\circ + (1 - x_B)^2 \Omega + RT(x_B \ln x_B). \tag{1.73}$$

**Table 1.9** Domains for invariant transformations in binary systems. Data from ref. 194.

| Liquid | $\Omega > 0$ | $\Omega < 0$ | $\Omega > 0$ | $\Omega < 0$ | $\Omega = 0$ | $\Omega = 0$ | $\Omega > 0$ | $\Omega < 0$ |
|--------|--------------|--------------|--------------|--------------|--------------|--------------|--------------|--------------|
| Solid  | $\Omega > 0$ | $\Omega < 0$ | $\Omega < 0$ | $\Omega > 0$ | $\Omega > 0$ | $\Omega < 0$ | $\Omega = 0$ | $\Omega = 0$ |

These equations might also be used to calculate the interaction co-efficient $\Omega$ from the experimental values of the activities or of the chemical potentials, determined by thermochemical or electro-chemical measurements on electrochemical cells.

On that basis it is also possible to forecast the equilibrium distri-bution of a chemical species among coexisting phases as a function of the temperature and composition and foresee the topological aspects of the corresponding phase diagrams, generally reported in tem-perature-composition $(T, x)$ coordinates.

Since in a vast number of semiconductor systems spontaneous decomposition occurs if the pressure of the environment is not taken into proper account, the phase diagrams are reported in $(T, x, P)$ diagrams or of their $T, x$ projections.

The equilibrium distribution of a component i among two coex-isting phases, of which one is typically liquid and the other is solid, could be given in terms of a distribution or segregation coefficient $k_i(x, T)$, defined as the ratio $k_i = \dfrac{x_i^s}{x_i^l}$, where $x_i^s$ is the equilibrium atomic fraction of the component i in the solid phase and $x_i^l$ is the equilibrium atomic fraction of i in the liquid phase, both of which could be ex-perimentally determined using the appropriate analytical techniques.

The knowledge of equilibrium segregation of impurities in semi-conductors is particularly critical, as will be seen in Chapter 2, be-cause it allows a proper handling of semiconductor doping and refining processes.

It is important, therefore, to show the relationships occurring between the segregation coefficient and the thermodynamic properties of the phases across which the repartition of the solution components occurs.

This can be done on the base of the following statement

$$\mu_i^{0,\alpha} + RT \ln \gamma_i^\alpha x_i^\alpha = \mu_i^{0,\beta} + RT \ln \gamma_i^\beta x_i^\beta \tag{1.74}$$

that defines the equilibrium conditions of the component i between two generic $\alpha$ and $\beta$ phases and allows deducing the equilibrium distribution of a generic solute i among the two coexisting phases as a function of the chemical potential $\mu_i^0$ of the pure solute i and on the activity coefficients $\gamma_i$ of i in the solid and liquid state, in the whole range of temperatures where the two phases coexist

$$\ln k_i = \ln \frac{x_i^s}{x_i^l} = -\frac{\mu_i^{0,s} - \mu_i^{0,l}}{RT} - \ln \frac{\gamma_i^s}{\gamma_i^l} = -\frac{\Delta G_i^{0,f}}{RT} - \ln \frac{\gamma_i^s}{\gamma_i^l} = \frac{\Delta S_i^{0,f}}{R} - \frac{\Delta H^{0,f}}{RT} - \ln \frac{\gamma_i^s}{\gamma_i^l}.$$

$$\tag{1.75}$$

The ratio $\frac{\gamma_i^s}{\gamma_i^l}$ depends on the specific interaction of i with the other components of the liquid and solid phases, and, thus, on the temperature and the composition of the phase, which in the case of multicomponent systems is, in fact, a solution, and the difference $\mu_i^{o,s} - \mu_i^{o,l}$ is the Gibbs energy of melting $\Delta G_i^{o,f} = \Delta H_i^{o,f} - T\Delta S_i^{o,f}$ of the pure component i at the melting temperature $T_i^f$.

The segregation coefficient is unitary $k_i = 1$ and $x_i^s = x_i^l$ when the following relationship does occur between the activity coefficients and the Gibbs energy of melting, for a solid/liquid equilibrium

$$\ln \frac{\gamma_i^s}{\gamma_i^l} = -\frac{\Delta G_i^{o,f}}{RT} \tag{1.76}$$

and $\Delta G_i^{o,f} \rightarrow 0$ or $\frac{\gamma_i^s}{\gamma_i^l} = 1$. In this case a distinction between the solid and liquid phase is thermodynamically impossible.

The segregation coefficient is also unitary for the segregation of i at the melting temperature $T_i^f$ of i, because in this case the free energy of melting $\Delta G_i^{o,f} = 0$ and $\mu_i^{o,s} = \mu_i^{o,l}$.

For the equilibrium of i among a couple of multicomponent phases behaving as ideal solutions, for which, by definition, $\gamma_i^s = 1$ and $\gamma_i^l = 1$, the segregation coefficient is given by the equation

$$\ln k_i = \ln \frac{x_i^s}{x_i^l} = -\frac{\Delta G_i^f}{RT} \tag{1.77}$$

where $\Delta G_i^{o,f}$ is again the free energy of fusion of the pure component i.

Except in these cases, the equilibrium distribution of a generic component i among two condensed phases or a condensed phase and a gaseous or vapour phase, given by eqn (1.71), and depends on the specific thermodynamic properties of the coexisting phases, which should be experimentally available or theoretically determined.

To arrive at the theoretical evaluation of a phase diagram, it is necessary to model the temperature, composition (and pressure) dependence of the free energy $G(T, P, x)$ of the whole system, consisting of two coexisting solid phases, two liquid and solid phases, and of the potentially stable intermediate compounds.

An example of the useful application of thermodynamical modelling to a semiconductor system is that reported in Figure 1.44,

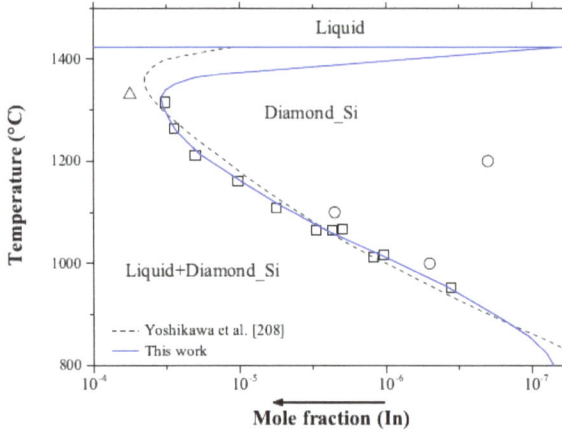

**Figure 1.44** Calculated section of the phase diagram of the In–Si system. Reproduced from ref. 195, https://doi.org/10.3390/ma10060676, under the terms of the CC BY 4.0 license, https://creativecommons.org/licenses/by/4.0/.

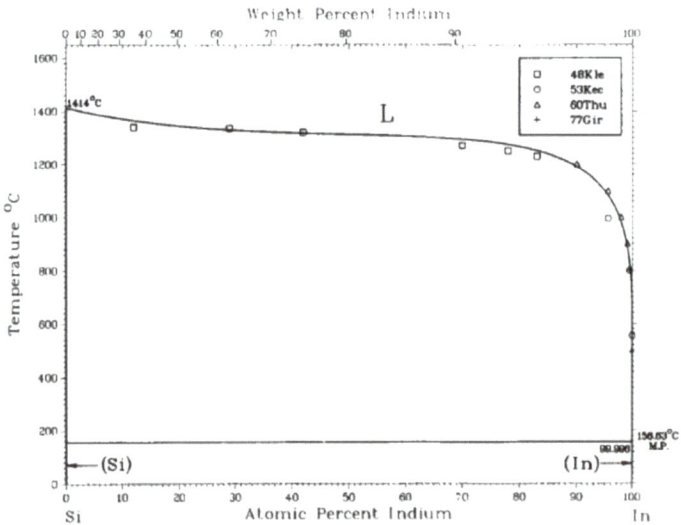

**Figure 1.45** Phase diagram of the binary In–Si. Reproduced from ref. 196 with permission from Springer Nature, Copyright 1985.

corresponding to the section of the In–Si system in the range of the extremely diluted solid solution of indium in silicon,[195] typical of In doped Si. It appears that this range is undetectable in the full phase diagram of Figure 1.45,[196] making this phase diagram useless for the selection of the conditions needed to grow In-doped Si.

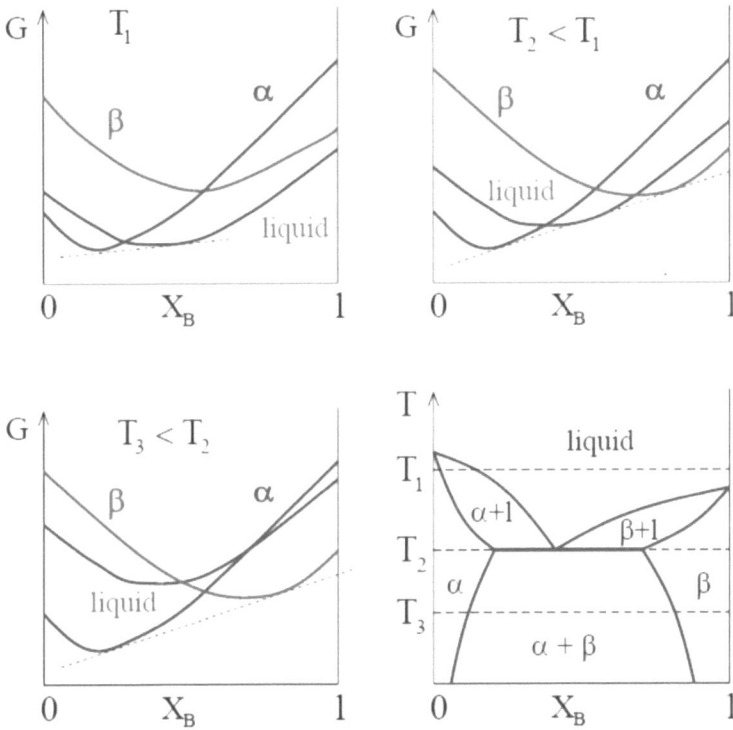

**Figure 1.46** Calculated eutectic type of phase diagram for the case of pure components having different structures and limited reciprocal solubility. Reproduced from ref. 197 with permission from Leonid V. Zhigilei.

Using the regular solution model, van Laar[187,188] demonstrated that many of the observed types of simple phase diagrams might be obtained by a systematic variation of the $\Omega$ values ($\Omega^\beta$ and $\Omega^\alpha$) for liquid and solid phases, respectively, in the positive domain ($\Omega > 0$) with $\Omega^\beta \leq \Omega^\alpha$.

By varying $\Omega^\alpha$ from zero to very large positive values (keeping the liquid solution as ideal with $\Omega^\beta = 0$), it was possible to obtain lens shaped isomorphous, simple eutectic, peritectic and isomorphous with congruent maximum or minimum diagrams, as it is possible to see in Figure 1.46 for eutectic-type phase diagrams.

It is apparent that here the $G(x)$ isotherms are given for three typical conditions, but the full diagram is, in fact, obtained by calculating the number of isotherms sufficient to give fully detailed features of the phase diagram.

# References

1. D. Dyatkin, While transistors slim down, microchip manufacturing challenges expand, *MRS Bull.*, 2021, **46**(1), 16.
2. S. L. Rumyantsev, M.-S. Shur and M. L. Levinshtein, Materials properties of nitrides, *Int. J. High Speed Electron. Syst.*, 2011, **14**(1), DOI: 10.1142/S012915640400220X.
3. T. R. Wei, *et al.*, Exceptional plasticity in the bulk single-crystalline van der Waals semiconductor InSe, *Science*, 2020, **369**(6503), 542.
4. C. J. Bartel, S. L. Millican, A. M. Deml, J. R. Rumptz, W. Tumas, A. W. Weimer, S. Lany, V. Stevanović, C. B. Musgrave and A. M. Holder, Physical descriptor for the Gibbs energy of inorganic crystalline solids and temperature dependent materials chemistry, *Nat. Commun.*, 2018, **9**, 4168.
5. E. A. Guggenheim, *Thermodynamics*, Elsevier Science Publishers B.V., 1967.
6. B. Himmetoglu, A. Floris, S. de Gironcoli and M. Cococcioni, Hubbard-corrected DFT energy functionals: the LDA+U description of correlated systems, *Int. J. Quantum Chem.*, 2014, **114**(1), 14.
7. D. W. Palmer, Properties of III-V semiconductors, www.semiconductors.co.uk, 2006.02.
8. C. Henager and J. R. Morris, Atomistic simulation of CdTe solid-liquid coexistence equilibria, *Phys. Rev. B: Condens. Matter Mater. Phys.*, 2009, **80**, 245309.
9. K. Yamaguki, K. Kameda, Y. Takeda and K. Itagaki, Measurements of high temperature heat content of the II-VI and IV-VI compounds, *Mater. Trans., JIM*, 1994, **35**, 118.
10. J. Wang and M. Isshiki, Wide-Bandgap II-VI semiconductors: Growth and Properties, *Springer Handbook of Electronic and Photonic Materials*, Springer-Verlag, 2007, ISBN: 978-0-387-26059-4.
11. S. I. Lopatin, V. L. Stolyarova, V. G. Sevast'yanov, Ya. Nosatenko, V. V. Gorskii, D. Sevast'yanov and N. T. Kuznetsov, Determination of the saturation vapor pressure of silicon by Knudsen cell mass spectrometry, *Russ. J. Inorg. Chem.*, 2012, **57**, 219.
12. A. W. Searcy and R. D. Freeman, Measurement of the Molecular Weights of Vapors at High Temperature. II. The Vapor Pressure of Germanium and the Molecular Weight of Germanium Vapor, *J. Chem. Phys.*, 1955, **23**, 88.
13. F. P. Bundy, Pressure-temperature phase diagram of elemental carbon, *Phys. A*, 1989, **156**, 169.
14. C. H. Bates, F. Dachille and R. Roy, High-Pressure Transitions of Germanium and a New High-Pressure Form of Germanium, *Science*, 1965, **147**(3660), 860.
15. *Properties of silicon*, EMIS Data Rewiew Series No. 4, INSPEC, London, 1988.
16. O. O. Kurakevych, Y. Le Godec, W. A. Crichton and T. A. Strobel, Silicon allotropy and chemistry at extreme conditions, *Energy Proc.*, 2016, **92**, 839.
17. C. C. Yang and Q. Jiang, Temperature–pressure phase diagram of germanium determined by Clapeyron equation, *Scr. Mater.*, 2004, **51**(11), 1081.
18. C. C. Yang and Q. Jiang, Temperature–pressure phase diagram of silicon determined by Clapeyron equation, *Solid State Commun.*, 2004, **129**(7), 437.
19. J. Y. Howe, L. E. Jones and A. N. Cormack, The oxidation of diamond, Oak Ridge National Laboratory Document, 2001.
20. A. Joshi, R. Nimmagadda and J. Herrington, Oxidation kinetics of diamond, graphite, and chemical vapor deposited diamond films by thermal gravimetry, *J. Vac. Sci. Technol., A*, 1990, **8**, 2137.
21. Y. Fedortchouk and D. Canil, Diamond oxidation at atmospheric pressure: development of surface features and the effect of oxygen fugacity, *Eur. J. Mineral.*, 2009, **21**(3), 623.

22. M. J. Kim and R. W. Carpenter, Composition and structure of native oxide on silicon by high resolution analytical electron microscopy, *J. Mater. Res.*, 1990, **5**, 347.

23. A. H. Al-Bayati, K. G. Orrman-Rossiter, J. A. van den Berg and D. G. Armour, Composition and structure of the native Si oxide by high depth resolution medium energy ion scattering, *Surf. Sci.*, 1991, **241**(1–2), 91.

24. B. E. Deal and A. S. Grove, General Relationship for the Thermal Oxidation of Silicon, *J. Appl. Phys.*, 1965, **36**(12), 3770.

25. K. Kajihara, T. Miura, H. Kamioka, A. Aiba, M. Uramoto, Y. Morimoto, M. Hirano, L. Skuja and H. Hosono, Diffusion and reactions of interstitial oxygen species in amorphous $SiO_2$: A review, *J. Non-Cryst. Solids*, 2008, **354**(2–9), 224.

26. Z. H. Lu and M. J. Graham, Effects of growth temperature on the SiO2/Si (100) interface structure, *J. Vac. Sci. Technol., B: Microelectron. Nanometer Struct.-Process., Meas., Phenom.*, 1995, **13**(4), 1626.

27. N. F. Mott, S. Rigo, F. Rochet and A. M. Stoneham, Oxidation of silicon, *Philos. Mag. B*, 1989, **60**(2), 198.

28. X. Wang, T. Nishimura, T. Yajima and A. Toriumi, Thermal oxidation kinetics of germanium, *Appl. Phys. Lett.*, 2017, **111**, 052101.

29. W. L. Jolly and W. M. Latimer, The Equilibrium $Ge(s) + GeO_2(s) = 2GeO(g)$. The Heat of Formation of Germanic Oxide, *J. Am. Chem. Soc.*, 1952, **74**(22), 5757.

30. K. Oniki, H. Koumo, Y. Iwazaki and T. Ueno, Evaluation of GeO desorption behavior in the metal/$GeO_2$/Ge structure and its improvement of the electrical characteristics, *J. Appl. Phys.*, 2010, **107**(12), 124113.

31. S. K. Wang, H.-G. Liu and A. Toriumi, Kinetic study of GeO disproportionation into a $GeO_2$/Ge system using X-Ray photoelectron spectroscopy, *Appl. Phys. Lett.*, 2012, **101**, 061907.

32. P. Ettmayer and W. Lengauer, Nitrides, *Ullmann's Encyclopedia of Industrial Chemistry*, Wiley-VCH Verlag GmbH & Co. KGaA, 2000, pp. 227–249.

33. S. L. Rumyantsev, M.-S. Shur and M. L. Levinshtein, Materials properties of nitrides, *J. High Speed Electron. Syst.*, 2011, **14**(1), DOI: 10.1142/S012915640400220X.

34. J. B. Mullin, W. R. Macewan, C. H. Holliday and A. E. V. Webb, Pressure balancing: A technique for suppressing dissociation during the melt-growth of compounds, *J. Cryst. Growth*, 1972, **13–14**, 629.

35. NMS Archive, Properties of semiconductors, Copyright 1998–2001 by Ioffe Institute.

36. W. Nakao, H. Fukuyama and K. Nagata, Gibbs Energy Change of Carbothermal Nitridation Reaction of $Al_2O_3$ to Form AlN and Reassessment of Thermochemical Properties of AlN, *J. Am. Ceram. Soc.*, 2002, **85**(4), 889.

37. W. Utsumi, H. Saitho, H. Kaneko, T. Watanuki, K. Aoki and O. Shimomura, Congruent melting of gallium nitride at 6 GPa and its application to single-crystal growth, *Nat. Mater.*, 2003, **2**, 735.

38. J. B. Mullin and A. Royle, A study on the relationship between growth technique and dopants on the electrical properties of GaAs with special reference to LEC growth, *J. Cryst. Growth*, 1980, **50**(3), 625.

39. J. F. Muth, J. H. Lee, I. K. Shmagin, R. M. Kolbas and H. C. Casey, Absorption coefficient, energy gap, exciton binding energy, and recombination lifetime of GaN obtained from transmission measurements, *Appl. Phys. Lett.*, 1997, **71**, 2572.

40. W. A. Bonner, InP synthesis and LEC growth of twin-free crystals, *J. Cryst. Growth*, 1981, **54**(1), 21.

41. R. Rudolph, Present State and Future Tasks of III-V Bulk Crystal Growth, *Proc. Int. Conf. Indium Phosphide Related Mater*, 2007, DOI: 10.1109/ICIPRM.2007.381191.

42. J. Karpinski, J. Jun and S. Porowski, Equilibrium pressure of $N_2$ over GaN and high-pressure solution growth of GaN, *J. Cryst. Growth*, 1984, **66**, 1.
43. R. Rudolph and M. Jurisch, Bulk growth of GaAs. An overview, *J. Cryst. Growth*, 1999, **198–199**, 325.
44. T. F. Kuech, III-V compound semiconductors: Growth and structures, *Prog. Cryst. Growth Charact. Mater.*, 2016, **62**(2), 352.
45. S. Porowski, B. Sadovyi, S. Gierlotka, S. J. Rzoska, I. Grzegory, I. Petrusha, V. Turkevich and D. Stratiichuk, The challenge of decomposition and melting of gallium nitride under high pressure and high temperature, *J. Phys. Chem. Solids*, 2015, **85**, 138.
46. I. Katayama, A. Iseda and K. Kemori Kozuka, Measurements of Standard Gibbs Energies of Formation of ZnO and $ZnGa_2O_4$ by E.M.F. Method, *Trans. Jpn. Inst. Met.*, 1982, **23**(9), 556.
47. *CRC Handbook of Chemistry and Physics*, CRC Press, 103rd edn, 2022.
48. S. Porowski, Growth and properties of single crystalline GaN substrates and homoepitaxial layers, *Mater. Sci. Eng., B*, 1997, **44**, 407.
49. R. Fornari, A. Brinciotti, A. Sentiri, T. Görög, M. Curti and G. Zuccalli, Pressure of phosphorus in equilibrium with solid InP at different temperatures, *J. Appl. Phys.*, 1994, **75**, 2406.
50. C. D. Thurmond, Phase equilibria in the GaAs and the GaP systems, *J. Phys. Chem. Solids*, 1965, **26**, 785.
51. T. Matsuoka, Current status of GaN and related compounds as wide-gap semiconductors, *J. Crystal Growth*, 1992, **124**, 433.
52. K. T. Jacob, S. Singh and Y. Waseda, Refinement of thermodynamic data on GaN, *J. Mater. Res.*, 2007, **22**(12), 3475.
53. K. T. Jacob and G. Rajitha, Discussion of enthalpy, entropy, and free energy of formation of GaN, *J. Cryst. Growth*, 2009, **311**, 3806.
54. J. A. Van Vechten, Quantum Dielectric Theory of Electronegativity in Covalent Systems. III. Pressure-Temperature Phase Diagrams, Heats of Mixing, and Distribution Coefficients, *Phys. Rev. B: Solid State*, 1973, **7**(4), 1479.
55. K. Harafuji, T. Tsuchiya and K. I. Kawamura, Molecular dynamics simulation for evaluating melting point of wurtzite-type GaN crystal, *J. Appl. Phys.*, 2004, **96**(5), 2501.
56. S. Krukowski, C. Skierbiszewski, P. Perlin, M. Leszcynski, M. Bokowski and S. Porowski, Blue and UV semiconductor lasers, *Acta Phys. Pol., B*, 2006, **37**, 1267.
57. G.-I. Grzegory, J. Jun, M. Bockowski, S. Krukowski, M. Wroblewski, B. Lucznik and S. Porowski, III–V Nitrides—thermodynamics and crystal growth at high $N_2$ pressure, *J. Phys. Chem. Solids*, 1995, **56**(314), 639–647.
58. T. J. Peshek, J. C. Angus and K. Kash, Experimental investigation of the enthalpy, entropy, and free energy of formation of GaN, *J. Cryst. Growth*, 2008, **311**, 185.
59. V. Davydov, W. J. Boettinger, U. R. Kattner and T. J. Anderson, Thermodynamic analysis of the Ga-N system, *Phys. Status Solidi A*, 2001, **188**, 407.
60. F. E. Simon and G. Glatzel, Bemerkungen zur Schmelzdruckkurve, *Z. Anorg. Allg. Chem.*, 1929, **178**, 309.
61. F. E. Simon, The Melting of Iron at High Pressures, *Nature*, 1953, **172**, 746.
62. J. Garai and J. Chen, Pressure effect on the melting temperature, *arXiv*, 2009, e-prints, arXiv:0906.3331, https://arxiv.org/abs/0906.333
63. A. Drozd-Rzoska, S. J. Rzoska and A. R. Imre, On the pressure evolution of the melting temperature and the glass transition temperature, *J. Non-Cryst. Solids*, 2007, **353**, 3915.
64. J. Nord, K. Albe, P. Erhartand and K. Nordlund, Modelling of compound semiconductors: Analytical bond-order potential for gallium, nitrogen, and gallium, *J. Phys.: Condens. Matter*, 2003, **15**, 5649.

65. C. Phillips, Dielectric definition of electronegativity, *Phys, Rev. Lett.*, 1968, **20**, 550.
66. A. G. Sokol, Y. N. Palyanov and N. V. Surovtsev, Incongruent melting of gallium nitride at 7.5 GPa, *Diamond Relat. Mater.*, 2007, **16**, 431.
67. W. Utsumi, *et al.*, High pressure science with multi-anvil apparatus at the SPring-8, *J. Phys.: Condens. Matter*, 2002, **14**, 10497.
68. V. B. Eltsov, J. Nissinen and G. E. Volovik, Safe and dangerous routes to anti-spacetime, *Europhys. News*, 2019, **50**(5–6), 34.
69. F. A. Kröger and H. J. Vink, Relations between the concentration of imperfections in crystalline solids, in *Solid State Physics*, ed. F. Seitz and D. Turnbull, Academic Press, New York, 1956, vol. 3.
70. H. Zimmermann, Accurate measurement of the vacancy equilibrium concentration in silicon, *Appl. Phys. Lett.*, 1991, **59**, 3133.
71. J. H. Crawford, L. M. Slifkin, *Point Defects in Solids*, Plenum Press, New York, 1972, vol. 1 and 2.
72. M. Lannoo and J. Burgoin, *Point defects in semiconductors*, Springer Verlag, Berlin, 1981, vol. I and II.
73. F. Agullo-Lopez, C. R. Catlow and P. D. Towsend, *Point defects in Materials*, Academic Press, London, 1988.
74. S. Pizzini, *Defect Interaction and clustering in semiconductors*, Scitec Publications Limited, 2002.
75. P. Pichler, *Intrinsic Point Defects, Impurities, and Their Diffusion in Silicon*, Springer Verlag, 2004.
76. V. I. Fistul, *Impurities in semiconductors: solubility, migration and interaction*, CRC Press, Boca Raton, 2004.
77. G. D. Watkins, Intrinsic Point Defects in Semiconductors, *Mater. Sci. Semicond. Process.*, 2000, **3**, 227.
78. D. A. Drabold and S. Estreicher, *Theory of defects in semiconductors*, Springer, 2007.
79. D. McCluskey and E. E. Haller, *Dopants and Defects in Semiconductors*, CRC Press, Boca Raton, 2012.
80. S. Pizzini, *Physical Chemistry of semiconductor materials and processes*, John Wiley & Sons, 2015.
81. S. Pizzini, *Point defects in group IV semiconductors*, CRC Press, 2017.
82. G. D. Watkins, An EPR study of the lattice vacancy in silicon, *J. Phys. Soc. Jpn.*, 1963, **18**(II), 22.
83. T. Südkamp and H. Bracht, Self-diffusion in crystalline silicon: A single diffusion activation enthalpy down to 755 °C, *Phys. Rev. B*, 2016, **94**, 125208.
84. K. T. Koga, M. J. Walter, E. Nakamura and K. Kobayashi, Carbon self-diffusion in a natural diamond, *Phys. Rev. B: Condens. Matter Mater. Phys.*, 2005, **72**, 024108.
85. H. Bracht, Self-and Dopant diffusion in silicon, germanium and their alloys, in *Silicon, Germanium and their alloys*, ed. G. Kissinger and S. Pizzini, CRC Press, Boca Raton, 2015.
86. G. D. Watkins, The vacancy in silicon: Identical diffusion properties at cryogenic and elevated temperatures, *J. Appl. Phys.*, 2008, **103**, 106106.
87. P. Rudolph, Fundamentals and engineering of defects, *Prog. Cryst. Growth Charact. Mater.*, 2016, **62**, 89.
88. E. Kamiyama, K. Sueoka and I. Vanhellemont, Vacancies in Si and Ge, in *Silicon, Germanium, and their alloys*, ed. G. Kissinger and S. Pizzini, CRC Press, 2015.
89. A. Carvalho and R. Jones, Self-Interstitials in silicon and germanium, in *Silicon, Germanium, and their alloys*, ed. G. Kissinger and S. Pizzini, CRC Press, 2015.
90. R. Car, P. J. Kelly, A. Oshiyama and S. T. Pantelides, Microscopic Theory of Atomic Diffusion Mechanisms in Silicon, *Phys. Rev. Lett.*, 1984, **52**, 1814.

91. G. B. Bachelet, Jacucci, G. R. Car and M. Parrinello, Free energy of formation of lattice vacancies in silicon, *Proceedings 18th International Conference on the Physics of Semiconductors. Stockholm (Sweden)*, ed. O. Engstrom, World Scientific, Singapore, 1987.

92. R. Car, P. E. Blöchl and E. Smargiassi, Ab-initio Molecular Dynamics of Semiconductor Defects, *Mater. Sci. Forum*, 1992, **83–87**, 433.

93. E. Smargiassi and R. Car, First-principles free-energy calculations on condensed-matter systems: *Lattice vacancy in silicon*, *Phys. Rev. B: Condens. Matter Mater. Phys.*, 1996, **53**, 9760.

94. T. Nishimatsu, M. Sluiter, H. Mizuseki, Y. Kawazoe, Y. Sato, M. Miyata and M. Uehara, Prediction of XPS spectra of silicon self-interstitials with the all-electron mixed-basis method, *Phys. B*, 2003, **340–342**, 570.

95. S. A. Centoni, B. Sadigh, G. H. Gilmer, T. J. Lenosky, T. Diaz de la Rubia and C. T. B. Musgrave, First-principle calculation on intrinsic defect volumes in silicon, *Phys. Rev. B: Condens. Matter Mater. Phys.*, 2005, **72**, 195206.

96. R. Jones, A. Carvalho, J. P. Goss and P. R. Briddon, The self-interstitial in silicon and germanium, *Mater. Sci. Eng., B*, 2008, **159–160**, 112.

97. A. Goyal, P. Gorai, E. S. Toberer and V. Stevanović, First-principles calculation of intrinsic defect chemistry and self-doping in PbTe, *npj Comput. Mater.*, 2017, **3**, 42.

98. T. Schick, C. G. Morgan and P. Papoulias, Gallium interstitial contributions to diffusion in gallium arsenide, *AIP Adv.*, 2011, **1**, 032161.

99. S. T. Murphy, M. W. D. Cooper and R. W. Grimes, Point defects and non-stoichiometry in thoria, *Solid State Ionics*, 2014, **267**, 80.

100. I. V. Horichok, H. Ya. Hurhula, V. V. Prokopiv and M. A. Plylyponiuk, Semi-empirical energies of vacancy formation in semiconductors, *Ukr. J. Phys.*, 2016, **61**(11), 992.

101. J. Vanhellemont and P. Spiewak, Intrinsic Point Defect Cluster Formation During Czochralski Crystal Growth, *Phys. Status Solidi C*, 2009, **6**(8), 1906.

102. J. Shim, E.-K. Lee, Y. J. Lee and R. M. Nieminen, Density-functional calculations of defect formation energies using supercell methods: Defects in diamond, *Phys. Rev. B: Condens. Matter Mater. Phys.*, 2005, **71**, 035206.

103. K. Laaksonen, M. G. Ganchenkova and R. M. Nieminen, Vacancies in wurtzite GaN and AlN, *J. Phys.: Condens. Matter*, 2009, **21**, 015803.

104. S. Komatsuda, W. Sato and Y. Ohkubo, Formation energy of oxygen vacancies in ZnO determined by investigating thermal behavior of Al and In impurities, *J. Appl. Phys.*, 2014, **116**, 183502.

105. J. K. Frodason, K. M. Johansen, T. S. Bjørheim, B. G. Svensson and A. Alkauskas, Zn vacancy-donor impurity complexes in ZnO, *Phys. Rev. B*, 2018, **97**, 104109.

106. R. J. Caveney, Chemical bonding and structure of the semiconductor compounds, *Philos. Mag.*, 1968, **17**(149), 943.

107. J. A. Van Vechten, Quantum Dielectric Theory of Electronegativity in Covalent Systems. I. Electronic Dielectric Constant, *Phys. Rev.*, 1969, **182**, 891.

108. J. C. Phillips, Bonds and Bands in Semiconductors, *Science*, 1970, **169**(3950), 1035.

109. H. Blank, Covalent and ionic crystal radii and Phillips ionicity, *Solid State Commun.*, 1974, **15**(5), 907.

110. T. Hidaka, Pauling's Ionicity and Phillips' Ionicity, *J. Phys. Soc. Jpn.*, 1978, **44**, 1204.

111. L. Q. Wu, Y. C. Li, S. Q. Li, Z. Z. Li, G. D. Tang, W. H. Qi, L. C. Xue, X. S. Ge and L. L. Ding, Method for estimating ionicities of oxides using O1s photoelectron spectra, *AIP Adv.*, 2015, **5**, 097210.

112. N. E. Christensen, S. Sapaty and Z. Pawlowska, Bonding and ionicity in semiconductors, *Phys. Rev. B*, 1987, **36**(2), 1032.

113. C. R. A. Catlow and A. M. Stoneham, Ionicity in solids, *J. Phys. C: Solid State Phys.*, 1983, **16**, 4321.
114. T. Mattila and A. Zunger, Predicted bond length variation in wurtzite and zinc-blende InGaN and AlGaN alloys, *J. Appl. Phys.*, 1999, **85**(1), 160.
115. A. Garcia and M. L. Cohen, First-principles ionicity scales. I. Charge asymmetry in the solid state, *Phys. Rev. B: Condens. Matter Mater. Phys.*, 1993, **47**, 4215.
116. D. Schmeißer, K. Henkel and C. Janowitz, Ionicity of ZnO—A key system for transparent conductive oxides, *Europhys. Lett.*, 2018, **123**, 27003.
117. W.-F. Li, C.-M. Fang, M. Dijkstra and M. A. van Huis, The role of point defects in PbS, PbSe and PbTe: a first principles study, *J. Phys.: Condens. Matter.*, 2015, **27**, 355801.
118. L. S. dos Santos, W. G. Schmidt and E. Rauls, Group-VII point defects in ZnSe, *Phys. Rev. B: Condens. Matter Mater. Phys.*, 2011, **84**, 115201.
119. I. V. Horichok, U. M. Pysklynets and V. V. Prokopiv, Formation Energies of Native Point Defects in II–VI Crystals, *Inorg. Mater.*, 2012, **48**(2), 119.
120. C.-Y. Yeh, Z. W. Lu, S. Froyen and A. Zunger, Zinkblende-wurtzite polytypism in semiconductors, *Phys. Rev. B: Condens. Matter Mater. Phys.*, 1992, **46**(16), 10086.
121. A. Ashrafi and C. Jagadish, Review of zincblende ZnO: Stability of metastable ZnO phases, *J. Appl. Phys.*, 2007, **102**, 071101.
122. J. E. Jaffe and A. C. Hess, Hartree-Fock study of phase changes in ZnO at high pressure, *Phys. Rev. B: Condens. Matter Mater. Phys.*, 1993, **48**, 7903.
123. P. Rudolph, Non-stoichiometry related defects at the melt growth of semi-conductor compound crystals – a review, *Crystal Res. Technol.*, 2003, **38**(7–8), 542.
124. J. Nishizawa, Stoichiometry control and point defects in compound semi-conductors, *Mater. Chem. Phys.*, 2000, **64**(2), 93.
125. S. Nakamura, GaN Growth Using GaN Buffer Layer, *Jpn. J. Appl. Phys.*, 1991, **30**, L1705.
126. S. Nakamura, T. Mukai and M. Senoh, Candela-Class High-Brightness InGaN/AlGaN Double-Heterostructure Blue-Light-Emitting-Diodes, *Appl. Phys. Lett.*, 1994, **64**, 1687.
127. J. M. Besson, J. P. Itié, A. Polian, G. Weill, J. L. Mansot and J. Gonzalez, High-pressure phase transition and phase diagram of gallium arsenide, *Phys. Rev. B: Condens. Matter Mater. Phys.*, 1991, **44**, 4214.
128. J. S. Blakemore, Semiconducting and other major properties of gallium ar-senide, *J. Appl. Phys.*, 1982, **53**, R123.
129. T. F. Kuech, III-V compound semiconductors: Growth and structures, *Prog. Cryst. Growth Charact. Mater.*, 2016, **62**(2), 352.
130. K. Laasonen, R. M. Nieminen and M. J. Puska, First-principles study of fully relaxed vacancies in GaAs, *Phys. Rev. B: Condens. Matter Mater. Phys.*, 1992, **45**(8), 4122.
131. F. El-Mellouhi and N. Mousseau, Self-vacancies in Gallium Arsenide: an *ab initio* calculation, *Phys. Rev. B: Condens. Matter Mater. Phys.*, 2005, **71**(12), 125207.
132. P. A. Schultz and O. Anatole von Lilienfeld, Simple intrinsic defects in gallium arsenide. Modelling and Simulation, *Mater. Sci. Eng.*, 2009, **17**(8), 084007.
133. D. T. J. Hurle, A thermodynamic analysis of native point defect and dopant solubilities in zinc-blende III–V semiconductors, *J. Appl. Phys.*, 2010, **107**, 121301.
134. S. B. Zhang and J. E. Northrup, Chemical potential dependence of defect for-mation energies in GaAs: Application to Ga self-diffusion, *Phys. Rev. Lett.*, 1991, **67**, 2339.
135. H. A. Tahini, A. Chroneos, S. T. Murphy, U. Schwingenschlögl and R. W. Grimes, Vacancies and defect levels in III–V semiconductors, *J. Appl. Phys.*, 2013, **114**, 063517.

136. C. G. Van de Walle and J. Neugebauer, First-principles calculations for defects and impurities: Applications to III-nitrides, *J. Appl. Phys.*, 2004, **95**, 3851.
137. J. Gebauer, M. Lausmann, F. Redmann, R. Krause-Rehberg, H. S. Leipner, E. R. Weber and Ph. Ebert, Determination of the Gibbs Free Energy of Formation of Ga Vacancies in GaAs by Positron Annihilation, *Phys. Rev. B: Condens. Matter Mater. Phys.*, 2003, **67**(23), 235207.
138. M. B. Panish and H. C. Casey Jr., Temperature Dependence of the Energy Gap in GaAs and GaP, *J. Appl. Phys.*, 1969, **40**, 163.
139. C. D. Thurmond, The Standard Thermodynamic Functions for the Formation of Electrons and Holes in Ge, Si, GaAs, and GaP, *J. Electrochem. Soc.*, 1975, **122**(8), 1133.
140. V. Bodnarenko, *Positron annihilation study of equilibrium point defects in GaAs*, PhD Dissertation Thesis, Martin-Luther-Universität Halle-Wittenberg, 2003.
141. S. Dannefear, P. Mascher and D. Kerr, Monovacancy formation enthalpy in silicon, *Phys. Rev. Lett.*, 1986, **56**, 2195.
142. S. Dannefear, V. Avalos and O. Andersen, Grown-in vacancy defects in poly-and single crystalline silicon investigated by positron annihilation, *Eur. Phys. J.: Appl. Phys.*, 2007, **37**, 213.
143. R. Krause-Rehberg, H. S. Leipner, T. Abgarjan and A. Polity, Review of defect investigations by means of positron annihilation in II–VI compound semiconductors, *Appl. Phys. A: Mater. Sci. Process.*, 1998, **66**, 599.
144. C. Rauch, I. Makkonen and F. Tuomisto, Identifying vacancy complexes in compound semiconductors with positron annihilation spectroscopy: A case study of InN, *Phys. Rev. B: Condens. Matter Mater. Phys.*, 2011, **84**, 125201.
145. T. Tuomisto and I. Makkonen, Defect identification in semiconductors with positron annihilation: Experiment and theory, *Rev. Mod. Phys.*, 2013, **85**(4), 1583.
146. V. Bondarenko, J. Gebauer, F. Redmann and R. Krause-Rehberg, Vacancy formation in GaAs under different equilibrium conditions, *Appl. Phys. Lett.*, 2005, **87**, 161906.
147. S. Abbaspour, B. Mahmoudian and J. P. Islamian, Cadmium Telluride Semiconductor Detector for Improved Spatial and Energy Resolution Radioisotopic Imaging, *World J. Nucl. Med.*, 2017, **16**(2), 101.
148. I. Dharmadasa, A. A. Ojo, H. I. Salim and R. Dharmadasa, Next generation solar cells based on graded bandgap device structures utilizing rod-type nanomaterials, *Energies*, 2015, **8**, 5440.
149. D. N. Krasikov, A. V. Scherbinin, A. A. Knizhnik, A. N. Vasiliev, B. V. Potapkin and T. J. Sommerer, Theoretical analysis of non-radiative multiphonon recombination activity of intrinsic defects in CdTe, *J. Appl. Phys.*, 2016, **119**, 085706.
150. M. A. Berding, Native defects in CdTe, *Phys. Rev. B: Condens. Matter Mater. Phys.*, 1999, **60**, 8943.
151. J. H. Greenberg, V. N. Guskov, V. B. Lazarev and O. V. Shebershneva, Vapor pressure scanning of non-stoichiometry in cadmium telluride, *Mater. Res. Bull.*, 1992, **27**, 847.
152. H. Greenberg, P-T-X phase equilibrium and vapor pressure scanning of non-stoichiometry in CdTe, *J. Cryst. Growth*, 1996, **161**(1–4), 1.
153. J. Greenberg, Experimental Data on P-T-X Phase Diagrams and Non-stoichiometry, *Thermodynamic Basis of Crystal Growth*, Springer Series in Materials Science, 2002, vol. 44.
154. Ch. Avetissov, E. N. Mozhevitina, A. V. Khomyakov, R. I. Avetisov, A. A. Davydov, V. P. Chegnov, O. I. Chegnova and V. Zhavoronkov, Homogeneity limits and nonstoichiometry of vapor grown ZnTe and CdTe crystals, *CrystEngComm*, 2015, **17**, 561.

155. L. Yujie, M. Guoli and J. Wanqi, Point defects in CdTe, *J. Cryst. Growth*, 2003, **256**(3–4), 266.
156. R. Grill and A. Zappettini, Point defects and diffusion in cadmium telluride, *Prog. Cryst. Growth Charact.*, 2004, **49**(1), 209.
157. P. Fochuk, R. Grill and O. Panchuk, The nature of point defects in CdTe, *J. Electron. Mater.*, 2006, **35**, 1354.
158. A. Lindström, *Defects and Impurities in CdTe. An ab Initio Study*. Acta Universitatis Upsaliensis, Assala University, 2015, ISBN: 978-91-554-9171-0.
159. M. R. M. Elsharkawy, G. S. Kanda, E. E. Abdel-Hady and D. J. Keeble, Characterization of point defects in CdTe by positron annihilation spectroscopy, *Appl. Phys. Lett.*, 2016, **108**(24), 242102.
160. J.-H. Yang, W.-J. Yin, J.-S. Park, J. Ma and S.-H. Wei, Review on first-principles study of defect properties of CdTe as a solar cell absorber, *Semicond. Sci. Technol.*, 2016, **31**(8), 083002.
161. T. Ablekim, S. K. Swain, W.-J. Yin, K. Zaunbrecher, J. Burst, T. M. Barnes, D. Kuciauskas, S.-H. Wei and K. G. Lynn, Self-compensation in arsenic doping of CdTe, *Sci. Rep.*, 2017, **7**, 4563.
162. M.-H. Du, H. Takenaka and D. J. Singh, Native defects and oxygen and hydrogen-related defect complexes in CdTe: Density functional calculations, *J. Appl. Phys.*, 2008, **104**(9), 093521.
163. A. Kolmakov, D. O. Klenov, Y. Lilach, S. Stemmer and M. Moskovits, Enhanced Gas Sensing by Individual $SnO_2$ Nanowires and Nanobelts Functionalized with Pd Catalyst Particles, *Nano Lett.*, 2005, **5**(4), 667.
164. L. Zhu and W. Zeng, Room-temperature gas sensing of ZnO-based gas sensor: A review, *Sens. Actuators*, 2017, **267**, 242.
165. A. M. Deml, A. M. Holder, R. P. O'Hayre, C. B. Musgrave and V. Stevanovic, Intrinsic Material Properties Dictating Oxygen Vacancy Formation Energetics in Metal Oxides, *J. Phys. Chem. Lett.*, 2015, **6**, 1948.
166. W. Ding, D. Liu, J. Liu and J. Zhang, Oxygen Defects in Nanostructured Metal-Oxide Gas Sensors: Recent Advances and Challenges, *Chin. J. Chem.*, 2020, **38**, 1832.
167. J. F. Muth, J. H. Lee, I. K. Shmagin and R. M. Kolbas, Absorption coefficient, energy gap, exciton binding energy, and recombination lifetime of GaN obtained from transmission measurements, *Appl. Phys. Lett.*, 1977, **71**, 2572.
168. I. Katayama, A. Iseda, N. Kemori and Z. Kozuka, Measurements of Standard Gibbs Energies of Formation of ZnO and $ZnGa_2O_4$ by E.M.F. Method, *Trans. Japan Instit. Metals*, 1982, **23**(9), 556.
169. G. M. Khan and M. S. Subkani, Standard free energy of formation of ZnO, *Z. Phys. Chem.*, 1982, **263**, 1034.
170. A. Janotti and C. G. Van de Walle, Fundamentals of zinc oxide as a semiconductor, *Rep. Prog. Phys.*, 2009, **72**, 126501.
171. N. S. Parmar, L. A. Boatner, K. G. Lynn and J.-W. Choi, Zn Vacancy Formation Energy and Diffusion Coefficient of CVT ZnO Crystals in the Sub-Surface Micron Region, *Sci. Rep.*, 2018, **8**, 13446.
172. H. J. Allsopp and J. P. Roberts, Non-stoichiometry of zinc oxide and its relation to sintering. Part 1. Determination of non-stoichiometry in zinc oxide, *Trans. Faraday Soc.*, 1959, **55**, 1386.
173. E. A. Secco, Decomposition of Zinc Oxide, *Can. J. Chem.*, 1960, **38**(4), 596.
174. T. Imoto, Y. Harano and Y. Nishi, The Thermal Decomposition of Zinc Oxide, *Bull. Chem. Soc. Jpn.*, 1964, **37**(8), 1181.
175. K. Lott, S. Shinkarenko, T. Kirsanova, L. Türn, E. Gorohova and A. Grebennik, Zinc nonstoichiometry in ZnO, *Solid State Ionics*, 2004, **173**(1–4), 29.
176. J. C. Greenbank and B. B. Argent, Vapour pressure of magnesium, zinc, and cadmium, *Trans. Faraday Soc.*, 1965, **61**, 655.

177. J. Shi, C. Chul and H. Yo, Study of the nonstoichiometric composition of zinc oxide, *J. Phys. Chem. Solids*, 1976, **37**(12), 1149.
178. P. Ehrart, A. Klein and K. Albe, First principles study of the structure and stability of oxygen defects in zinc oxide, *Phys. Rev. B: Condens. Matter Mater. Phys.*, 2005, **72**, 085213.
179. A. Pöppl and G. Völkel, ESR investigation of the oxygen vacancy in pure and $Bi_2O_3$-doped ZnO ceramics, *Phys. Status Solidi A*, 1989, **115**(1), 247.
180. F. Morazzoni, R. Scotti and S. Volontè, Electron paramagnetic Resonance investigation of paramagnetic point defects in ZnO and ZnO-supported ruthenium, *J. Chem. Faraday Trans.*, 1990, **86**(9), 1587.
181. W. J. Moore and E. L. We, Diffusion of zinc and oxygen in zinc oxide, *Discuss. Faraday Soc.*, 1959, **28**, 86.
182. S. Pizzini, N. Buttà, D. Narducci and M. Palladino, Thick Film ZnO resistive gas sensors, *J. Electrochem. Soc.*, 1989, **136**(7), 1945.
183. Q. Ou, T. Matsuda, M. Measko, A. Ogino and M. Nagatsu, Cathodoluminescence property of ZnO nanophosphors prepared by laser ablation, *Jpn. J. Appl. Phys.*, 2008, **47**(1), 389.
184. D. Gaspar, L. Pereira, K. Gehrke, B. Galler, E. Fortunato and R. Martins, High mobility hydrogenated zinc oxide thin films, *Sol. Energy Mater. Sol. Cells*, 2017, **163**, 255.
185. C. van de Walle, Hydrogen as cause of doping in Zinc oxide, *Phys. Rev. Lett.*, 2000, **85**, 1012.
186. E. V. Lavrov, Hydrogen in ZnO, *Phys. B*, 2009, **404**(23), 5075.
187. J. J. van Laar, Die Schmelz- oder Erstarrungskurven bei binären Systemen, wenn die feste Phase ein Gemisch (amorphe feste Lösnng oder Mischkristalle) der beiden Komponenten ist, *Z. Phys. Chem.*, 1908, **63**, 216.
188. J. J. van Laar, Über Dampfspannungen von binären Gemischen, *Z. Phys. Chem.*, 1908, **64**, 257.
189. J. H. Hildebrand, Solubility. XII. Regular solutions, *J. Am. Chem. Soc.*, 1929, **51**, 66.
190. R. H. Fowler and E. A. Guggenheim, *Statistical Thermodynamics*, University Press, Cambridge, 1939.
191. E. A. Guggenheim, *Mixtures*, Oxford University Press, 1952.
192. G. B. Stringfellow, The calculation of regular solution interaction parameters between elements from groups III, IV and V of the periodic table, *J. Phys. Chem. Solids*, 1973, **34**, 1749.
193. A. Baldereschi and M. Peressi, Atomic scale structure of ionic and semiconducting solid solutions, *J. Phys.: Condens. Matter.*, 1993, **5**, 837.
194. B. N. Sarma, S. S. Prasad, S. Vijayvergiva, V. B. Kumar and S. Lele, Existence domains for invariant reactions in binary regular solution phase diagrams exhibiting two phases, *Bull. Mater. Sci.*, 2003, **26**(4), 423.
195. A. Mostafa and M. Medraj, Binary phase diagrams and thermodynamic properties of silicon and essential doping elements (Al, As, B, Bi, Ga, In, N, P, Sb and Tl), *Materials*, 2017, **10**, 676.
196. R. W. Olesinski, N. Kanani and G. J. Abbaschian, The In-Si (Indium-silicon) System, *Bull. Alloy Phase Diagrams*, 1985, **6**(2), 128.
197. L. Zhigilei, Phase Diagrams and Kinetics, Lecture Notes, Web reference MSE 3050, 2010, http://virginia.edu/.
198. A. Liang, Y. Liu, L. Shi, L. Lei, F. Zhang, Q. Hu and D. He, Melting temperature of diamond and cubic boron nitride at 15 gigapascals, *Phys. Rev. Research*, 2019, **1**, 033090.

# 2 Role of Defects, Impurities and Deviations from the Stoichiometry in the Optoelectronic Properties of Semiconductors

## 2.1 Introduction

Elemental and compound semiconductors are the materials that allowed the birth and the further growth of information, lighting, and solar light-harvesting technologies.

The revolution started at the Bell Labs in the years 1947 and 1948, when the Nobel laureates Bardeen, Brattain and Shockley succeeded with the fabrication of the first transistors, which worked as a solid-state amplifier, using germanium and silicon single crystals, giving rise also to single crystal growth technologies.[1]

The second revolutionary event was the realization of the first, viable, red-light emitting diode (LED) in the 1960s,[†][2] followed, in 1993, by the fabrication of the first green and blue LEDs and lasers by Nakamura *et al.*,[3–7] using single and multiple InGaN multiquantum well structures, grown using a metal–organic phase deposition process, after having also found a way to dope p-type GaN using Mg as the dopant.[8]

---

[†] Although the capability of compound semiconductors to emit light was known already in the 19th century.

---

Chemistry of Semiconductors
By Sergio Pizzini
© Sergio Pizzini 2024
Published by the Royal Society of Chemistry, www.rsc.org

The third revolutionary event was the start-up in China of giant production plants of polycrystalline silicon and of GW-capacity factories for the mass production of silicon solar cells, with a record decrease in the selling cost of polysilicon,[‡] silicon wafers and solar cells.

Simultaneously with the development of these new technologies, research focused interest on the role of impurities, defects and crystal size in the opto-electronic properties of semiconductors, with the development of novel experimental and theoretical techniques, which led not only to the improvement of the already existing growth technologies but also to the birth of semiconductor nanotechnologies.

The present chapter deals with the physico-chemical grounds of the detrimental role of metallic impurities and of various types of defects on the performances of semiconductor devices, including their dopability, this last is shown to be critical in several compound semiconductors like GaN and ZnO, and is specifically addressed at the understanding of the physico-chemical nature of interactions involving defects and impurities, without neglecting their purely physical aspects.

We will also deal with chemical gettering technologies based on oxide precipitates, addressed at the removal of device-degrading impurities from the active circuit regions, which is an important step for enhancing the yield of VLSI manufacturing.

## 2.2 Semiconductors for Micro-electronic Applications

Microelectronic and photonic applications of semiconductors are based, as is well known, on the use of homo- or hetero-junctions, transistors, CMOS, FETs and MOSFETs, which represent the building blocks of all semiconductor devices, including solar cells, light emitting devices (LEDs) and lasers.

For microelectronic applications and solar cells, indirect gap semiconductors are appropriate, though not ideal, while for photonic applications (light emitting diodes and lasers) direct gap semiconductors are, instead, necessary, as we will discuss in the next section.

The main requirements of semiconductor materials for their microelectronic applications are large energy gap ($E_G$) values, low values

---

[‡] High purity polysilicon is the material used to grow silicon crystals for solar cell uses.

**Table 2.1** Physical and thermal properties of semiconductors used in microelectronic applications,[9] where $E_G$ is the energy gap, $\varepsilon$ is the dielectric constant, $\mu_e$ is the electron mobility, $k$ is the thermal conductivity and $E_c$ is the critical electric field for breakdown.

| | $E_G$ (eV) | $\varepsilon$ (F m$^{-1}$) | $\mu$ (cm$^2$ V$^{-1}$ s$^{-1}$) | $\kappa$ (W K$^{-1}$ cm$^{-1}$) | $E_c$ (V cm$^{-1}$) |
|---|---|---|---|---|---|
| Ge | 0.66 | 16.2 | $3.9 \times 10^3$ | 0.58 | $10^5$ |
| Si | 1.11 | 11.9 | $1 \times 10^3$ | 1.5 | $3 \times 10^5$ |
| InP | 1.34 | 12.4 | $6.6 \times 10^3$ | 0.67 | $4.5 \times 10^5$ |
| GaAs | 1.43 | 12.5 | $11.5 \times 10^3$ | 0.54 | $4 \times 10^5$ |
| | 3.2 | 10 | $0.9 \times 10^3$ | 4 | $3.5 \times 10^6$ |
| GaN | 3.4 | 9.5 | $0.44 \times 10^3$ | 1.3 | $2 \times 10^6$ |
| Diamond | 5.6 | 5.5 | $0.9 \times 10^3$ | 20–30 | $5 \times 10^6$ |

of the dielectric constant $\varepsilon$, high thermal conductivities $\kappa$, high electron mobilities $\mu_e$ and high critical electric fields for breakdown $E_c$, that translate into the ability of a semiconductor device to allow operation in harsh thermal environments, in the high radio frequency range (RF) typical of 5G applications and to support high critical fields before breakdown. High electric breakdown values favour, in fact, the high frequency and high-power operation of RF devices.[9]

Table 2.1 displays these properties for several semiconductors of common and less common use in semiconductor devices, from which one already sees the potentialities of compound semiconductors with respect to silicon and germanium. It is well known, however, that the microelectronic world is for today and will be for the future based on high purity, single crystal silicon substrates,[§] which grant excellent performance at ambient temperatures of a variety of devices, including, as a non-conventional example, the Si-strip radiation detectors installed and operating in the high luminosity LHC supercollider operating at CERN in Geneva,[12] whose compatibility with the radiation damage arising from fast neutrons irradiation was studied, among others, by Acciarri *et al.*[13]

An advantage of Si *vs.* Ge[¶] is its larger energy gap ($E_G$), which grants an intrinsic carrier concentration at room temperature ($1.45 \times 10^{10}$ cm$^{-3}$) three orders of magnitude lower than that of Ge ($2 \times 10^{13}$ cm$^{-3}$), favouring the onset of lower thermal leakage currents,[‖] although the higher carrier mobility of Ge and its quasi direct gap in strained

---

[§] Today we see an emerging role of GaN also in microelectronic devices.
[¶] This is, historically, the first semiconductor used for a solid-state diode.
[‖] A thermal leakage current in a diode is the current flowing to ground without a bias applied.

quantum wells favours its use in advanced applications. This is the case of SiGe heterojunctions, bipolar transistors, and strained Si metal–oxide–semiconductor (MOS) transistors for advanced complementary metal–oxide–semiconductor (CMOS) and BiCMOS (bipolar CMOS) technologies.[10,11]

The main advantage of silicon is, however, the availability of a full set of advanced crystal growth technologies, strongly improved in the last several decades,[14–16] making of it the highest pure and almost defect-free material, produced in estimated 8.5 million tonnes quantities worldwide[**] at very low production and selling costs.

Silicon carbide (SiC) came into the game few decades ago because of its suitability for electronic and optoelectronic devices operating under high temperature, high power, high frequency and strong radiation conditions, a field where conventional semiconductor materials like Si and GaAs have reached their limits.[17]

SiC belongs to a family of more than 200 polytypes, with band gaps ranging between 2.3 and 3.3 eV, of which the most common are the cubic SiC-3C, and the hexagonal SiC-4H and SiC-6H ones.

SiC claims many superior properties, such as its excellent thermal stability and chemical inertia, a larger operating temperature range in comparison to silicon, a high critical breakdown field, and high resistance to radiation damage. Owing to these outstanding properties, silicon carbide already plays a significant role in high temperature applications, power devices and microwave devices, although its high temperature synthesis, and the problems arising from its polymorphism and defectivity, have been for years a limit to its employment.[18]

Direct-gap InP and GaAs[††] have an advantage due to their high electron mobilities for high frequency (RF) applications, but their low thermal conductivities is a problem in high power RF operations.[9] More recently, GaN found applications in power-RF devices, in advanced HEMTs (high electron mobility transistors) for 5G devices,[19] for small and efficient battery chargers[9] and dc–dc controllers, thanks to improved methods of its growth on low-cost foreign substrates, which we will deal with in Chapter 3, making GaN is a real competitor to silicon in these fields of use. Its main application, however, has been and still is as the standard material for LED and lasers in the blue and UV range,[20] as we will see in the next section.

---

[**] In the year 2021.
[††] GaAs is systematically used for optical applications (LED and lasers).

Both silicon and SiC take, eventually, advantage of the excellent thermodynamic stability of the silicon oxide $SiO_2$, which can be thermally grown on Si and SiC[21] surfaces, with the formation of the insulation layer, which enables the fabrication of metal/oxide/semi-conductor (MOS) structures.

## 2.3 Semiconductors for Photonic Applications

Intrinsic optical absorption, with the associated generation and transport of free carriers, and the light generation from polarized direct gap semiconductor diodes, are at the background of all photonic applications of semiconductors, which include light emission, solar light harvesting, and radiation detection.[22]

For light-emission applications, most direct gap compound semiconductors present the required potentialities, when coupled to efficient and stable inorganic phosphors used as wavelength emission converters, but the success is strictly bound to their easy deposition as thin crystalline layers on a convenient substrate, to their dopability and, then, to the effectiveness of p–n homo- or hetero-junctions fabricated with them. Today, GaN-based light emitting diodes (LEDs) are in mass production for lighting purposes for homes, cars, streets, and many other applications. Since laser lighting has the advantage of high luminosity and easier optical management of the light, it is also conceivable that high intensity lighting applications will be operated, in the near future, by GaN laser sources.

Although Ge and Si are very inefficient light emitters, silicon nanocrystal might fulfil this function,[24,25] as well as hexagonal Si–Ge alloys, which behave as direct gap semiconductors,[26] and whose emission wavelength might be tuned by changing their composition.

These findings open novel applications to silicon and its alloys, with SiGe alloys that could represent an ideal material system, capable of combining electronic and optoelectronic functionalities on a single chip, opening the way towards integrated device concepts and information-processing technologies.[‡‡]

For solar cell applications, both indirect gap and direct gap semiconductors with energy gaps in the range 1–2 eV are appropriate, see Figure 2.1, since their energy gaps fit well with the wavelength range of solar light. These semiconducting materials include Ge, Si, GaAs, InP and CdTe, but many others present the required potentialities,

---

[‡‡] Original sentence from ref. 26.

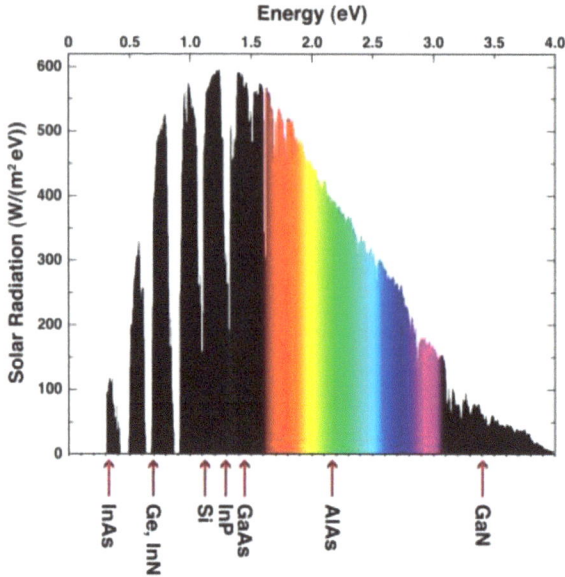

**Figure 2.1** Solar radiation spectrum and energy gaps of semiconductors of potential use in solar cell applications. Reproduced from ref. 27 with permission from John Wiley & Sons, Copyright © 2009 Wiley-VCH Verlag GmbH & Co. KGaA, Weinheim.

including ternary III–V nitrides,[§§] and the inorganic and mixed organic-inorganic perovskites, these last offering efficiencies close to 30%, but with some degradation problems under sunlight illumination.

Silicon remains, however, the material that offers the best compromise between efficiency and cost, with efficiencies close to 25% and with hundreds of GWatt of material available, at a silicon cost of US$0.2 per Watt (NREL 2020), that support the large-scale development of the photovoltaic industry.

The last, but not least, application of semiconductors is in the field of radiation detectors, which are today employed in a wide range of duties, extending from astrophysical research to subnuclear physics and to medicine.

While silicon and germanium have been the standard materials for laboratory spectroscopies under cryogenic operation, needed to reduce thermal noise (see details in Section 2.6), all wide-gap compound semiconductors could potentially support continuous spectroscopic coverage from the far infrared through to γ-ray wavelengths, including neutrons,[29] without the problem of cryogenic

---

[§§] With their high open circuit voltages.

operation. However, while they could be routinely used at infrared and optical wavelengths, in high frequency bands their use is hindered by poor radiation hardness, defectivity and material fabrication problems. This is particularly true at hard X- and $\gamma$-ray wavelengths, except for a few compounds like GaAs, CdZnTe and HgI$_2$.[30]

Since an excellent radiation hardness is the prerequisite for the operation of detectors in the High Luminosity Hadron Collider (HLHC) at CERN (Geneva), devoted to advanced subnuclear research; extensive research efforts and defect engineering attempts have been applied to both Si and SiC, in view of optimizing the operation of float zone (FZ) silicon-based strip detectors used in the ATLAS experiment,[31] and extending to SiC detectors the spectroscopic investigation of subnuclear events, given that SiC offers excellent potentialities for radiation detector ($\alpha$ and $\beta$ particles, ions and X-rays) applications.[32–36]

A successful example of chemical engineering applied silicon detectors has been the oxygen enrichment of high resistivity FZ-silicon, that succeeded in achieving a strong improvement of the radiation hardness under irradiation with charged, high energy hadrons. This beneficial effect was recently found to be even more pronounced under gamma irradiation.[37]

## 2.4 Equilibrium Thermodynamics of Free Carriers in a Semiconductor and in a p–n Junction

The practical performance of semiconductor devices depends not only on the intrinsic electronic properties of the material used, but on the thermophysical conditions of carrier generation and recombination processes, which occur in the presence of dopant impurities, and of metallic impurities and defects, the latter behaving as carrier recombination centres, leading to the degradation of the device performance.

The equilibrium properties of free carriers in an intrinsic semiconductor could be quantitatively discussed for a system consisting of $N$ particle states in thermal equilibrium,[38] each occupied by a single particle, according to the Pauli exclusion principle, and each defined by an energy $E$ and a weight $P_N(E)$

$$P_N(E) = \frac{\exp - E/k_B T}{\sum\limits_i^N \exp - E_i^N/k_B T} \tag{2.1}$$

where $k_B$ is the Boltzmann constant, $T$ is the absolute temperature, $E_i^N$ is the energy of the $i$th state, and the sum $\sum$ is extended to all the $N$ states.

Given that the sum $Q = \sum_i^N \exp - E_i^N/k_B T$ at the denominator of eqn (2.1) is the partition function, which is directly related to the Helmholtz free energy $F = U - TS$ of the system[39]

$$F(T) = -k_B T \ln Q \tag{2.2}$$

we have also that

$$\sum_i^N \exp - E_i^N/k_B T = \exp - F_N/k_B T \tag{2.3}$$

which allows us to rewrite eqn (2.1) as a function of the difference $E - F_N$

$$P_N(E) = \exp - (E - F_N)/k_B T. \tag{2.4}$$

If we further consider that the energy of a system with $N+1$ states $E^{N+1}$ differ from the energy $E^N$ of a system with $N$ states of an energy $\varepsilon = E^{N+1} - E^N$, we can also see that the weight $P_N$ of a system with $N$ states differs from that of a system with $N+1$ states by a term $\exp - \dfrac{(\varepsilon - \mu)}{k_B T}$

$$P_N(E^N) = \exp - \frac{(\varepsilon - \mu)}{k_B T} P_{N+1}(E^{N+1}) \tag{2.5}$$

where $\mu = F_{N+1} - F_N$ is the chemical potential of the system, which, in fact, accounts for the change in the Helmholtz energy of the system arising from the addition of one electron. Since $N$ is large, the addition of one electron to the system does not macroscopically change the value of $E$.

This conclusion fits with the thermodynamic nature of the chemical potential $\mu$, which applies to an open phase of non-constant composition, and thus also to a phase with a variable concentration of electrons, for which a variation $\delta F$ of the Helmholtz free energy $F = U - TS$[¶¶] associated with a change $\delta n_e$ in the

---

[¶¶] Where $U$ is the internal energy and $S$ is the entropy.

concentration of electrons is, in fact, represented by the chemical potential

$$\left(\frac{\delta F}{\delta n_e}\right)_S = \left(\frac{\delta U}{\delta n_e}\right)_S = \left(\frac{\delta u}{\delta n_e}\right)_S = \mu \qquad (2.6)$$

where $u = U/V$ is the internal energy density per unit volume.

In such a system the probability $f_i^N$ that a one-electron level of energy $\varepsilon_i$ is occupied is given by the following equation

$$f_i^N = \frac{1}{[\exp(\varepsilon_i - \mu)/k_B T] + 1} \qquad (2.7)$$

which is an expression of the Fermi–Dirac distribution. It is easy to see that eqn (2.7) reduces to the Boltzmann distribution if $\exp(\varepsilon_i - \mu)/k_B T \gg 1$.

It can be further shown, with some simplifications of the statistical factors,[38] that the intrinsic concentrations of electrons $n_e(T) = n$ and holes $n_h(T) = p$ in an impurity-free semiconductor are given by the following equations

$$n_e(T) = n = N_C(T)\exp - (E_C - \mu)/k_B T \qquad (2.8)$$

$$n_h(T) = p = P_V(T)\exp - (\mu - E_V)/k_B T \qquad (2.9)$$

where $N_C(T)$ and $P_V(T)$ are the densities of the one-electron levels in the conduction and valence bands,[‖] $E_C$ and $E_V$ are the energies of the bottom of the conduction band and of the top of the valence band, respectively, and $\mu$ is, again, the chemical potential of the system, or the Fermi level $E_F$ of the semiconductor.

It is apparent that in conditions of thermal and electroneutrality equilibrium $n_e(T) = n_h(T)$, the terms $(E_C - \mu)$ and $(\mu - E_V)$ in eqn (2.8) and (2.9) represent the energies needed to generate electrons in the conduction band or holes in the valence band and that $\mu$ lies exactly in the middle of the energy gap at $T = 0$, since

$$P_V(T)\exp - (\mu - E_V)/k_B T = N_C(T)\exp - (E_C - \mu)/k_B T \qquad (2.10)$$

and

$$\mu = 1/2kT \, \ln \frac{P_V(T)}{N_C(T)} + 1/2(E_C + E_V) = \frac{1}{2}kT \, \ln \frac{P_V(T)}{N_C(T)} + \frac{1}{2}E_G. \qquad (2.11)$$

---

[‖] We recall here that by the Pauli exclusion principle only one electron could be placed in a single electron level.

In thermodynamic equilibrium conditions, and in the absence of impurities or defects that could behave as electron donors or acceptors, the product of the electron and holes concentrations in a semiconductor phase is given, eventually, by the following equation derived by a product of terms of eqn (2.8) and (2.9)

$$np = P_V(T)N_C(T)\exp - \frac{\exp - (E_C - E_V)}{k_B T} = A \ \exp - (E_G)/k_B T \qquad (2.12)$$

which is a form of the mass of action law, where the term $A \exp - (E_G)/k_B T$ represents the value of the equilibrium constant.

The intrinsic equilibrium concentration of free carriers $n_i = \sqrt{np}$ depends, therefore, on the energy gap $E_G$ of the semiconductor, and increases with temperature, and is ruled by carrier generation and recombination processes written with the following mass law equilibrium equation

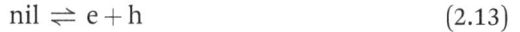

$$\text{nil} \rightleftharpoons e + h \qquad (2.13)$$

where nil means an electron in the valence band, e is an electron in the conduction band and h is a hole in the valence band.

Their equilibrium generation/recombination rates $(R_G/R_R)$ should, necessarily, satisfy the electroneutrality condition, and thus

$$R_G = R_R \qquad (2.14)$$

to maintain equal and constant, at constant temperature, the concentrations of free electrons and holes.

Thermal energy is not the only source of free-carrier generation, according eqn (2.12), since the generation of an excess carrier concentration in a semiconductor, and their recombination to bring the system again to thermodynamic equilibrium, might occur with the absorption or the emission of photons of energy $h\nu \geq E_G$, where $h$ is the Planck constant, $\nu$ is the frequency of the absorbed or emitted photons, and $E_G = E_c - E_V$.

A straight emission–recombination process of this kind, however, occurs only in direct gap semiconductors, where electrons and holes are in the same $k$ position in the $E - k$ space, see Figures 2.2 and 2.3, enabling the conservation of crystal momentum during the $\Delta k = 0$ transition. In the case of indirect semiconductors, the minimum energy of the conduction band and the maximum of the valence band do not correspond in the $E - k$ space, see Figure 2.3b and,

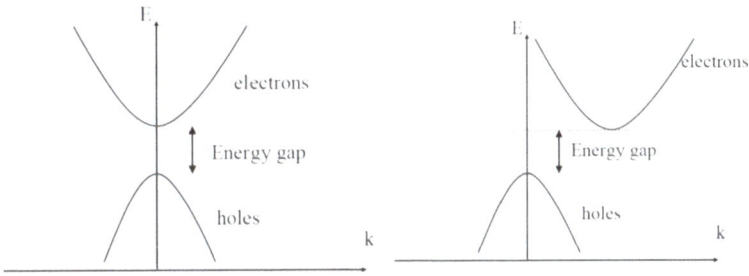

**Figure 2.2** Band structures in the $E-k$ space of a direct gap (left) and an indirect gap (right) semiconductor.

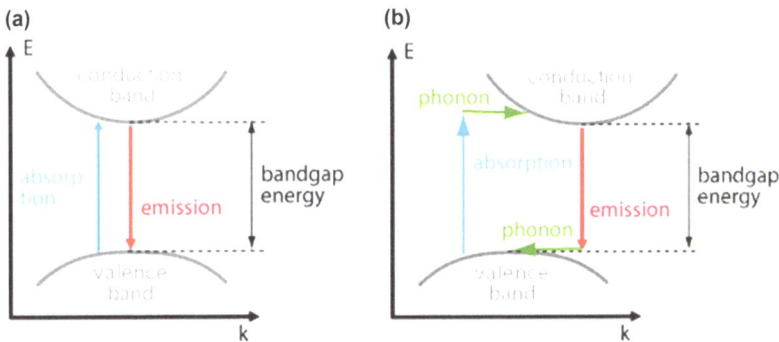

**Figure 2.3** (a) Direct e–h recombination in a direct gap semiconductor with the emission of a photon of energy $hv = E_G$. (b) Phonon-assisted e–h recombination in an indirect gap semiconductor. Reproduced from ref. 39 with permission from RP Photonics Encyclopedia.

therefore, generation–recombination processes occur with the assistance of phonons, which provide the conditions of conservation of energy and crystal momentum. A process involving three bodies, of which one is a phonon with the required energy, is, however, less efficient, and indirect gap semiconductors are, therefore, poor light emitter materials.

## 2.4.1 Role of Dopant Impurities on the Carrier Equilibrium Concentration in Elemental Semiconductors

For all semiconductor devices, impurities, atomic defects and operation temperature necessarily influence their performance, with consequences that depend on the properties of the semiconductor used, *i.e.*, on its direct or indirect energy gap and ionicity, and on the nature of the defect centres involved, which could be metallic and non-metallic impurities, atomic defects, and extended defects (dislocations, grain boundaries and stacking faults, to mention the most important).

We know,*** in fact, that metallic impurities and all kinds of defects (atomic, extended and precipitates)[40,41] dissolved or dispersed[†††] in the semiconductor matrix, generate levels in the semiconductor gap, which might behave as shallow[‡‡‡] levels, with very extended wave functions, in the case of dopant impurities, or as deep levels with localized wave functions, in the case of non-dopant metallic impurities.

Dopants are heterovalent impurities, which are intentionally introduced in the semiconductor lattice during the growth process from a melt containing the dopant impurity, or after the growth process using diffusion or ion-implantation techniques from suitable sources.

Doping of elemental semiconductors (silicon, germanium and diamond) is a relatively simple affair, since, different from compound semiconductors, both donors and acceptors can only substitute equivalent lattice sites (silicon, germanium, or carbon sites), as is the case of B and P in silicon, with the following formal reactions

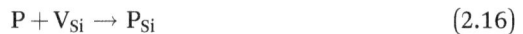

$$B + V_{Si} \rightarrow B_{Si} \tag{2.15}$$

$$P + V_{Si} \rightarrow P_{Si} \tag{2.16}$$

which occur with the assistance of a vacancy mechanism.

The thermal ionization of dopants generates an excess of electrons or holes over their intrinsic concentration, with an increase in the electrical conductivity of the material, as is the case, again, of phosphorus (P), that behaves as an electron donor

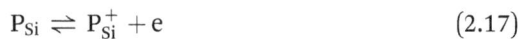

$$P_{Si} \rightleftharpoons P_{Si}^{+} + e \tag{2.17}$$

and boron (B) which behaves as an electron acceptor

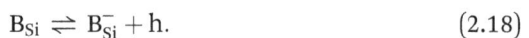

$$B_{Si} \rightleftharpoons B_{Si}^{-} + h. \tag{2.18}$$

In both single cases we assume that the product of electron and hole concentration $np$ remains constant as defined by eqn (2.12), under

---

*** See Chapter 1 for atomic defects thermodynamics.
[†††] Extended defects and precipitates behave like secondary phases dispersed in a solid matrix.
[‡‡‡] Above the top of the valence band or below the bottom of the conduction band.

fulfilment of equilibrium electroneutrality conditions, provided the dopant concentration remains in the limit of diluted solutions, which is normally the case, given that their concentrations are in the range $10^{15}$–$10^{19}$ cm$^{-3}$ in most practical uses.

Since silicon semiconductor devices are planned to work at or close to ambient temperature, the ionization energies of good dopants of silicon,[42–44] see Table 2.2, and of compound semiconductors, see Tables 2.3 and 2.4, should have a good fit with the value of the thermal energy $k_BT$ (298 K), to ensure at least their partial ionization at room temperature, on the purely thermodynamic assumption that

**Table 2.2** Properties of some typical donor and acceptor impurities in silicon. The covalent radius of Si = 110 pm.

| Impurity | Type | Ionization energy (meV) | Covalent radius (pm) | Ref. |
|---|---|---|---|---|
| P | Donor | 60.7 | 111 | 43 |
| As | Donor | 53.77 | 82 | 43 |
| Sb | Donor | 44.00 | 124 | 43 |
| Bi | Donor | 78.55 | 148 | 43 |
| B | Acceptor | 45 | 82 | 42 |
| Al | Acceptor | 57 | 118 | 42 |
| Ga | Acceptor | 65 | 126 | 42 |
| In | Acceptor | 160 | 142 | 42 |
| Tl | Acceptor | 255 | 145 | 44 |

**Table 2.3** Ionization energies (in meV) of donor and acceptor impurities in GaAs.[42] The covalent radius of Ga = 126 pm and that of As = 118 pm.

| Impurity | Type | Ionization energy (meV) | Covalent radius (pm) |
|---|---|---|---|
| Si | Donor | 5.854 | 110 |
| Ge | Donor | 5.908 | 122 |
| S | Donor | 5.87 | 102 |
| Be | Acceptor | 30 | 90 |
| Mg | Acceptor | 30 | 130 |
| Zn | Acceptor | 31.4 | 125 |
| C | Acceptor | 26.7 | 75 |

**Table 2.4** Ionization energies of dopants in GaN.[58]

| Donors | Ionization energy (meV) | Acceptors | Ionization energy (meV) |
|---|---|---|---|
| N | $53 \pm 8$ | Be | $150 \pm 10$ |
| O | 28.7 | Mg | 240 |
| Si | 12–15 | Zn | 580 |
| Te | $50 \pm 20$ | Ca | $169 \pm 12$ |

the concentration of electrons or holes increases exponentially with the temperature according to the following equations

$$n = (N_C N_D)^{1/2} \exp - \frac{E_C - E_D}{2k_B T} = (N_C N_D)^{1/2} \exp - \frac{\Delta E_D}{2k_B T} \qquad (2.19)$$

$$p = (N_V N_A)^{1/2} \exp - \frac{E_A - E_V}{2k_B T} = (N_V N_A)^{1/2} \exp - \frac{\Delta E_A}{2k_B T} \qquad (2.20)$$

where $n$ is the electron concentration $(cm^{-3})$, $p$ is the concentration of holes $(cm^{-3})$, $N_C$ is the density of energy levels in the conduction band, $N_D$ is the density of the donor levels, $E_C$ is the density of the energy levels in the conduction band, $N_A$ is the density of donor levels, $E_A$ is the energy of the acceptor level, $E_V$ is the energy of the top of the valence band and $\Delta E_D = E_C - E_D$ is the ionization energy of the donor, while $\Delta E_A = E_A - E_V$ is the ionization energy of the acceptor.

It is obvious that conditions (2.19) and (2.20) are only satisfied when formation of dopant–impurity complexes or dopant–point defect complexes can be excluded, which otherwise would affect the equilibrium concentration of electrons and holes.

In the absence of interactions of any type involving dopant species in the semiconductor lattice, it could be also expected that a linear relationship would occur between the electrical conductivity $\sigma$, the electrical resistivity $\rho$, the dopant concentration $x_D$, and the carrier density $n$ of electrons or holes

$$\sigma = \mu \theta x_D = \mu n = \frac{1}{\rho} \qquad (2.21)$$

where $\mu$ is the carrier mobility and $\theta$ is the ionization ratio of the dopant impurity.

The experimental resistivity *vs.* dopant density relationships in B-doped and P-doped silicon have been reported, among many others, by Thurber *et al.*,[45,46] see Figure 2.4, where one could remark that a linear relationship is followed in the range $10^{13}$–$10^{16}$ at $cm^{-3}$, whereas deviations are observed at higher dopant concentrations.

The deviation of the resistivity from the ideal behaviour at higher dopant concentrations might be associated with the non-ideality of the solution and, thus, to interactions of a chemical nature between the dopant species and the atoms of the silicon lattice, which could also influence the mobility of the carriers.

From the phase diagrams of the systems B–Si and P–Si[47] displayed in Figures 2.5 and 2.6, and from the historical results of Olesinski and

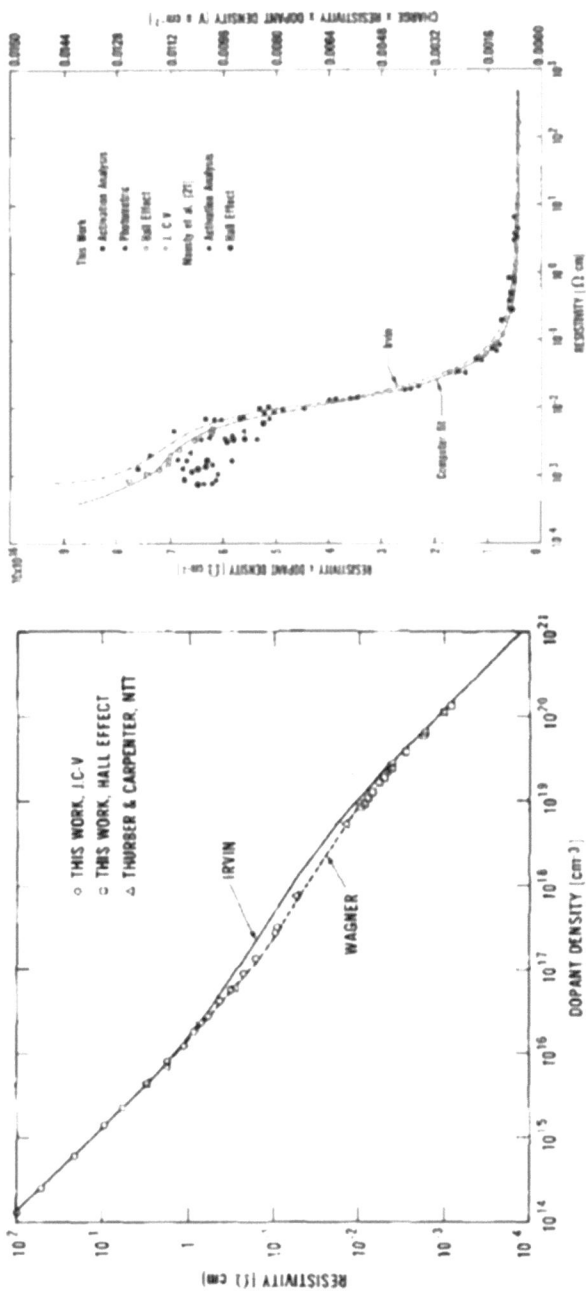

**Figure 2.4** The relationships between resistivity and dopant density for (left) boron doped and (right) phosphorous doped silicon (the atomic density of pure silicon holds $5 \times 10^{22}$ cm$^{-3}$). Reproduced from ref. 45 and 46 with permission from IOP, Copyright 1980.

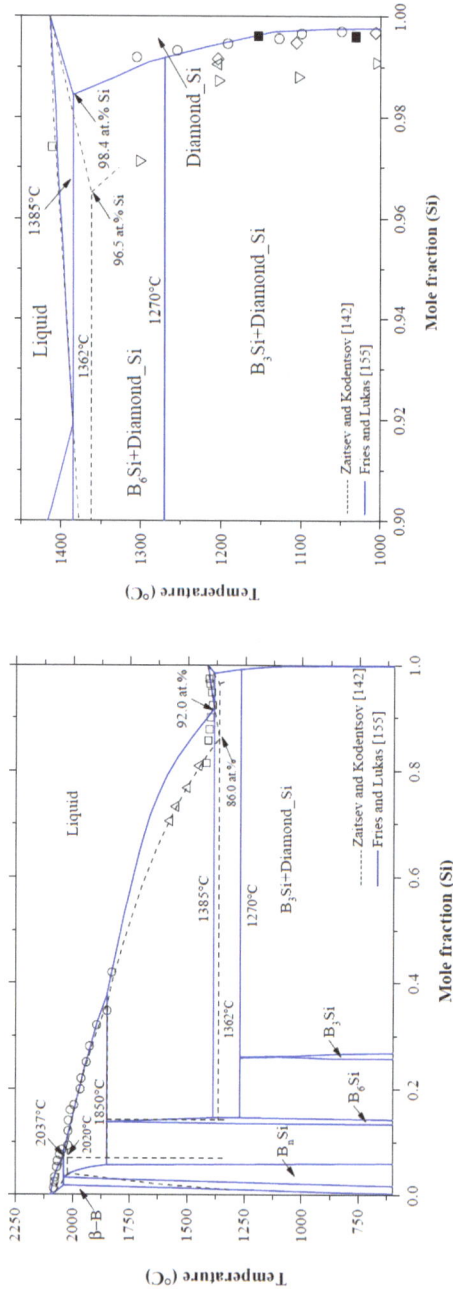

**Figure 2.5** Phase diagram of the B–Si system. Reproduced from ref. 47, https://doi.org/10.3390/ma10060676, under the terms of the CC BY 4.0 license, https://creativecommons.org/licenses/by/4.0/.

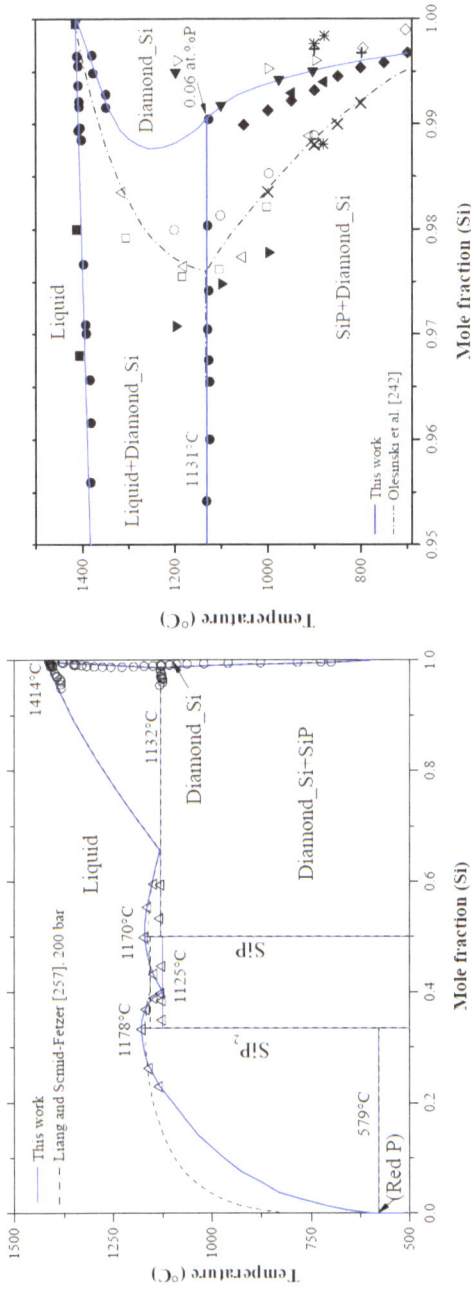

**Figure 2.6** Phase diagram of the P–Si system. Reproduced from ref. 47, https://doi.org/10.3390/ma10060676, under the terms of the CC BY 4.0 license, https://creativecommons.org/licenses/by/4.0/.

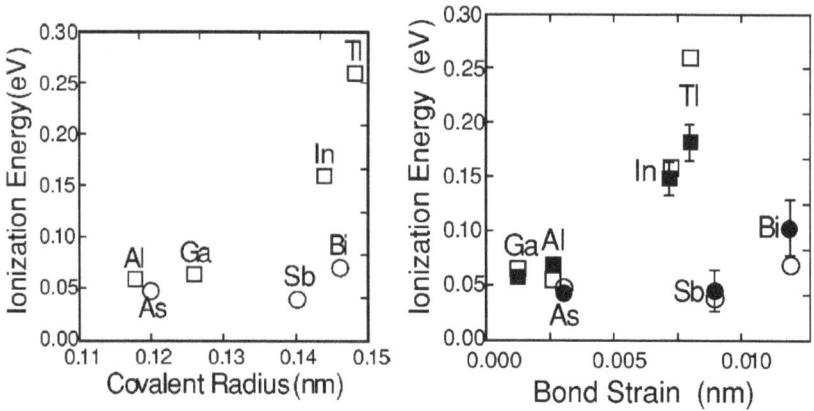

**Figure 2.7**  Ionization energies of monovalent dopant impurities *vs.* their covalent radii and DFT-calculated bond strain. Reproduced from ref. 50 with permission from American Physical Society, Copyright 2003.

Abbaschian[48,49] we could infer, or at least not exclude, the occurrence of interactions of a chemical nature§§§ between the solute impurities and the solvent atoms, with formation of complexes and, thus, deviation from the ideal, since in both cases the formation of secondary phases $B_nSi$ and $P_nSi$ occurs at higher dopant concentrations.

The bond strain arising from differences of the covalent radii of the host and of the dopant impurity has shown, instead, to have a relevant influence on the ionization energies of dopants. It is, in fact, apparent from Figure 2.7, which displays the ionization energies of some donor and acceptor impurities in silicon as a function of their covalent radii (see also Table 2.2) and of the DFT-calculated bond strain, that the ionization energies of acceptors and donors increase with increases in the covalent radius and bond strain, but that the increase is faster for acceptors, leading In and Tl to behave more as deep levels than as shallow levels, despite their similar electronic configurations.

In the attempt to explain this behaviour, we must recall that the covalent radii used by Rockett *et al.*[50] are those of the un-ionized impurities, not of ionized donor and acceptors. Therefore, while the good behaviour of donors could be explained by a covalent radius contraction due ionization, which leads to a better fit with the covalent radius of silicon and to an absolute reduction of strain, the notable deviations observed with acceptor impurities could be

---

§§§ As would be the case of the formation of thermodynamically stable complexes.

explained by the covalent radius expansion associated with ionization, which is particularly relevant for In and Tl.

## 2.4.2 Doping of III–V and II–VI Compound Semiconductors

Efficient doping of II–VI and III–V compound semiconductors is, often, a difficult task,[51] and the main limiting issues, common to all compound semiconductors, are:

- Temperature and composition dependent substitutional incorporation of dopants in both sublattices.
- Self-compensation effects by amphoteric doping, when a single dopant enters into both sublattices, or into one sublattice and in interstitial positions.
- The deactivation of dopants by lattice relaxation or by complex formation with oppositely charged atomic defects.

The case of GaAs doping with Si could be used as an illustrative example of these effects.[52–57]

In fact, silicon behaves as a donor, with an exceptionally low ionization energy (see Table 2.3), when it enters into a substitutional Ga position as a $Si_{Ga}$ species

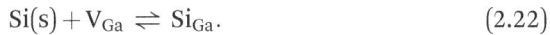

$$Si(s) + V_{Ga} \rightleftharpoons Si_{Ga}. \tag{2.22}$$

Stoichiometry conditions are, then, restored with the formation of a vacancy in the As-sublattice

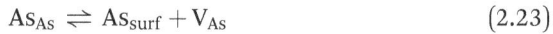

$$As_{As} \rightleftharpoons As_{surf} + V_{As} \tag{2.23}$$

leading to the final equilibrium condition

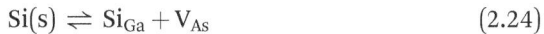

$$Si(s) \rightleftharpoons Si_{Ga} + V_{As} \tag{2.24}$$

and to the $Si_{Ga}$ donor ionization

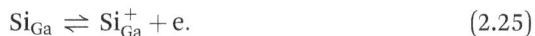

$$Si_{Ga} \rightleftharpoons Si_{Ga}^{+} + e. \tag{2.25}$$

Given the good fit of the covalent radius of Si (110 pm) with that of As (118 pm), Si could also enter in the As-sublattice

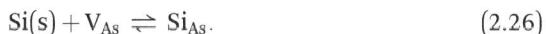

$$Si(s) + V_{As} \rightleftharpoons Si_{As}. \tag{2.26}$$

Stoichiometry conditions are, then, restored with the formation of a vacancy in the Ga-sublattice

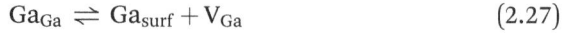

$$Ga_{Ga} \rightleftharpoons Ga_{surf} + V_{Ga} \qquad (2.27)$$

leading to the final equilibrium condition

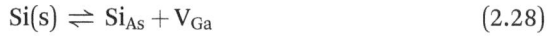

$$Si(s) \rightleftharpoons Si_{As} + V_{Ga} \qquad (2.28)$$

and to the ionization of $Si_{AS}$, which behaves as an acceptor

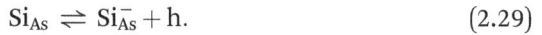

$$Si_{As} \rightleftharpoons Si_{As}^- + h. \qquad (2.29)$$

It could be supposed that the processes running in eqn (2.25) and (2.29) would occur simultaneously, with relative rates, which should depend on temperature, leading to temperature-dependent self-compensation effects.

It has been, in fact, shown by Vazquez-Cortas *et al.*[55] that in thermally annealed GaAs samples, Si is mainly incorporated in Ga sites at 450 °C with donor behaviour, mostly in As sites with acceptor behaviour at 500 °C and in both sites above 500 °C, leading to self-compensated samples with a temperature-dependent degree of self-compensation.

Ahlgren *et al.*[54] showed, additionally, that above 800 °C Si sits only in As positions, and that the solubility of Si in substitutional As positions $x(Si_{AS})$, deduced from diffusion experiments, follows an exponential relationship

$$x(Si_{As}) = \left(A + B \times 10^7 \ \exp - \frac{\Delta H_s}{kT}\right) 10^{18} \text{ at cm}^{-3} \qquad (2.30)$$

where $\Delta H_s = -138.9 \text{ kJ mol}^{-1}$ is the enthalpy of solution of silicon in substitutional As sites of GaAs, leading to a Si solubility in As sites of $10^{19} \text{ cm}^{-3}$ at 800 °C and of $10^{20} \text{ cm}^{-3}$ at 1000 °C, in good agreement with the results reported by Northrup and Zhang[57] displayed in Figure 2.8.

The same authors calculated the concentration of silicon and of gallium vacancies in GaAs at 940 °C, see Figure 2.9, from which one can see that the concentration of $Si_{AS}$ equals the concentration of $Si_{Ga}$ at the higher silicon concentrations and that at the highest silicon concentrations the neutral $(Si_{Ga}-Si_{AS})$ pairs dominate, with full self-compensation.

A second example of the difficulties encountered in doping epitaxial films of compound semiconductors is given by the case of GaN,

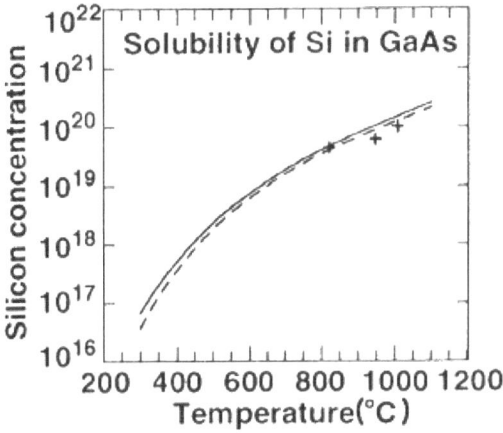

**Figure 2.8**  Temperature dependence of the solubility of Si in GaAs. The solid line is the calculated solubility, the dashed line is the experimental solubility. Reproduced from ref. 57 with permission from American Physical Society, Copyright 1993.

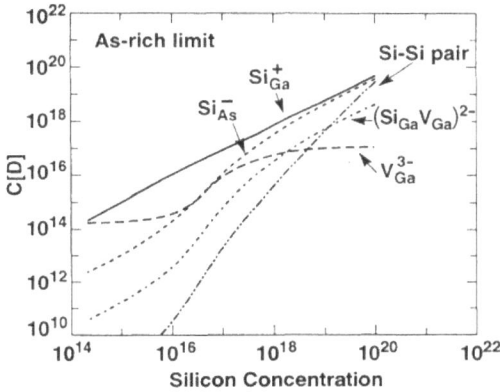

**Figure 2.9**  Calculated concentration of Si-defect species in As-rich GaAs ($T = 940$ °C). Reproduced from ref. 57 with permission from American Physical Society, Copyright 1993.

which is invariably found n-type doped, probably due to the presence of nitrogen vacancies.[58] n-type doping of GaN is favoured by the low ionization energies of donors (see Table 2.4), particularly low for Si, which is readily incorporated in substitutional Ga-sites, behaving as a shallow donor with an ionization energy of 12–15 meV, leading to its nearly complete ionization at room temperature. Si doping works well over a wide range of densities from low-$10^{17}$ cm$^{-3}$ to mid-$10^{19}$ cm$^{-3}$, although some structural problems occur in thick and heavily doped film. While doping with Si is the most common procedure for n-type doping of GaN, other elements such as O, Te, S, Se[58] are also employed.

p-type doping of GaN, instead, has been considered for years almost impossible, due to the high values of the ionization energies of acceptors (see Table 2.4), which would imply the use of high concentrations of dopants, although doping efficiency is limited at high dopant concentrations, including the case of Mg, which is known as being the only effective acceptor dopant.

The break-out of the doping efficiency at high Mg concentrations is not the only problem with Mg-doping. In fact, only a fraction of the Mg in the as-grown Mg-doped GaN films, independently of the growth process used, is electrically active.

Nakamura *et al.*[59] gave a solution to the problem showing that the activation of Mg acceptors in GaN can be obtained by dedicated post-growth thermal annealing, but that their activation could be suppressed by annealing in a hydrogen ambient, enlightening the crucial role of hydrogen in the Mg-doping process.

Similar results were obtained by Eiting *et al.*[60] on Mg-doped GaN and AlGaN epitaxial films deposited with a metal–organic chemical vapor deposition (MOCVD) technique using MgCP2 (bis(cyclopentadienyl) magnesium) as the Mg precursor and hydrogen as carrier gas. They confirmed that the as-grown Mg-doped films are semi-insulators, suggesting that the ineffectiveness of Mg p-type doping is due to a compensation from ionized donor defects.

This conclusion agrees with the arguments of Neugebauer and Van de Walle,[61] who demonstrated, using first principles calculations, that hydrogen dissolves in GaN and that the formation energy of hydrogen interstitials in n-type GaN is larger than that in p-type GaN, leading to larger solubilities of hydrogen in p-type GaN. Furthermore, they also showed, see Figure 2.10, that hydrogen interstitials $H_i$ behave as ionized donor species $H_i^+$ in p-type GaN and as a ionized acceptor species $H_i^-$ in n-type GaN, establishing for hydrogen the role of a compensating species for both Mg and Si dopants.

This is well evidenced in Figure 2.10 (left) for the case of Mg, where one can observe that at the crossing point in the p-type region, where the energies of formation and the concentrations are equal, the compensation of the ionized donor $H_i^+$ with the $M_{Ga}^-$ acceptor occurs. Similar conditions occur for silicon in the n-type region, see Figure 2.10 (right).

Compensation establishes equilibrium conditions of the Fermi level, as well as the formation of the $[Mg–H]^x$ neutral complex

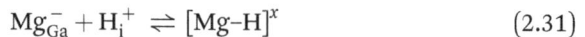

$$Mg_{Ga}^- + H_i^+ \rightleftharpoons [Mg–H]^x \qquad (2.31)$$

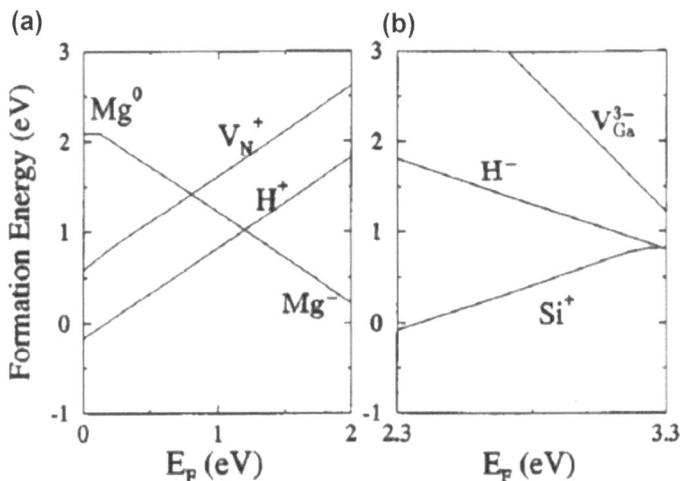

**Figure 2.10** First principles calculated values of the formation energy of defects in (a) p-type doped and (b) n-type doped GaN. Reproduced from ref. 61 with permission from AIP Publishing, Copyright 1996.

with a binding energy estimated to hold 0.7 eV, which might explain the electrical inactivity of Mg dopant in MOCVD deposited GaN and AlGaN found by Eiting *et al.*,[60] due to the presence of residual hydrogen.

Compensation of a couple of ionized donors and acceptors, with the outcome of a fixed value of the Fermi level, is the analogous case of a buffer solution, which maintains constant the pH of an aqueous solution, with hydrogen playing in both cases the key role.

According to Miceli and Pasquarello,[62] hybrid functional calculations of formation energies of point defects and point defect complexes in GaN suggest that the formation energy of Mg interstitials $Mg_i$, of nitrogen vacancies $V_N$, and of $Mg_{Ga}$–$V_N$ complexes are lower than that of Mg in Ga-substitutional positions $Mg_{Ga}$ in p-type conditions, see Figure 2.11.

Since Mg behaves as a double donor***** when sitting in interstitial positions $Mg_i^{2+}$, while the complex $Mg_{Ga}$–$V_N$ behaves as a deep donor||||[63,64] it is also suggested that heavy Mg-doping of GaN leads to self-compensation by amphoteric doping.

The last example concerns the case of ZnO, already considered in Chapter 1, that presents doping problems not yet completely set-off, due to the n-type self-doping of ZnO crystals and of ZnO

---

***** The fully ionized $Mg_i^{2+}$ is a double donor.
|||| Suggested to be the level responsible of the luminescence at 2.9 eV in Mg-doped GaN.

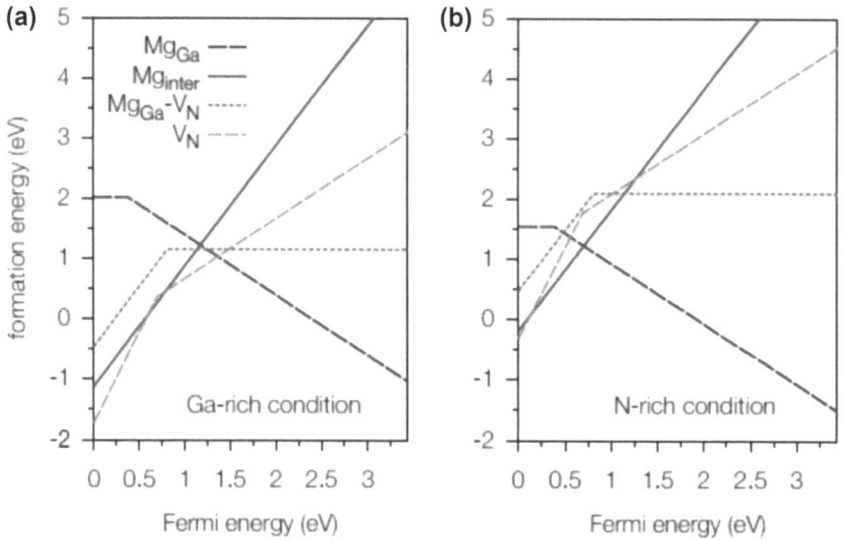

**Figure 2.11**   Calculated formation energies of relevant defects in GaN (a) Ga-rich conditions (b) N-rich conditions. Reproduced from ref. 62 with permission from American Physical Society, Copyright 2016.

microcrystalline and nanocrystalline films, and to the few candidates available as acceptors.[65]

Depending on its high ionicity, the conventional density functional theory in the local density approximation (DFT-LDA)[66] severely underestimates the band gap of ZnO and, consequently, the energies of formation of defects species. Implementation is obtained by accounting for additional Coulomb interactions among lattice species, although not with entirely satisfactory results.

For a long time ZnO self-doping has been associated with oxygen vacancies ($V_O$) or with zinc interstitials ($Zn_i$), but the role of $V_O$ could be excluded because it behaves as a deep acceptor[67,68] and the role of $Zn_i$ is still under debate, although it is suggested that $Zn_i$ in Zn-rich ZnO is responsible for the near band edge luminescence at 375 nm (3.3 eV).[69]

According to Van de Walle[70,71] and Koßmann and Hättig,[72] interstitial hydrogen $H_i$ in various configurations could be considered instead the donor responsible for the n-type self-doping of ZnO, because the calculated values of the $H^\circ/H^+$ level stays very close to the conduction band minimum (CBM), see Figure 1.42, with almost complete ionization.

Hydrogen, in fact, and environmental water vapours, with dissociative adsorption, as observed with MD simulations in the case of MgO,[70] may easily contaminate ZnO during its preparation.

Concerning potential p-type dopants of ZnO, the elements of the first group (Li, Na, K) in Zn positions are shown to behave as deep acceptors, like the elements of the V group (P, As, Sb) in oxygen lattice positions. Substitutional $Ag_{Zn}$, among IB elements, with a calculated acceptor level at 400 meV, might be, instead, a good candidate.[73]

Nitrogen appears to be the natural substituent for oxygen in ZnO, and dissolved nitrogen allows p-type doping of ZnO,[67,74–76] with an experimental acceptor level at 165 meV, but the question still open is about the nature of the defect species or of the defect position.

The assignment to nitrogen in substitutional oxygen positions of $N_O$ to the role of shallow acceptor is, in fact, excluded by Jannotti *et al.*,[77] Boonchun and Lambrecht[67] and Tarun *et al.*,[78] who showed that $N_O$ behaves as a deep acceptor, with a 0/−1 level at $\sim E_V + 1.3$ eV, confirmed by a PL emission band peaking near 1.7 eV in N-doped ZnO.

Boonchun and Lambrecht[67] showed, instead, that molecular nitrogen on a Zn site $(N_2)_{Zn}$ or the $N_O$-$V_{Zn}$ complex, with one of the oxygens surrounding the oxygen vacancy substituted by nitrogen, can behave as a shallow acceptor.

It is interesting to note that the process chemistry adopted to dope ZnO with nitrogen is also critical. Tarun *et al.*[78] showed, as an example, that using a chemical vapour transport (CVT) process in ammonia ambient for the preparation of nitrogen-doped ZnO, ammonia provides the source not only of nitrogen but also of hydrogen, which leads to the formation of a N–H complex and of hydrogen interstitials $H_i$, that could be removed by thermal annealing at 675 °C, whereas an annealing at 775 °C is needed to dissociate the N–H pairs, leaving nitrogen in oxygen sites $N_O$, which behaves as a deep acceptor.

Yao *et al.* showed, eventually, that in ZnO prepared using magnetron sputtering at 510 K of a ZnO target, and nitrogen as the sputtering gas, nitrogen was found present in two different configurations, as substitutional nitrogen in oxygen sites $N_O$ and as molecular nitrogen on an oxygen position $(N_2)_O$ from XPS analysis.

Depending on the process used, therefore, several nitrogen containing species are present in nitrogen-doped ZnO. In oxidizing environments, the $N_O$ and the $(N_2)_O$ species are stable, while in neutral environments the $(N_2)_{Zn}$ and $N_O$-$V_{Zn}$ complexes are the species present in the deposited material.

A summary of the properties of donor and acceptor species in ZnO is displayed in Table 2.5, which shows that only the nitrogen species $(N_2)_{Zn}$ and $N_O$-$V_{Zn}$ behave as effective p-type dopants of ZnO,

**Table 2.5**  Experimental (*) and calculated (**) values of the energy levels of defects and defect complexes in ZnO.

| Defect | Defect level (eV) | Properties | Ref. |
|---|---|---|---|
| $V_O$ (2+/0) | $E_V + 2.91$ (*) | Deep donor | 67 |
| $Zn_i$ (1+/0) | $E_V + 3.3$ (*) | Shallow donor | 69 |
| $H_i$ (+/0) | $E_V + 3.3$ (**) | Shallow donor | 70 |
| $N_O$ (0/−1) | $E_V + 1.3$ (**) | Deep acceptor | 77 |
| $N_O$ (0/−1) | $E_V + 1.7$ (**) | Deep acceptor | 78 |
| $(N_2)_{Zn}$ (0/−) | $E_V + 0.37$ (**) | Shallow acceptor | 67 |
| $V_{Zn}$ (0/−) | $E_V + 0.95$ (**) | Deep acceptor | 67 |
| $N_O$–$V_{Zn}$ | $E_V + 0.21$ (**) | Shallow acceptor | 67 |

although with relatively deep levels, which explain why high concentration of dopants are necessary.

### 2.4.3  Impact of Metallic Impurities on Carrier Recombination Phenomena: Background Concepts

While dopant impurities influence the equilibrium carrier concentration, and are deliberately introduced in the semiconducting material either during the growth process or the device fabrication process, metallic impurities present as undesirable hosts of the semiconductor crystals used as the substrate and unwanted metallic impurities arising from metallurgical steps might contaminate the device and behave as carrier trapping and recombination centres, and, thus, degrade its *I/V* characteristics, as well as the light generation efficiency.

Carrier recombination processes, independently of the physical origin of an excess carrier concentration in the semiconductor material, operate through the intermediate of a trap, consisting of deep level impurity or of a lattice defect, which might exchange electrons and holes with the conduction band and valence band of the semiconductor, giving rise to carrier trapping and, thus, to carrier recombination processes.

Following Shockley and Reed,[79] the elementary processes associated with a recombination event at a trap lying at an energy level $E_t$ in the gap and stable in only two charge states,**** differing by one electronic unit of charge, being either neutral or negatively charged, are schematically displayed in Figure 2.12.

---

**** A real trap might present several oxidation states, as is the case of transition metals.

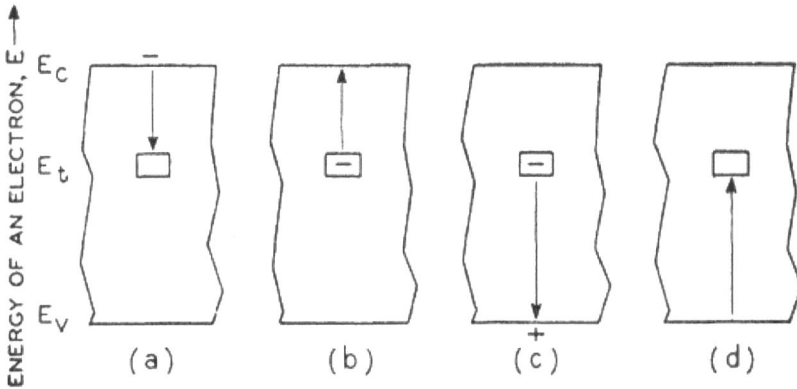

**Figure 2.12** The basic processes involved in the recombination of an electron with a hole at a trap level $E_t$: (a) capture of an electron from the CB, (b) emission of an electron from the negatively charged trap, (c) hole capture from the valence band and e–h recombination, (d) capture of an electron from the valence band. Reproduced from ref. 79 with permission from American Physical Society, Copyright 1952.

If the trap is neutral it may capture an electron from the conduction band (step (a) in the figure), acquiring a negative charge, allowing an electron–hole, eh, recombination to occur by injection of an electron in the valence band orbit the capture of a hole from the valence band (step (c) in the figure). The cases (b) and (d) represent the conditions of electron and hole re-emission. It is assumed that the overall recombination rate arises from the combined statistics of only these redox processes and that the rate limitation would be associated only to the availability of electrons and holes capable of entering the trap.

On that basis, these processes could be discussed using the Shockley–Read–Hall (SRH) statistics,[79–81] which show that the steady-state recombination rate $R = \dfrac{\delta x}{\delta t}$ is given by the following equation[80]

$$\frac{\delta x}{\delta t} = \frac{(np - n_i^2)}{[\tau_p(n + n_o) + \tau_n(p + p_o)]} \tag{2.32}$$

where $n$ and $p$ are the actual electron and hole concentrations, $n_i^2$ is the equilibrium carrier contribution due thermal generation, $\tau_n$ is the recombination lifetime of electrons, $\tau_p$ is the corresponding holes recombination lifetime, and $n_o$ and $p_o$ are auxiliary SRH parameters, that coincide with the equilibrium electron and hole concentrations in a sample whose Fermi level coincides with the position of the recombination centre[80] and hold

$$n_{\mathrm{o}} = N_{\mathrm{C}} \ \exp - \frac{(E_{\mathrm{C}} - E_{\mathrm{T}})}{kT} \qquad (2.33)$$

$$h_{\mathrm{o}} = N_{\mathrm{V}} \ \exp - \frac{(E_{\mathrm{T}} - E_{\mathrm{V}})}{kT} \qquad (2.34)$$

where $N_{\mathrm{V}}$ and $N_{\mathrm{C}}$ are the effective densities of states in the valence and conduction bands, respectively, $E_{\mathrm{V}}$ is the valence band maximum, $E_{\mathrm{C}}$ is the conduction band minimum, $E_{\mathrm{T}}$ is the energy of the trap level, and the terms $(E_{\mathrm{T}} - E_{\mathrm{V}})$ and $(E_{\mathrm{C}} - E_{\mathrm{F}})$ are the values of the kinetic barriers to be overcome by holes and electrons in a recombination event.

Apparently, the recombination rate $R$ is an inverse function of the recombination lifetimes and takes its maximum value for a trap $T$ located at $E_{\mathrm{T}}$ halfway from the valence band maximum $E_{\mathrm{V}}$ and the conduction band minimum $E_{\mathrm{C}}$.

It can be demonstrated that the recombination lifetime $\tau_{\mathrm{e,h}}$ is inversely correlated to the capture cross sections $\sigma_{\mathrm{h,e}}$ for the electrons and holes of the trap centre and to their concentration $N_{\mathrm{T}}$

$$\tau_{\mathrm{e,h}} = \frac{1}{N_{\mathrm{T}} \nu_{\mathrm{th}} \sigma_{\mathrm{e,h}}} \qquad (2.35)$$

where $\nu_{\mathrm{th}}$ is the thermal velocity of electrons or holes, and decreases with the increase in the concentration $N_{\mathrm{T}}$ of traps, *i.e.* of the impurities or defects density in the semiconductor matrix.

The capture cross section $\sigma$ is the figure of merit of a trap, consisting of an impurity or of a lattice defect, which behaves as a preferential sink for electrons or for holes, leading to the establishment of a stable, electrically charged defect. In the experimental practice, the capture cross section $\sigma$ values could be experimentally measured with deep level transient spectroscopy (DLTS) measurements.[82,83]

### 2.4.4   Recombination Features of Metallic Impurities

Transition metal (TM) impurities are known to behave as traps and recombination centres in semiconductors, with the formation of deep levels in their gaps.[84] Their role on the performance of silicon microelectronic and optoelectronic devices, extensively studied in the last decades, might be taken as an illustrative example for the whole semiconductor family, although with quantitative differences in the case of compound semiconductors, arising from the different chemical composition, structure, electronic configuration and ionicity.

As shown among others by Istratov *et al.*,[85] Fe, like other TMs, dissolves in tetrahedral interstitial positions of undoped silicon, with the formation of a deep level $E_V + 0.38$ eV, while its orbitals are hybridized with the silicon ligands, leading to the electronic structure of the neutral $Fe_i^0$ and ionized iron $Fe_i^+$ displayed in Figure 2.13.[86] It is apparent that the redox character of Fe (like that of the other transition metals) is at the background of the carrier trapping and recombination properties exhibited by transition metal impurities, which in the case of Fe might by formalized by the following equilibrium

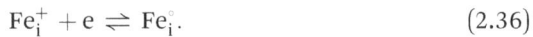

$$Fe_i^+ + e \rightleftharpoons Fe_i^0. \tag{2.36}$$

In p-doped silicon, iron forms stable complexes with acceptors (B, Al, Ga, In)[87] with experimental pair-binding energies holding about 0.6 eV, and a deep level lying at $E_C - 0.23$ eV for the $Fe_i$–$B_{Si}$ pairs.

Recombination lifetime or capture cross sections of TMs in silicon have been experimentally measured, as an example, by Itsumi[88] with a photoconductivity decay method, or by Lang *et al.*[89,90] with DLTS spectroscopy.

Using DLTS spectroscopy, Macdonald and Geerligs[91] measured the recombination activity of interstitial iron and iron–boron pairs, see Table 2.6, as well as that of other transition metals, see Table 2.7, in term of their capture cross sections in undoped silicon.

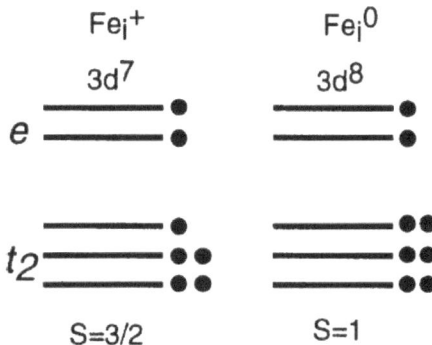

**Figure 2.13**  The electronic structure of ionized and neutral iron in silicon. Reproduced from ref. 86 with permission American Physical Society, Copyright 1960.

**Table 2.6** Recombination properties of iron and iron–boron pairs in silicon[91] ($n_o$ and $h_o$ are the SRH parameters of eqn (2.28) and (2.31).

| Recombination centre | Energy level (eV) | $\sigma_e$ $(\mathrm{cm}^{-2})$ | $\sigma_h$ $(\mathrm{cm}^{-2})$ | $n_o$ $(\mathrm{cm}^{-3})$ | $h_o$ $(\mathrm{cm}^{-3})$ |
|---|---|---|---|---|---|
| $Fe_i$ | $E_V + 0.38$ | $5 \times 10^{-14}$ | $7 \times 10^{-17}$ | $6.9 \times 10^6$ | $1.1 \times 10^{13}$ |
| $Fe_i$–$B_{Si}$ | $E_c - 0.23$ | $3 \times 10^{-14}$ | $2 \times 10^{-15}$ | $3.6 \times 10^{15}$ | $2.1 \times 10^4$ |

**Table 2.7** Room temperature energy levels and recombination properties of transition metal impurities in undoped silicon.[91]

| Impurity | Type | Lattice site | Energy level (eV) | $\sigma_n$ $(\mathrm{cm}^{-2})$ | $\sigma_h$ $(\mathrm{cm}^{-2})$ | $k = \sigma_n/\sigma_p$ |
|---|---|---|---|---|---|---|
| Au | Acceptor | Substitutional | $E_c - 0.55$ | $1.4 \times 10^{-16}$ | $7.6 \times 10^{-15}$ | 0.02 |
| Zn | Acceptor | Substitutional | $E_v + 0.33$ | $1.5 \times 10^{-15}$ | $4.4 \times 10^{-15}$ | 0.34 |
| Ti | Donor | Interstitial | $E_c - 0.27$ | $3.1 \times 10^{-14}$ | $1.4 \times 10^{-15}$ | 22 |
| V | Donor | Interstitial | $E_c - 0.27$ | $5 \times 10^{-14}$ | $3 \times 10^{-18}$ | $1.7 \times 10^4$ |
| Cr | Donor | Interstitial | $E_c - 0.22$ | $2 \times 10^{-13}$ | $1.1 \times 10^{-13}$ | 2 |
| Fe | Acceptor | Interstitial | $E_v + 0.38$ | $5 \times 10^{-14}$ | $7 \times 10^{-17}$ | 700 |
| Mo | Acceptor | Interstitial | $E_v + 0.28$ | $1.6 \times 10^{-14}$ | $6.0 \times 10^{-16}$ | 30 |

One can see from Table 2.6 that the formation of the $Fe_i$–$B_{Si}$ pair shifts the $Fe_1$ level at $E_V + 0.38$ upward of 0.56 eV, corresponding well to the value of the binding energy of $Fe_i$–$B_{Si}$ pairs (0.6 eV) given by Zhao et al.,[87] changing the character of the impurity from acceptor-like to donor-like. This changes the height of the recombination barriers and, indeed, the recombination rate.

From Table 2.7 one can remark, instead, that in undoped silicon the capture cross sections for the holes $\sigma_h$ are smaller than those of the electrons $\sigma_e$ for all TM interstitials, except for gold and zinc, which behave as substitutional impurities. If we account for the values of the ratio $k = \sigma_n/\sigma_p$, which is a measure of deviation from equal values of capture cross section of electrons and holes, we can see that vanadium presents the maximum deviation, with $k = \sigma_n/\sigma_p$ values of the order of $10^5$, which qualitatively seem to depend on the number (4) of the stable oxidation states of this element. We can also see that the interstitial impurities, $Ti_i$, $V_i$, $Cr_i$ and the $Fe_i$–$B_{Si}$ pairs behave as donor-like impurities, while $Fe_i$ behaves as acceptor-like impurities.

It is, eventually, expected and known[88] that the recombination lifetime of TM impurities would depend on doping and that an approximate relationship exists between the lifetimes $\tau$ in p-doped and n-doped silicon in the presence of Cu or Fe impurities

$$\tau_n = A\tau_p^{1.5} \tag{2.37}$$

where $A$ is a constant that depends on the nature of the impurity.

Not only metallic impurities, but also native atomic defects (vacancies and self-interstitials), which have been dealt in Chapter 1 as equilibrium chemical species in semiconductors, can behave as traps or recombination centres. They could be, in fact, indirectly involved in recombination processes if their interaction with impurities leads to the formation of stable, atomic defect-impurity complexes behaving as deep centres.[92]

In the presence of an impurity I, the formation of a stable vacancy–impurity complex $V_{Si}$–I could, in fact, occur

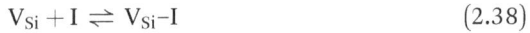

$$V_{Si} + I \rightleftharpoons V_{Si}\text{–I} \tag{2.38}$$

as is the case of the vacancy-oxygen complex in silicon, the A centre, see Figure 2.14[93,94] or the O–B complex,[95–97] this last strongly influences the photovoltaic efficiency of silicon solar cells with the so-called light induced degradation (LID) effect.[98]

Both complexes are thermodynamic stable. The stability of the A-centre arises from the saturation of two of the four dangling bonds in the core of the vacancy with oxygen, while that of the B–O complex $B_{Si}$–$O_{2i}$, consisting of a substitutional boron $B_{Si}$ bonded to an interstitial oxygen dimer $O_{2i}$, depends on the strong O–B binding energy ($787 \pm 42$ kJ mol$^{-1}$).[99]

From what we have discussed so far, it is clearly apparent that impurities, depending on their individual chemical character, on their electronic properties and on the thermodynamic stability of defect–impurity complexes, have a critical influence on the semiconductor properties and on the device performances. And it is already understandable that semiconductor growth and further device processing require delicate attention to impurity chemistry in all the steps of the semiconductor growth and processing.

Therefore, independent of the details of the physics of recombination processes, structural perfection and chemical purity are the prerequisites for a semiconductor that should be used in microelectronic (and photonic) applications. We will also see that non-stoichiometry and ionicity are additional hindrances in the case of compound semiconductors, as well as surface defects in the case of semiconductor nanostructures, where the large surface/volume ratio enhances the role of surface defects in the recombination activity.

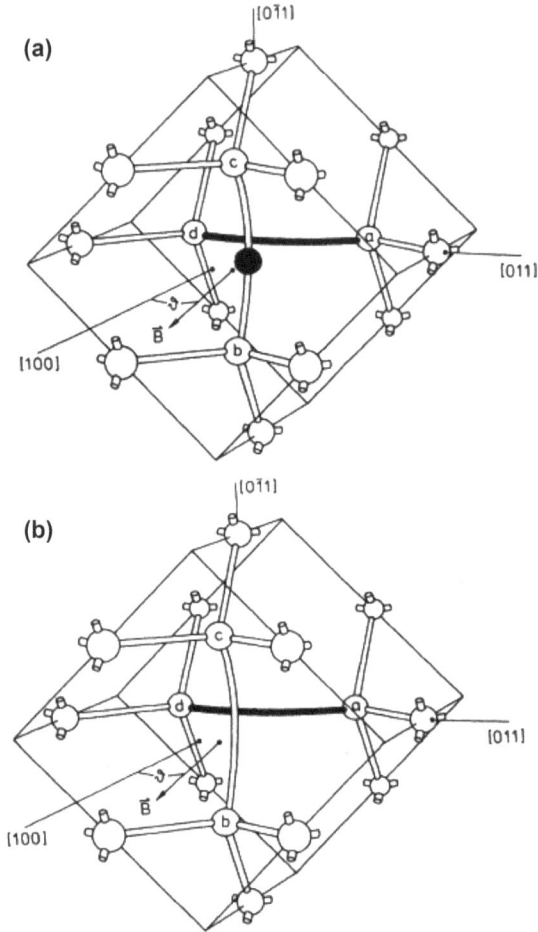

**Figure 2.14**  Model of the negative oxygen-silicon vacancy complex (a) and of the negatively charged silicon vacancy (b). The oxygen atom in (a) is represented by a black sphere. Reproduced from ref. 93 with permission from American Physical Society, Copyright 1989.

## 2.4.5 Impact of Structural Defects on Carrier Recombination Phenomena

Not only atomic defects, but also extended defects (ED) with their different dimensionalities (dislocations, stacking faults and precipitates,[††††] to mention only the most common) behave as sinks for impurities, reducing their absolute concentration with an impurity

---

[††††] Dislocations are 1D defects, stacking faults are 2D defects and precipitates are 3D defects.

segregation mechanism, but also behaving as free-carrier recombination centres.

Interaction of impurities with EDs is a thermodynamically favourite process if, as result of the interaction, there is an overall reduction of the internal energy of the system, with a free energy of interaction $\Delta G_r < 0$, and this is a good reason for suggesting that a system containing extended defects is thermodynamically stable when these defects are "reacted".

Interaction processes of impurities with EDs could be treated as a chemical reaction with the unsaturated bonds at the interface of 2D and 3D defects or in the core region of 1D defects, with the formation of true chemical bonds, or as a repartition (segregation) of impurities between two phases,[‡‡‡‡] of which one is the crystal matrix and the other is the extended defect, this last is considered as a low dimensionality Gibbs type of phase.

In the last case, as a gross approximation, equilibrium conditions require that both the thermodynamic potential of the impurity i and the internal pressure $P$ are the same in both "phases", labelled for convenience $\alpha$ and $\beta$

$$\mu_i^\alpha = \mu_i^\beta \qquad (2.39)$$

$$P^\alpha = P^\beta \qquad (2.40)$$

neglecting the stress induced by the segregation of an impurity at an ED.

While it is experimentally known that both clean[§§§§] and impurity-reacted extended defects behave as recombination centres for free carriers, the actual configuration of a carrier recombination event occurring at lattice defects depends on the nature of the defect itself and whether it is a clean- or an impurity-reacted defect.

Here we will discuss only the case of dislocations, as a unique but an illustrative example, in view of their ubiquitous presence in the whole family of semiconductors, where they behave as traps for the majority carriers (electrons or holes, depending on doping) at intrinsic electronic states $E_{Dh}$ or $E_{De}$ of their 1D bands,[¶¶¶¶] see Figure 2.15,[100] and induce the onset of an electrostatic Coulomb

---

[‡‡‡‡] These conditions hold since we assume that the impurity segregated at the extended defect behaves as a solute in a low-dimensionality phase.
[§§§§] In the absence of impurities segregated on them.
[¶¶¶¶] Dislocations are one-dimensional defects, behaving like a 1D phase of silicon.

potential, $e\Phi$, which behaves as a screening barrier and holds, approximately[101]

$$e\Phi \approx \frac{e^2}{2\pi\varepsilon\varepsilon_0} N_{tot} \left[ (\ln(N_{tot}L_{scr}) - \frac{1}{2} \right] \qquad (2.41)$$

where $\varepsilon$ is the dielectric constant of silicon, $\varepsilon_0$ is the dielectric constant of vacuum and $N_{tot}$ is the total charge of the dislocation per unit length.

In turn $L_{scr}$ is the screening radius of the electrostatic barrier around the dislocation

$$L_{scr} = [a\pi(N_D - N_A)]^{-1/2} \qquad (2.42)$$

where $a$ is the interatomic distance of Si atoms inside the dislocation core, and $N_D$ and $N_A$ are the donor and acceptor concentrations in the crystal matrix,[100–105] which reduce the capture rate of a factor $\exp\left(-\dfrac{e\Phi}{kT}\right)$ and limit the equilibrium occupancy of the dislocation states by electrons, but favour the capture rate of minority carriers (holes), and, thus, enhance their recombination.

Recombination might occur by direct eh recombination, see Figure 2.15, (path 1), if the dislocation is clean or, following path 2, when the dislocation is decorated by impurity atoms segregated in the core of the dislocation, or by some "native" core defects, caused by vacancies or impurity atoms incorporated into dislocation core, leading to the onset of a cluster of gap levels at $E_{oo}$ in the figure.

It is reasonable to suppose that at a reacted dislocation, recombination would work with a SRH mechanism, in full analogy with the case of TM impurities in a semiconductor crystal, with differences associated, however, with the different electronic configurations of dislocations.

The electronic configuration of dislocations in silicon, taken again as a virtuous example, might be inferred from the low temperature (12 K) photoluminescence (PL) spectrum of clean dislocations generated by plastic deformation at 750 °C of a (111) oriented, high purity FZ silicon sample,‖‖‖ displayed in Figure 2.16.[106] One can see a quartet of emission lines at 0.807 eV, 0.877 eV, 0.945 eV and 0.999 eV, conventionally called D1, D2, D3 and D4, plus the band to band

---

‖‖‖ Low temperature is needed to enhance the intensity of the emission that at room temperature is negligible, given the indirect gap of silicon.

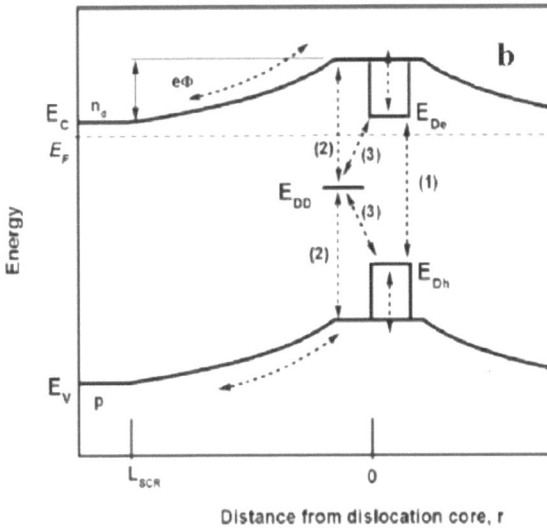

**Figure 2.15** Schematic view of recombination processes at a clean dislocation, path (1) where a direct recombination occurs among the dislocation levels $E_{De}$ and $E_{Dh}$ and path (2) where indirect recombination does occur at a contaminated dislocation with an impurity level at $E_{DD}$. The onset of the electrostatic Coulomb potential $e\Phi$ is supposed to result from electron trapping at the dislocation level $E_{De}$. Reproduced from ref. 100, https://doi.org/10.3390/cryst6070074, under the terms of the CC BY 4.0 license, https://creativecommons.org/licenses/by/4.0/.

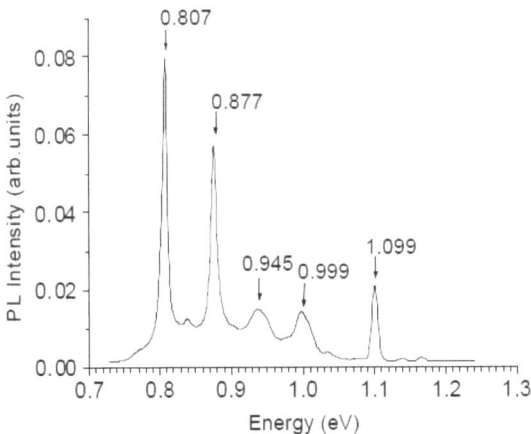

**Figure 2.16** Photoluminescence spectrum at 12 K of a plastically deformed, (111)-oriented FZ silicon sample. The emission at 1.099 eV is the signature of the band to band emission of the silicon matrix. Reproduced from ref. 106 with permission from Trans Tech Publications, Copyright 2004.

emission of the silicon matrix at 1.099 eV, which in fact corresponds to the energy gap of silicon at 12 K.

According to Kveder and Kittler[104] and Lightowlers and Higgs,[107] the emissions at 0.999 eV and 0.945 eV are features that originate from the dislocation core, where the D4 line at 0.999 eV originates from the direct eh recombination, path 1 in Figure 2.15, and the D3 line at 0.945 eV is its phonon replica. It is apparent that the energy gap of the dislocation, displaying properties of a 1D system, takes a value of 0.999 eV, slightly lower than the energy gap of the 3D silicon matrix.

It is, further, supposed that the other two emissions originate from point defects clusters trapped in the in the dislocation core, although the role of traces of impurities segregated at the dislocation core can also not be excluded.[104]

In fact, even if ultra-pure FZ silicon is used as the base material, plastic deformation is carried out at 750 °C, a temperature at which environmental impurities could easily diffuse and contaminate the sample.

The recombination rate in the presence of an impurity in the energy gap of the dislocation core (step 2 of Figure 2.15) could be calculated considering that the total rate $R_{\text{C-M}}$ of electron capture from the conduction band of silicon into the impurity level in the dislocation is given by the KKS (Kveder, Kittler, Schröter) equation[103]

$$R_{\text{C-M}} \approx n\sigma_e v_{\text{th}}(N_M - n_M)\exp - \frac{e\Phi}{kT} - N_C\sigma_e v_{\text{th}}n_M \exp - \frac{E_C - E_M}{kT} \qquad (2.43)$$

where the first term represents the free-electron capture rate and the second the re-emission rate, $n$ is the equilibrium concentration of electrons arising from the dopant concentration, $\sigma_e$ is the capture cross section for electrons at the impurity segregated at the dislocation core,***** $v_{\text{th}}$ is the thermal velocity of electrons, $N_M$ is the concentration of the impurity per unit length of dislocation, $n_M$ is the concentration of electrons captured at the impurity level, $N_C$ is the effective density of states in the conduction band, and $E_C - E_M$ is the energy of the impurity level with respect to the bottom of the conduction band.

The corresponding capture rate of holes at the impurity level, following the mechanism illustrated in Figure 2.15, neglecting their

---

***** The core of a dislocation is the region of radius $r$ around the dislocation, above which the elastic field of the dislocation satisfy linear elasticity properties.

re-emission because they will react immediately with electrons, is given by the following equation

$$R_{\text{V-M}} \approx p_{\text{D}} \sigma_{\text{h}} v_{\text{thh}} n_{\text{M}} \tag{2.44}$$

where $p_{\text{D}}$ is the concentration of holes at the reaction boundary, $\sigma_{\text{h}}$ is the capture cross section of holes at the impurity segregated at the dislocation core and $v_{\text{thh}}$ is the thermal velocity of holes.

The equilibrium recombination rate is therefore, given by the value $R^\circ = R_{\text{V-M}} = R_{\text{C-M}}$.

It is apparent that the recombination rate at an impurity-contaminated dislocation could be computed if the impurity nature and its concentration is known. Unfortunately, traces of multiple impurities are systematically present in semiconductor materials, and dislocations are efficient impurity gettering sites, making a quantitative evaluation of the recombination rate difficult.

It could, therefore, be concluded that the individual chemical character of impurities, the electronic properties of dislocations, the thermodynamic stability of defect–impurity complexes and their segregation rate at extended defects, have a critical influence on the semiconductor material properties and on the device performance. And it is already understandable that semiconductor growth and further processing would require a critical attention to impurity contamination in all their steps.

## 2.4.6 Recombination Processes in a p–n Diode

As a final case, we discuss the recombination processes occurring at an un-polarized and polarized semiconductor diode, in condition of thermal equilibrium, which ensures conditions of constant and equal thermal generation rates of carriers.[108]

A schematic view of a unpolarized (zero bias) p–n diode is displayed in Figure 2.17a. Here, the junction between the p and n regions of the diode is a metallurgical boundary, normally prepared with an epitaxial growth process of the p-region on an n-type single crystal substrate, that allows it to maintain the crystal order across the boundary and a virtual[†††††] absence of interface defects.

The thermodynamic and electrical equilibrium across the p and n regions of the diode are accomplished by a constant value of the

---

[†††††] Entropic factors would always favour the onset of local disorder.

(a) p-n junction under zero bias

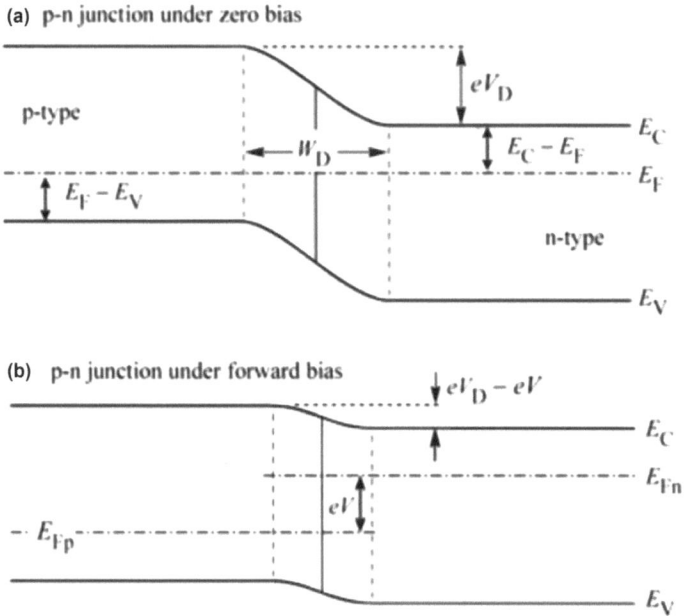

(b) p-n junction under forward bias

**Figure 2.17** Schematic diagram of an unpolarized p-n junction (a) and of a polarized p-n junction under forward bias (b).

Fermi energy $E_F$ across the entire diode, realized thanks to the set-up of a charged double layer of width $W_D = W_p + W_n$, in correspondence with the physical position of the junction, consisting of ionized acceptors at the p-side edge and of ionized donors at the n-side edge, and then of a built-in voltage $eV_D$.

For a width $W_D$ the junction is virtually[†††††] completely depleted of electrons or holes, since those arising by the thermal ionization of dopants, disappear by eh recombination.

Electron–hole recombination might occur by electrons dropping from the conduction band into in an unoccupied state in the valence band (direct eh recombination) or by electrons dropping into a defect deep level in the gap, from where it could capture a hole from the valence band (indirect eh recombination)

The built-in voltage $eV_D$ and the corresponding strong electric field $\varphi_i$ inhibits the flow of charge carriers across the junction in the absence of an externally applied voltage at the p and n sides of the diode.

---

[†††††] A residual electron and holes density in these regions occurs, arising from carriers thermally generated that survive to carrier recombination events.

Under an applied forward polarization,$^{\S\S\S\S\S}$ leading to a decrease in $eV$ of the built-in voltage, which is set at a value $eV_D - eV$, see Figure 2.17b, a current of holes and electrons $I_D = I_e + I_h$ is allowed to flow in the diode, whose absolute value depends on the absolute resistance of the diode, and also on the carriers lost in the junction region of the diode, where they may, in fact, recombine non-radiatively$^{\P\P\P\P\P}$ at defects and impurities (deep traps), with a rate $R_t$ per unit volume

$$R_t \approx \frac{np}{\tau(n+p)} \tag{2.45}$$

which depends on the total carrier density, and on the recombination time $\tau$

$$\tau = \frac{1}{N_t v_{th} \sigma} \tag{2.46}$$

depending on the density of traps $N_t$, on their capture cross section $\sigma$ for electrons or the holes and on the thermal velocity of electrons $v_{th}$, supposed equal to that of holes.[108–112]

Under reverse polarization, leading to an increase in $eV$ of the built-in voltage, which is set at a value $eV_D - eV$, the current that flows across the junction, called the saturation current, is again that allowed only by carriers thermally generated in the junctions that do not recombine and by carriers thermally excited from defect levels, so far the applied field has not overcome the limit at which a diode breakdown occurs.

The ideal diode current–voltage characteristics are displayed in Figure 2.18.

We can expect, however, a decrease in the ideal (dark)$^{\|\|\|\|\|\|}$ diode current under forward bias, due to the increase in the carrier recombination rate in the junction region, when the np diode is manufactured with an impure semiconductor or inadvertently contaminated during the fabrication process.

---

$^{\S\S\S\S\S}$ A forward polarization occurs when the diode is polarized when a positive voltage is applied to the n-type region of the diode.

$^{\P\P\P\P\P}$ Electrons and holes can, in fact, recombine radiatively with the emission of photons. In silicon or germanium semiconductors, however, the recombination time for a non-radiative recombination is orders of magnitude higher than that for a direct recombination.

$^{\|\|\|\|\|\|}$ Light induces the generation of additional carriers when $h\nu > E_G$.

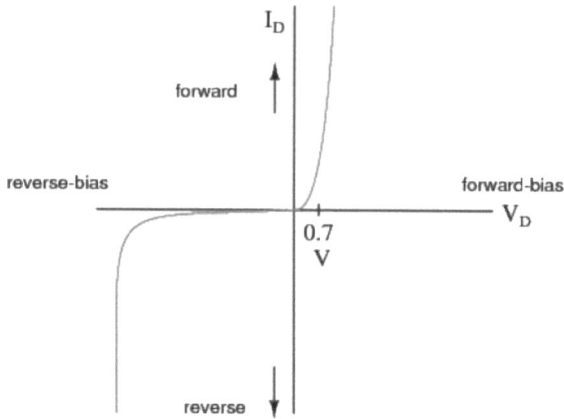

**Figure 2.18** *I–V* characteristics of an ideal diode.

## 2.5  Physico-chemical Aspects of Segregation of Impurities in Semiconductors

Undesired impurities****** in a semiconductor crystal are the remains of impurities contaminating the precursors from which the semiconductor crystal is grown, and still present, despite the occurrence of an intrinsic purification process during the growth process.

Depending on the concentration of the impurities dissolved in a semiconductor crystal, they may remain in solution after any kind of thermal processes operated during the device fabrication, or segregate as a second phase when supersaturation conditions might occur.

Thermodynamic equilibrium considerations predict, in fact, that an impurity $Me(x_\alpha)$ dissolved in solid solution conditions in a phase $\alpha$ of an elemental or compound semiconductor at the concentration $x_\alpha$, segregates as a second phase $\beta$ at the temperature $T_{\alpha-\beta}$ when its concentration $x_\alpha$ exceeds its solubility $x_\alpha^0$ in the $\alpha$ phase, with $\Delta G_{\alpha \to \beta} = 0$ and $\mu_M^\alpha = \mu_M^\beta$.

The second phase $\beta$ might be the solid crystalline†††††† phase of the impurity Me, or that of a compound MX arising by a solid state reaction of the impurity M with a component X of the phase $\alpha$ or with impurities present in the phase $\alpha$, with a process ruled kinetically by

---

****** Impurities dissolved substitutionally or interstitially in the lattice of the semiconductor form a solid solution as long as the solubility limits are not overcome, *i.e.* when the segregation of the excess concentration does occur under form of a second phase (a precipitate).

†††††† Or amorphous.

an homogenous or heterogeneous nucleation, with a very rich family of different behaviours.

Among the myriad of examples available in the literature concerning the segregation of impurities in semiconductors, we intend to discuss here the case of copper[113–115] and of oxygen precipitation in silicon, since they represent iconic examples of the possible segregation features of metallic and non-metallic impurities in semiconductor silicon.

## 2.5.1 Copper Segregation in Silicon

Copper impurities are virtually absent in as-grown semiconductor silicon crystals and wafers, but are ubiquitous in silicon devices, due to the ease of Cu contamination occurring during the fabrication processes, favoured by the high Cu diffusivity in silicon.[115]

From X-ray measurements, Cu in $Cu_xSi_{1-x}$ solid solutions is known to sit in interstitial positions, and from electrical measurements it is also known to behave as a shallow donor, and is therefore present as a thermally ionized positively charged $Cu_i^+$ species, with only benign recombination activity.

The equilibrium solubility of Cu in silicon within 840 and 300 °C, measured using Cu-saturated intrinsic and B-doped samples at these temperatures, is shown to depend on doping, see Figure 2.19.[116] Apparently, the copper solubility values are equal for intrinsic and p-doped samples, while the (calculated) solubility of Cu in silicon is lower in n-type samples.

From the phase diagram of the Cu–Si system, Figure 2.20,[118] one can observe that at temperatures below the eutectic at 802 °C, Cu segregates from a supersaturated Cu–Si solid solution, as the intermetallic η phase consisting of the silicide $Cu_{19}Si_6$, often referred to in the literature as a $Cu_3Si$ phase.[‡‡‡‡‡‡]

Different from interstitial $Cu_i$, intermetallic $Cu_3Si$ precipitates lead to the set-up of deep levels and are, therefore, very harmful recombination centres.

According to Istratov *et al.*[115] and Buonassisi *et al.*[114] it is also experimentally proven that the intermetallic $Cu_3Si$ phase is the only copper-containing phase that segregates from Cu-rich silicon crystals, since oxygen impurities, always present, react preferentially with Si leading to the precipitation of $SiO_2$ instead of that with Cu, leading to the precipitation of $Cu_2O$. This conclusion is, apparently, rather

---

‡‡‡‡‡‡ Similar conclusions could be inferred from the phase diagram of ref. 117.

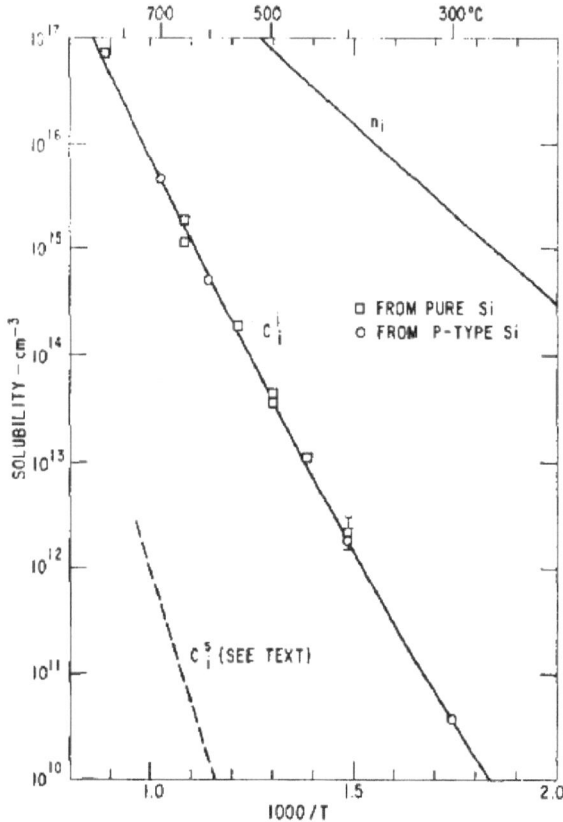

**Figure 2.19**  Copper solubility in silicon. □ Cu-solubilities in undoped silicon, ○ Cu-solubilities in p-type samples. $C_i^s$ are the calculated Cu solubilities in n-type samples. Reproduced from ref. 116 with permission from AIP Publishing, Copyright 1964.

obvious, considering that the standard molar Gibbs formation energy $\Delta G_f^\circ$ of $SiO_2$ holds, $-856.4$ kJ mol$^{-1}$ *vs.* $-291.6$ kJ mol$^{-1}$ for the $Cu_2O$ phase.

Not only the Cu-solubility in silicon is influenced by doping, but also the segregation kinetics of the $Cu_3Si$ phase. In fact, its segregation is hindered in B-doped p-type silicon but favoured in n-type silicon. A qualitative explanation of this behaviour, according to Istratov *et al.*[115] and Buonassisi *et al.*,[114] can be found in the amphoteric character of the precipitation nuclei of copper silicide, which is positively charged in p-type Si, and neutral or negatively charged in n-type silicon.

This could allow us to suppose that positively charged $Cu_i^+$, present in supersaturated conditions in p-type silicon, finds an electrostatic

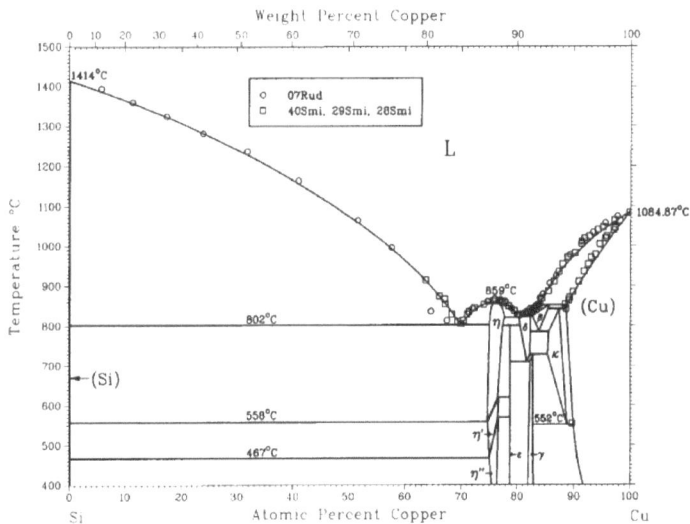

**Figure 2.20**    Phase diagram of the Cu–Si system. Reproduced from ref. 118 with permission from Springer Nature, Copyright 1986.

barrier in correspondence with the precipitation nuclei, which would prevent further precipitate growth.[115] In addition, positively charged $Cu_i^+$ easily pairs with negatively charged boron $B_{Si}^-$, leading to stable $[B_{Si}^- - Cu_i^+]$ neutral pairs, subtracting reactive $Cu_i^+$ ions in the precipitation process.

Eventually, different from copper in p-type silicon, copper present in supersaturated conditions in n-type Si easily segregates in the absence of electrostatic barriers, allowing the growth of platelets of the copper silicide phase $Cu_{19}Si_6$ with an [111] oriented habit. These platelets are metastable at room temperature, and change morphology after thermal annealing, taking the habit of spherical clusters.

## 2.5.2   Oxygen Segregation in Silicon

The study of the physico-chemical properties of oxygen in silicon, of the process of oxygen precipitation in dislocation-free and dislocated silicon, and of the role of oxygen precipitates as effective getters (*i.e.* chemical and physical traps) for metallic impurities in silicon wafers has been the subject of thousands of papers and hundreds of books in the last 70 years, given the technological and scientific importance of the subject, of which we will give here a few critical ref. 119–140.

Oxygen impurities in silicon critically affect the features and the yield of the microelectronic devices manufactured on top of CZ silicon wafers. It is, in fact, known that dissolved oxygen improves the mechanical properties of silicon, but that supersaturated solutions of oxygen in silicon segregate the excess oxygen as amorphous or crystalline $SiO_2$ precipitates at the process temperatures used for integrated device fabrication (800–1000 °C). These precipitates, when present in the volume of the wafers, behave as beneficial getters[§§§§§§] for residual transition metal (TM) impurities, or for unwanted TM impurities introduced in the wafer during the device manufacture process, and their proper use is at the background of the internal gettering processes, which we will discuss in the next section.

When oxygen precipitates are, instead, present at the surface of the wafer, at the p-region of a p–n diode, which is the active region of a microelectronic device or of a silicon solar cell,[132] they can severely degrade the device performance due to their strong recombination activity.[122]

It has been, therefore, mandatory not only to understand the behaviour of oxygen in silicon from the electrical, chemical and structural point of view, but also to study and model the nucleation and the oxygen precipitation processes in silicon[121] to be able to manage the oxygen precipitation processes as well as the impurity segregation (gettering) processes at oxygen precipitates.

Oxygen contamination of silicon arises either from the partial dissolution of the silica crucible during the Czochralski (CZ) growth of silicon ingots, or from a vapour/solid phase reaction between Si vapours sublimating from the molten charge and the silica deposited on the hot steel walls of the growth furnace during the float zone (FZ) growth process (see Chapter 3)

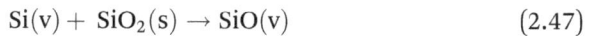

$$Si(v) + SiO_2(s) \rightarrow SiO(v) \tag{2.47}$$

followed by the dissolution of $SiO(v)$ vapours in the liquid silicon phase.

From the earliest IR investigations on oxygen in silicon,[133] and from most recent *ab initio* or density functional theory (DFT) investigations[134] it is known that oxygen in silicon sits in symmetrical bond-centred interstitial positions, see Figure 2.21a, under low hydrostatic pressures, but adopt a buckled configuration, see Figure 2.21b, at higher pressures.

There is also the experimental evidence that the as-grown CZ crystals contain oxygen in supersaturated conditions, with room temperature concentrations around $10^{18}$ at $cm^{-3}$ (against a value less

---

[§§§§§§] The gettering process is a physico-chemical process that immobilizes and renders electrically inactive the impurities at the surface of the getter material.

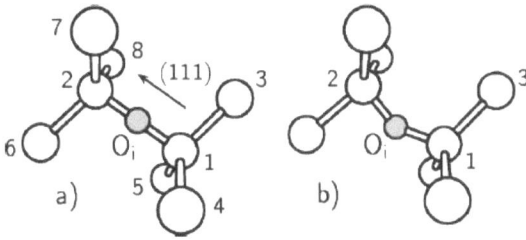

**Figure 2.21** Symmetrical and buckled configuration of interstitial oxygen in silicon. Reproduced from ref. 134 with permission from American Physical Society, Copyright 2000.

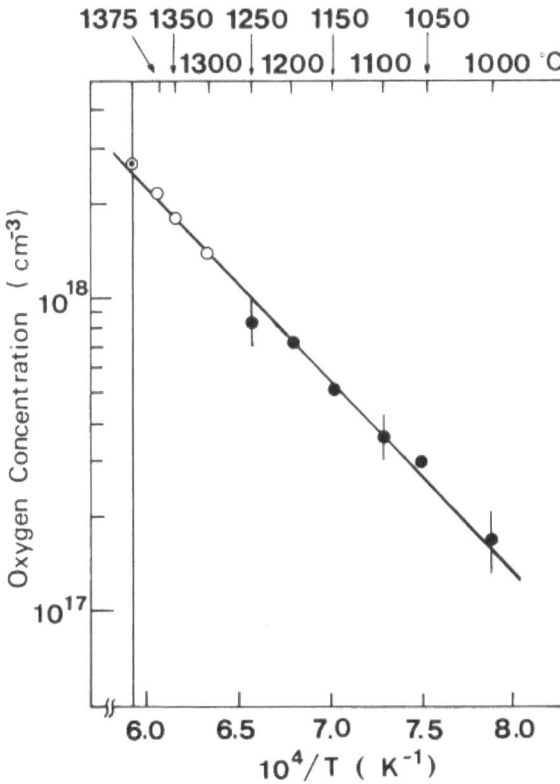

**Figure 2.22** Solubility of oxygen in silicon. The vertical solid line corresponds to the melting temperature of silicon. Reproduced from ref. 135 with permission from IOP, Copyright 1985.

than $10^{16}$ at cm$^{-3}$ in FZ-silicon), values close to the experimental values of oxygen solubility in silicon at the growth temperature,[135] see Figure 2.22. There is, therefore, the evidence that metastable, supersaturated solid solutions of oxygen, whose concentration stays inside the two-phase region of the diagram of Figure 2.23, survive down to

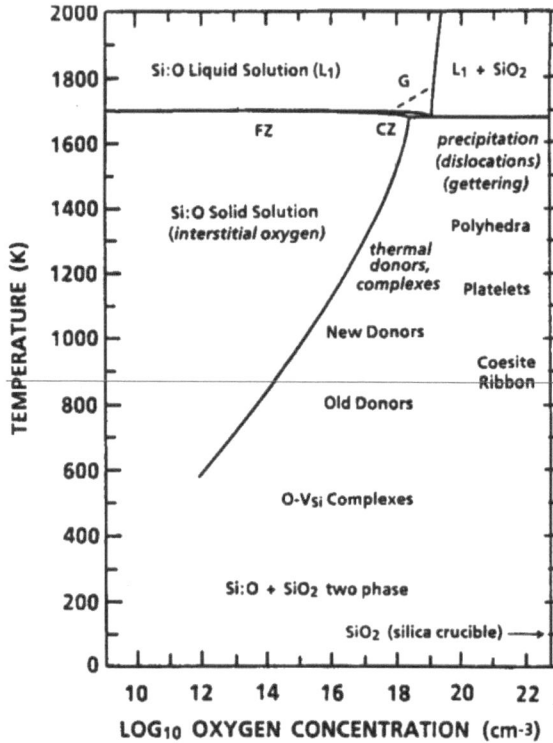

**Figure 2.23** Phase diagram of the Si–O system. Reproduced from ref. 119 with permission from Springer Nature, Copyright 1987.

room temperature with supersaturation ratios close to $10^4$ when the silicon ingots cool down during the CZ growth process.[136,137]

During the after-growth cooling of the CZ ingots in the growth furnace, as will be seen in Chapter 3, or by thermal annealing at 800 K of CZ wafers, the formation of oxygen clusters, the thermal donors (TDs), occurs, mentioned as old donors in Figure 2.23, while the precipitation of oxygen as amorphous $SiO_2$ arises at higher temperatures (1000–1600 K), with the intermediate formation of several precursors of α-quartz with different morphologies that depend on temperature, as seen in Figure 2.23.

From careful studies dedicated to their formation, structure and properties, we know that TDs are multi-atomic oxygen clusters, with a core consisting of four oxygen atoms, and with a size that increases with the annealing temperature, with the annealing time and with the initial concentration of oxygen in the ingot. Their stability reaches its maximum at 450 °C, while above 500 °C they tend to annihilate.[138]

TDs behave as double donors with ionization energies of 70 and 150 meV for the first and second ionization, which slightly depend on their size, and, therefore, ionize at room temperature, adding an unwanted carrier contribution to that arising from dopants deliberately added. For this reason, it is industrial practice to anneal at 500 °C the CZ silicon wafer to annihilate residual TDs and grant appropriate resistivity properties.

The nucleation and growth of oxide precipitates from these supersaturated O–Si solid solutions occurs by thermal annealing in the absence or in the presence of heterogeneous nucleation sources as dislocations and other extended defects in the silicon matrix.

In the case of dislocated samples, the nucleation is heterogeneous, as can be seen in Figure 2.24, and occurs in correspondence with a hexagonal dislocations half loop[128] with the growth of an octahedral oxide precipitate.

In the case of the dislocation-free CZ silicon wafers used for microelectronic applications, the nucleation stage is a homogeneous and energetically unfavourable process, occurring *via* the absorption of vacancies from equilibrium Frenkel pairs $V_{Si} + Si_i$ and the emission of self-interstitials.

In fact, the oxide segregation implies the generation of a large compressive stress in the silicon lattice, due to the large difference

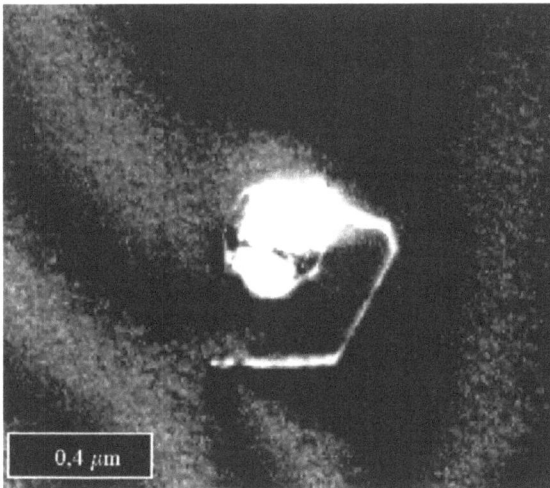

Figure 2.24    TEM micrograph of an oxide precipitate associated with a dislocation half-loop. Reproduced from ref. 128 with permission from AIP Publishing, Copyright 2002.

between the molar volume of the silicon dioxide as amorphous α-quartz (22.67 cm$^3$) and that of silicon (12.05 cm$^3$), which will strongly impair its segregation thermodynamics and kinetics.

The molar volume difference and the resulting stress could be, however, partially or entirely accommodated (a) by morphological and structural changes of the precipitates, which, depending on the process temperature, could be thermodynamically stable as amorphous precipitates or as crystalline needles, platelets or polyhedra with different aspect ratios, or (b) by a local volume compensation process associated with the absorption of silicon vacancies, which will deliver the requested excess volume and then reduce the thermodynamic cost of forming the silicon dioxide phase.[139,140]

The growth process occurring with the assistance of vacancies as the point-defects mediating the stress-relief mechanism, could be formally described by the following equation[141]

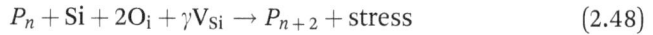

$$P_n + \text{Si} + 2\text{O}_i + \gamma V_{\text{Si}} \rightarrow P_{n+2} + \text{stress} \tag{2.48}$$

where $P_n$ is an oxide precipitate containing $n$ oxygen atoms, $\gamma$ is the number of silicon vacancies $V_{\text{Si}}$ per oxygen atom involved in the process, and $P_{n+2}$ is an oxide precipitate with $n + 2$ oxygen atoms and stress is the amount of stress not compensated for by vacancy absorption.

Since the equilibrium vacancy concentration is ruled by a Frenkel mechanism, eqn (1.19), the absorption of a vacancy is compensated for by the emission of a Si self-interstitial.

The thermodynamic driving force $\Delta G_{\text{segr}}$ for the oxide segregation is the sum of two terms

$$\frac{\Delta G_{\text{segr}}}{nk_{\text{B}}T} = -\ln \frac{C_{\text{O}}}{C_{\text{O}}^{\text{eq}}} - \gamma V_{\text{Si}} \ln \frac{C_{\text{V}}}{C_{\text{V}}^{\text{eq}}} \tag{2.49}$$

of which the first represents the contribution of the supersaturation ratio of oxygen $\dfrac{C_{\text{O}}}{C_{\text{O}}^{\text{eq}}}$ and the second is the contribution of the actual vacancy concentration $C_{\text{V}}$ with respect to the equilibrium vacancy concentration $C_{\text{V}}^{\text{eq}}$.

The strain energy $G_{\text{S}}$, in turn, is given by the following equation, using a continuum mechanical method[140]

$$G_{\text{s}} = \frac{1}{2}\Delta V^2 \frac{K^* \varphi(\beta)}{K^* + \varphi(\beta)} V_{\text{p}} \tag{2.50}$$

where $K^*$ is the bulk modulus[¶¶¶¶¶¶] of the oxide precipitate phase, $\varphi\beta$ is a shape function, depending on the modulus $K$ of the silicon matrix phase, $\beta$ is the aspect ratio of the precipitate, $V_p$ is the volume of the precipitate and

$$\Delta V = \frac{V_{SiO_2} - (V_{Si} + 2\gamma V_{Si})}{V_{Si} + 2\gamma V_{Si}}. \tag{2.51}$$

Eventually, the matrix/precipitate interfacial free energy $G_{i,n}$ is given by the following expression

$$G_{i,n} = \pi R_p^2 f(\beta^2) \beta^{-2/3} \left(\frac{1+e_C}{1+Xe_C}\right)^2 \sigma_O(n) \tag{2.52}$$

where $R_p$ is the radius of the precipitate, $e_C$ is the constrained strain, $X = 1 + \dfrac{4\mu}{3K}$, $\mu$ is the shear modulus and $\sigma_O(n)$ is the effective interfacial energy, which depends on the precipitate size and on other variables.

Equilibrium conditions are obtained by numerical minimization of the $G$ function with respect to $\gamma V$

$$\frac{\delta G(n, \gamma V, \beta)}{\delta \gamma V} = 0 \tag{2.53}$$

and to $\beta$

$$\frac{\delta G(n, \gamma V, \beta)}{\delta \beta} = 0. \tag{2.54}$$

A reasonably good fit within model values and experimental values of precipitate density of 13 samples nucleated at different temperatures and duration times was obtained by Yang *et al.*,[139] as can be seen in Figure 2.25, showing the reliability of the model.

In the industrial practice, however, oxygen precipitation processes have been carried out for years with empirical and often unsatisfactory schemes, aiming at obtaining:

- The appropriate control of the silicon vacancy concentration, which, according to eqn (2.49), drives the oxygen segregation process.

---

[¶¶¶¶¶¶] Measuring the material's resistance to uniform compression.

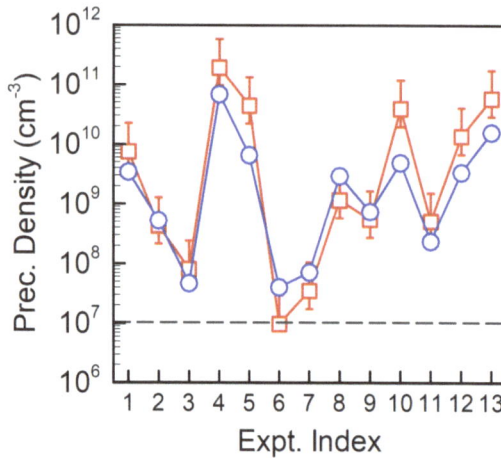

**Figure 2.25** Comparison of the experimental oxide precipitate density with that predicted using the optimized interface free energy model in eqn (2.52) for the 13-experimental samples. Red squares—experimental measurements), blue circles—simulation. The horizontal dashed line represents an estimate of the experimental detection threshold. Reproduced from ref. 140 with permission from AIP Publishing, Copyright 2019.

- A subsurface region fully denuded by precipitates, to avoid carrier recombination losses in the active region of the device.

A successful solution to this problem was found by Falster, Voronkov *et al.*[119,122,124,125,130] who succeeded in the realization of the so-called magic denuded zone (MDZ) process, by the installation of a proper vacancy profile inside a silicon wafer, *via* a three step process, which is carried out using rapid thermal annealing (RTA) conditions.

The first step of the process is the spontaneous formation of the equilibrium concentration of Frenkel pairs $V_{Si} + Si_i$, whose actual concentration depends on the temperature and on the Gibbs formation energy of Frenkel pairs, which occurs when the silicon sample is brought to high temperatures. Due to the presence of a surface, a coupled flux of the fast diffuser self-interstitials towards the surface of the wafer, where they are annealed, and of vacancies towards the bulk then occurs as the second step of the process. Upon subsequent fast cooling, the vacancy profile is frozen-on, given that vacancies are slow diffusers, leading to the final vacancy profile given in Figure 2.26. One can see that the vacancy concentration is very low at the wafer surface, so as to suppress the oxygen precipitation in the surface region during

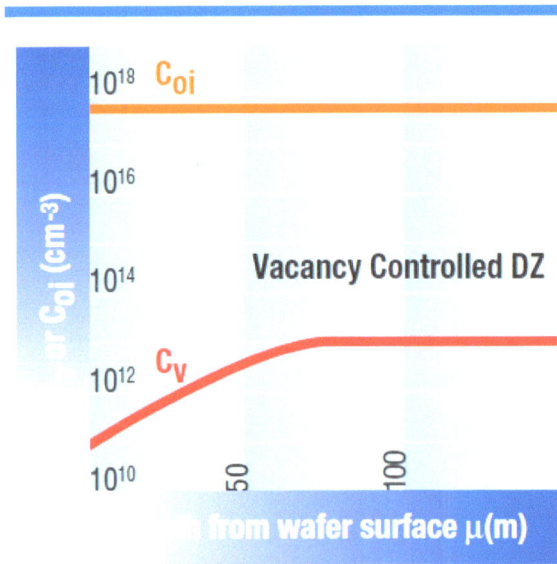

**Figure 2.26** Schematic illustration of the vacancy concentration profile in the MDZ process. Reproduced from https://www.yumpu.com/en/ document/read/14985070/the-control-of-oxygen-precipitation-and-the-impact-of-memc with permission from MEMC, Applications Note AE-008.

a further thermal annealing of the silicon sample, which is the final step of the process.

Figure 2.27 displays the distribution of microprecipitates obtained with the MDZ process, with a surface layer which is fully depleted of them. This oxygen precipitates distribution is ideal, since it preserves the structural quality of the wafer surface where the microelectronic devices are manufactured, while the homogeneous distribution of precipitates in the bulk of the wafer allows the internal gettering (IG) and electrical inactivation of residual dangerous metallic impurities present in the wafer or introduced during the device manufacturing process.

According to Falster *et al.*[130] the critical density of precipitates for effective IG is $10^7$ cm$^{-3}$, and a prolonged heating at 700 °C is needed to induce a morphological transition from unstrained (and inactive) precipitates to strained amorphous platelets, which are the effective IG particles.

The chemical backgrounds of the internal gettering processes will be discussed in the next section to arrive at a full understanding of these processes.

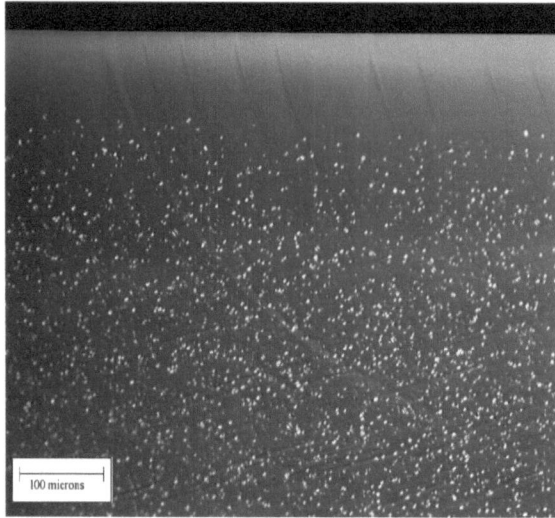

**Figure 2.27** Distribution of oxygen precipitates with a surface depleted layer obtained with the MDZ process Reproduced from ref. 131 with permission from Trans Tech Publications, Copyright 2008.

## 2.6 Chemistry of Impurity Gettering Phenomena at Oxide Precipitates

The chemistry and physics of impurity segregation at oxygen precipitates in semiconductor silicon, known as internal gettering (IG),[142] can be discussed considering the process as the irreversible transfer of a metallic impurity from a phase consisting of a solution of one or more TM impurities in silicon to an ensemble of nanoparticles distributed in a predetermined region of the semiconductor sample, which behaves as the gettering substrate.

Different from the case of external gettering, occurring for impurities in supersaturation conditions, which will be discussed in Section 4.2, here the segregation occurs for impurities present at concentrations lower than their solubility.

The understanding and the further modelling of the thermodynamic and kinetic features of a segregation process of impurities at a gettering substrate, consisting of a distribution of oxide precipitates in a silicon matrix, is based on the consideration that the impurity segregation might either occur with the formation of chemical bonds at the surface of the gettering phase, *i.e.* with a chemisorption process, or with the formation of a surface compound of the metallic impurity, *via* a surface reaction mediated by atomic defects.

Additionally, it is based on the experimental knowledge that the surface of oxide nanoparticles consists of a nanometrically thin ($\sim$3 nm) sub-stoichiometric $SiO_x$ phase.

The thermodynamics of a chemisorption process could be treated as the repartition of an impurity M among the silicon matrix (phase $\alpha$) and an ensemble of nanoparticles (phase $\beta$), both behaving as truly Gibbs-type of phases, given by the following ratio

$$n_M^\beta / n_M^\alpha = \Gamma\left(n_M^\beta, n_M^\alpha\right) \exp - \frac{\Delta G_r}{kT} \tag{2.55}$$

where $n_M^\beta$ is the average concentration of the impurity at the surface of the phase $\beta$, $n_M^\alpha$ is the concentration of the impurity in the bulk semiconductor phase, $\Gamma(n_M^\beta, n_M^\alpha)$ is a constant that accounts for the interaction coefficients of the impurity in the bulk and in the surface phase, and $\Delta G_r$ is the Gibbs energy for the chemisorption of the impurity at a silicon oxide phase, with the formation of M–O or M–Si bonds.

The thermodynamics of a segregation process occurring with the formation of a new surface phase depends, instead, on the specific reaction occurring at the surface of the oxide nanoparticles. We can, in fact, foresee a chemical reaction of the impurity $M_i$ dissolved in the interstices of the silicon phase, either with point defects (self-interstitials emitted to compensate the strain) and the formation of a metal silicide phase

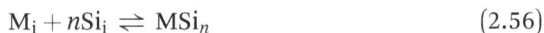

$$M_i + nSi_i \rightleftharpoons MSi_n \tag{2.56}$$

or with oxygenated species (*e.g.* $O_2^i$ molecules) at the sub-stoichiometric surface of the oxide nanoparticles, with the formation of an oxide or a metal silicate phase.

In both cases it is assumed that the establishment of an equilibrium process does occur, ruled by the Gibbs energies of formation of the silicides or of the oxidized metal phases.

The kinetics of the segregation process of a metallic impurity at a gettering phase, consisting of a distribution of oxide nanoparticles depends, as expected, on the rate of the slowest step of the interaction process, which includes the delivery of the reactive species at the physical reaction site.

If the reaction rate is fast, the kinetics is dominated by the rate of impurity delivery, *i.e.* by metal diffusion in the silicon phase, and its activation energy would correspond to the activation energy $E_D$ of the diffusion process

$$D_i = D_i^\circ \exp - \frac{E_D}{RT}. \tag{2.57}$$

When the reaction itself is, instead, a strongly activated process, the gettering process is reaction-limited. A truly reaction-limited process occurring at the interface between an oxide nanoparticle and an impurity dissolved in a solid matrix implies that the process of delivery of chemical species at the reaction interface is fast and that the slow rate step is the formation of a chemical bond between the gettering substrate and the impurity, or the formation of a solid phase with the participation of point defects or secondary impurities.

In these cases, the reaction kinetics are quantitatively described by an absolute rate constant $k$

$$k \ (s^{-1}) = A \exp - \frac{E_r}{kT} \tag{2.58}$$

where $E_r$ is the activation energy and thus the height of the reaction barrier.

If the reaction is mediated by point defects, behaving as catalysts or intermediate species, the activation rate of the process is a direct measure of the defect-mediated activation barrier. Details about the success of IG on several metallic impurities in silicon are discussed in the following section.

### 2.6.1 Internal Gettering of Ni, Fe and Cu: Experimental Evidence

Experimental and theoretical studies on IG of metallic impurities started in the 1990s[143–147] and the IG features of nickel, iron, and copper[§] at oxide precipitates in silicon will be reported here as examples of general interest.[148–172]

The gettering of Ni impurities at oxide precipitates, discussed by Ourmazd and Schröter,[143] was carried out at 752 °C, and is a typical case of gettering mediated by point defects. Because of the large molar volume difference ($\sim 150\%$) between (amorphous) silicon oxide

---

[§] These are among the most detrimental impurities for silicon devices.

and silicon, the precipitation of silicon oxide from an oxygen-rich silicon matrix occurs with the injection of silicon self-interstitials, leading to their local supersaturation, and to the formation of nickel disilicide

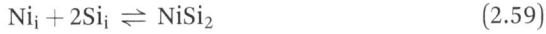

$$Ni_i + 2Si_i \rightleftharpoons NiSi_2 \qquad (2.59)$$

at the surface of oxide nanoparticles, experimentally revealed by bright field TEM micrographs.

Subsequent work carried out by Seibt and Graff,[158] and by Sueoka[159] and Nakamura,[160] confirmed that the phase segregated at the surface of oxide precipitates is nickel disilicide, that the gettering process is reaction-rate limited, thus not depending on the diffusion of Ni in the silicon matrix, and that the residual Ni concentration after the IG process is below the limits of Ni detection.

The success of Ni gettering at oxide precipitates is directly associated with the thermodynamics of the reaction (2.59), which is ruled by the Gibbs energy of formation of the NiSi$_2$ phase

$$K = \frac{a_{NiSi}}{a_{Ni_i} a_{Si_i}^2} = A \exp - \frac{\Delta G_{NiSi_2}}{RT} \qquad (2.60)$$

where $a_{NiSi}$ is the activity of the NiSi$_2$ phase, taken equal to 1, and $a_{Ni_i}$ and $a_{Si_i}$ are the activities of the Ni impurities and self-interstitials, taken equal to their concentrations in a regime of dilute solutions, and $\Delta G_{NiSi_2}$ [$-(88 \pm 12)$ kJ mol$^{-1}$ at 298 K][161] is the Gibbs energy of formation of nickel disilicide.

From eqn (2.60) we expect, however, that a redissolution of Ni-impurities would occur after a thermal annealing of the gettered system at temperatures higher than the IG process, in close analogy with the case of iron and copper impurities segregated at oxide precipitates,[150,154] which will be discussed below.

The case of iron was discussed, among many others, by Zang *et al.*,[153] who demonstrated that intentionally iron-contaminated, B-doped samples, submitted to an internal, two step, gettering anneal at 700 °C and 450 °C, followed by a fast quench to room temperature, are found to be virtually iron-free ($\sim 10^{10}$ cm$^{-3}$).

When the sample is further annealed at 800 °C for different times, see Figure 2.28, iron redissolution occurs and equilibrium conditions are obtained with an iron concentration of $\sim 5 \times 10^{12}$ cm$^{-3}$, which corresponds to the iron solubility at 800 °C. The iron concentration

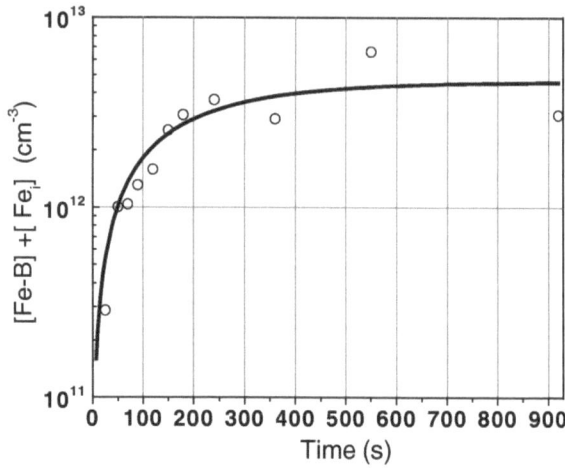

**Figure 2.28**   Kinetics of iron dissolution at 800 °C. Reproduced from ref. 153
with permission from Elsevier, Copyright 2003.

dependence on process time could be fitted by an exponential
function

$$C_t = C^0 \exp - \frac{t}{\tau_{\text{diss}}} \tag{2.61}$$

where $C^0$ is the iron solubility at 800 °C $(4.58 \times 10^{12} \text{ cm}^{-3})$ and
$\tau_{\text{diss}} = 198 \pm 83$ s is the time needed to redissolve completely the
segregated iron.

Since the experimental value of $\tau_{\text{diss}}$ is much greater than its calculated value $(\tau = 2.53 \text{ s})$ for a process whose rate is determined by the
metal diffusion of Fe in Si at 800 °C, the process is actually rate
limited by an activation barrier $E_{\text{act}}$ for redissolution, which was
experimentally found to be equal to $1.47 \pm 0.10$ eV, see Figure 2.29.
This value is greater than the activation energy $E_D = 0.67$ eV for the
diffusion of iron in silicon, and greater than the Gibbs energy of
formation of $FeSi_2$, which is 0.30 eV[162] and excludes, therefore, the
dissolution of a $FeSi_2$ phase as the rate determining step, as well the
formation of $FeSi_2$ in the IG process.

Considering that the oxygen precipitates consist of amorphous
silica $(\alpha\text{-SiO}_2)$ platelets and that the main defects in $\alpha\text{-SiO}_2$ are the $P_b$
defects, identified in the literature[164] as unsaturated Si dangling
bonds lying very close to the SiO/Si interface, and interstitial oxygen
molecules $O_2^i$, which are the species accounting for the excess oxygen
present in thermally grown bulk amorphous oxide and are also responsible for the oxygen diffusion in the oxide layer during the

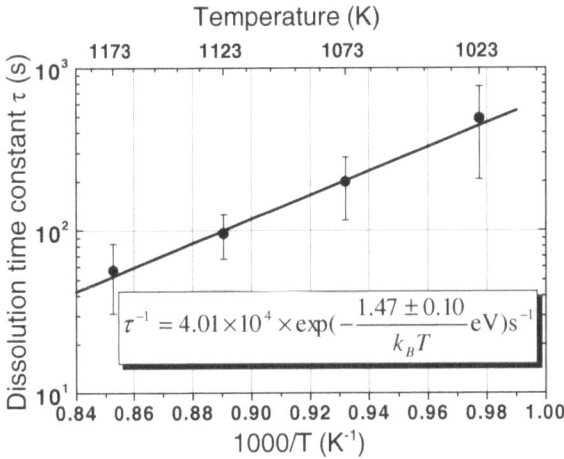

**Figure 2.29**     Temperature dependence of the dissolution constant of iron in silicon. Reproduced from ref. 153 with permission from Elsevier, Copyright 2003.

oxidation of silicon,[165] we could suppose that iron interacts either with $P_b$ defects or with interstitial oxygen molecules $O_2^i$.

In the first case the surface phase resulting from the chemical interaction would be, again, an iron disilicide $\gamma$-$FeSi_2$ phase, whose formation is experimentally excluded, despite being the thermodynamically stable silicide phase formed by solid state reaction of iron with silicon.[166]

In the second case we could suppose that interstitial oxygen molecules would react with iron forming a thermodynamically stable surface phase of FeO

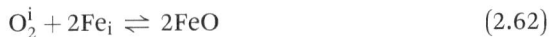

$$O_2^i + 2Fe_i \rightleftarrows 2FeO \qquad (2.62)$$

driven by the Gibbs energy of formation, which is $-171.384$ kJ mol$^{-1}$ at 1000 °C[167] corresponding to 1.78 eV, not so far from the value of the experimental activation barrier $E_{act}$ for iron redissolution, which is $1.47 \pm 0.10$ eV.

Given that the formation of FeO occurs at the surface of the precipitate as a nanophase, a deviation from the values of the Gibbs energy of formation of the bulk phase is well understandable, since the formation enthalpy depends on size, and increases (in absolute value) with the decrease of the size, as we will see in detail in Chapter 5.

Looking now to the IG of Cu at oxygen precipitates, which has been investigated, among others, by Kissinger *et al.*,[168] who showed that the surface of oxide precipitates consists of a thin, 3 nm thick, layer of $SiO_x$, in good agreement with other literature results[142,169–171] and that the sensitivity of the spectroscopic techniques with sufficient spatial resolution is too low to identify the chemical nature of the segregated impurities at a low density of oxide precipitates dispersed in a silicon matrix.

To overcome the analytical problem, already observed in the case of nickel and iron, they used a thermal oxide layer, 14.5 nm thick, at the surface of a silicon wafer, as the research substrate, and were able to show, using ToF-SIMS******* measurements, that copper segregation and its thermal dissolution occurs at the surface of the oxide precipitates.

If we assume, like in the iron case, that copper segregates at the $SiO_x$ surface as an oxidized Cu species, possibly with the support of interstitial oxygen molecules $O_2^i$,[165] gettering of copper would occur with the formation of CuO

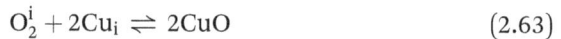

$$O_2^i + 2Cu_i \rightleftharpoons 2CuO \tag{2.63}$$

with a thermodynamically allowed surface reaction, driven by the Gibbs energy of formation of CuO, which is $-124.245$ kJ mol$^{-1}$ at 1000 K.

A problem, however, arises from the microstructural investigations carried out on the reacted oxide surface,[169] using scanning TEM, since no surface precipitates were found, leading to the conclusion that the gettered copper was dissolved in the interface $SiO_x$ layer with the formation of the Cu-silicate phase.

Extrapolating this conclusion to the case of Fe, it could be suggested that, depending on the nature of the metal, internal gettering of TM metals might occur with the formation of a surface phase of a non-stoichiometric metal silicate.

Though still speculative, this hypothesis would also explain the internal gettering of other metallic impurities at silicon dioxide precipitates.

# Appendix 2.1 Thermal Noise Effects

For all semiconductors, including silicon, there is also a temperature limit to their valuable microelectronic (and optoelectronic)

---

******* Time of flight SIMS.

applications due to the temperature dependence of their intrinsic electronic properties and to the set-up of intrinsic noise problems.

The intrinsic noise is a spontaneous fluctuation of current or voltage in a semiconductor device[172] whose main sources are the thermal noise and the generation–recombination noise.

The first, *i.e.*, the unwanted noise in an electronic component, arising from the increase in the dark current[††††††] or from the thermal fluctuations of electrons by heat, is unavoidable at non-zero temperatures and originates from the temperature dependence of the energy gap $E_G$ of semiconductors, which decreases with temperature with a law that for most semiconductors could be fitted by the following equation[173]

$$E_G(T) = E_G(0) - \frac{\alpha T^2}{T - \beta} \qquad (2.64)$$

where $E_G(0)$ is the energy gap at 0 K, $\alpha$ and $\beta$ are constants, and $T$ is the absolute temperature. We expect, therefore, an increase in the concentration of the intrinsic current carriers with the increase in the temperature, leading, in turn, to an increase in the leakage current.

More explicitly, the thermal noise depends on the random thermal fluctuation of charge inside an electrical conductor, affects every physical resistor in a circuit, including a p–n junction, and induces a fluctuating voltage at the resistor terminals, which increases with the temperature whose mean square $\overline{V_{th}^2}$ value is given by the equation

$$\overline{V_{th}^2} = 4k_B T R \Delta f \qquad (2.65)$$

where $k_B$ is the Boltzmann constant, $T$ is the absolute temperature, $R$ is the resistivity and $\Delta f$ is the frequency band width of the measurement system.

The generation–recombination noise depends, instead, on the fluctuation $\overline{(\Delta N)^2}$ of the number of current carriers due to the existence of carrier recombination centres, impurities, or structural defects, inducing the presence of deep levels in the gap of the semiconductor and leads to conductance changes of the device.

The spectral density of the generation–recombination noise $\dfrac{S_{g-r}(f)}{N^2}$ is given by the following equation

$$\frac{S_{g-r}(f)}{N^2} = \frac{\overline{(\Delta N)^2}}{N^2} \frac{4\tau}{1 + (2\pi f \tau)^2} \qquad (2.66)$$

---

†††††† Dark current is the unwanted leakage current of a pn junction.

where $N$ is the carrier density, $\tau$ is the carrier lifetime, and $f$ is the frequency. A temperature dependence is expected, given that $N = A\exp - \dfrac{E_G}{kT}$, and $E_G$ depends on temperature *via* eqn (2.64).

When several recombination centres of different nature are present, the overall spectral density is the superposition of several distributions, each described by an equation which takes the form of eqn (2.66).

# References

1. A. B. Garrett, The discovery of the transistor: W. Shockley, J. Bardeen and W. Brattain, *J. Chem. Educ.*, 1963, **40**(6), 302.
2. N. Zheludev, The life and times of the LED, a 100-year history, *Nat. Photonics*, 2007, **1**, 189.
3. S. Nakamura, M. Senoh, S.-I. Nagahama, N. Iwasa, T. Yamada, T. Matsushita, H. Kiyoku and Y. Sugimoto, Superbright Green InGaN single-quantum-well structure light emitting diodes, *Jpn. J. Appl. Phys.*, 1995, **34**, L1332.
4. S. Nakamura, M. Senoh, S.-I. Nagahama, N. Iwasa, T. Yamada, T. Matsushita, H. Kiyoku and Y. Sugimoto, Characteristics of InGaN multi-quantum well-structure laser diodes, *Appl. Phys. Lett.*, 1996, **68**, 3269.
5. S. Nakamura, M. Senoh, S.-I. Nagahama, N. Iwasa, T. Yamada, T. Matsushita, H. Kiyoku and Y. Sugimoto, InGaN-Based Multi-Quantum-Well-Structure Laser, *Jpn. J. Appl. Phys.*, 1996, **35**, L74.
6. S. Chichibu, T. Azuhata, T. Sota and S. Nakamura, Recombination of Localized Excitons in InGaN Single- and Multiquantum-Well Structures, *MRS Online Proc. Libr.*, 1996, **449**, 653.
7. S. Nakamura, InGaN/GaN/AIGaN-Based Laser Diodes with an estimated lifetime longer than 10000 hours, *MRS Bull.*, 1988, **23**(5), 37.
8. Y. Liao, C. Thomidis, C.-K. Kao and T. D. Moustakas, AlGaN based deep ultra-violet light emitting diodes with high internal quantum efficiency grown by molecular beam epitaxy, *Appl. Phys. Lett.*, 2011, **98**, 081110.
9. R. J. Trew, High frequency solid state electronic devices, *IEEE Trans. Electron Devices*, 2005, **52**(5), 638.
10. D. Christina, Germanium on silicon, Epitaxy and applications, in *Silicon, Germanium, and Their Alloys: Growth, Defects, Impurities, and Nanocrystals*, ed. G. Kissinger and S. Pizzini, CRC Press, 2014.
11. Y. M. Haddara, P. Ashburn and D. M. Agnall, Silicon-Germanium: Properties, Growth and Applications, in *Springer Handbook of Electronic and Photonic Materials*, ed. S. Kasap and P. Capper, Springer Handbooks, Springer, 2017.
12. The LHCf Collaboration *et al.*, The LHCf detector at the CERN Large Hadron Collider, *J. Instrum.*, 2008, **3**, S08006.
13. M. Acciarri, S. Acerboni, S. Binetti, S. Ferrari, S. Pizzini, M. Bosetti, G. P. Rancoita, M. Rattaggi and G. Terzi, A novel method for the determination of the radiation damage effects in silicon detectors, *Nucl. Phys. B, Proc. Suppl.*, 1993, **32**, 410.
14. M. Balkanski and R. F. Wallis, *Semiconductor Physics and Applications*, Oxford University Press, 2000.
15. J. Singh, *Semiconductor Devices, Basic Principles*, John Wiley& Sons, 2001.
16. S. Pizzini, *Physical Chemistry of semiconductor materials and processes*, John Wiley & Sons, 2015.

17. G. Dhanaraj, X. R. Huang, M. Dudley, V. Prasad and R.-H. Ma, *Silicon Carbide Crystals — Part I: Growth and Characterization in Crystal Growth Technology*, ed. K. Byrappa, W. Michaeli, H. Waarlimont and E. Weber, William Andrew Inc, 2003.
18. A. Kumar and M. S. Aspalli, SIC: An advanced semiconductor material for power devices, *Int. J. Res. Eng. Technol.*, 2014, **03**, 248.
19. *Compound Semiconductor Magazine*, 2021, Volume 27, Issue 1.
20. N. Lu and I. Ferguson, III-nitrides for energy production: photovoltaic and thermoelectric applications, *Semicond. Sci. Technol.*, 2013, **28**, 074023.
21. Y. Hijikata, S. Yagi, H. Yaguchi, S. Yoshida and S. Thermal, Oxidation Mechanism of Silicon Carbide, in *Physics and Technology of Silicon Carbide Devices*, Intech, 2013.
22. R. Bube, *Photoelectronic properties of semiconductors*, Cambridge University Press, 1993.
23. N. Trivellin, M. Yushchenko, M. Buffolo, C. De Santi, M. Meneghini, G. Meneghesso and E. Zanoni, Laser-Based Lighting: Experimental Analysis and Perspectives, *Materials*, 2017, **10**, 1166.
24. L. Mangolini, E. Thimsen and U. Kortshagen, High-yield plasma synthesis of luminescent silicon nanocrystals, *Nano Lett.*, 2005, **5**(4), 655.
25. D. Jurbergs, E. Rogojina, L. Mangolini and U. Kortshagen, Silicon nanocrystals with ensemble quantum yields exceeding 60%, *Appl. Phys. Lett.*, 2006, **88**, 233116.
26. E. M. T. Fadaly, *et al.*, Direct-bandgap emission from hexagonal Ge and SiGe alloys, *Nature*, 2020, **580**, 205.
27. G. F. Brown and J. Wu, Third generation photovoltaics, *Laser Photonics Rev.*, 2009, **3**, 394.
28. J. Y. Kim, J.-W. Lee, H. S. Jung, H. Shin and N.-G. Park, High-Efficiency Perovskite Solar Cells, *Chem. Rev.*, 2020, **120**(15), 7867.
29. A. G. Melton, E. Burgett, T. Xu and N. Hertel, Comparison of neutron conversion layers for GaN-based scintillators, *Phys. Status Solidi C*, 2012, **9**, 957.
30. A. Owens and A. Peacock, Compound semiconductor radiation detectors, *Nucl. Instrum. Methods Phys. Res.*, 2004, **531**, 18.
31. J. Lange, E. Cavallaro, S. Grinstein and I. López Paz, 3D silicon pixel detectors for the ATLAS Forward Physics experiment, *J. Instrum.*, 2015, **10**, C03031.
32. A. LeDonne, S. Binetti, M. Acciarri and S. Pizzini, Electrical characterization of electron irradiated X ray detectors based on 4H-SiC epitaxial layers, *Diamond Relat. Mater.*, 2004, **13**, 414.
33. A. LeDonne, S. Binetti, M. Acciarri, A. Castaldini, F. Nava, A. Cavallini and S. Pizzini, Electrical and optical characterization of electron irradiated X ray detectors based on 4H-SiC epitaxial layers, *Mater. Sci. Forum*, 2004, **457–469**, 1503.
34. G. Bertuccio, S. Binetti, S. Caccia, R. Casiraghi, A. Castaldini, A. Cavallini, C. Lanzieri, A. Le Donne, F. Nava, S. Pizzini, L. Rigutti, G. Verzellesi and E. Vittone, Silicon Carbide for Alpha, Beta, Ion and Soft X-Ray High Performance Detectors, *Mater. Sci. Forum*, 2005, **483–485**, 1015.
35. A. Cavallini, *et al.*, Recent advancements in the development of radiation hard semiconductor detectors for S-LHC, *Nucl. Instrum. Methods Phys. Res., Sect. A*, 2005, **552**, 7.
36. A. LeDonne, S. Binetti and S. Pizzini, Electrical and optical Characterization of electron-irradiated 4H-SiC epitaxial layers annealed at low temperature, *Diamond Relat. Mater.*, 2005, **14**, 1150.
37. E. Fretwurst, *et al.*, Recent advancements in the development of radiation hard semiconductor detectors for S-LHC, *Nucl. Instrum. Methods Phys. Res., Sect. A*, 2005, **552**, 7.
38. N. W. Ashcroft and N. D. Mermin, *Solid state physics*, Saunders College Publishing, New York, 1976.

39. E. A. Guggenheim, *Thermodynamics. An Advanced Treatment for Chemists and Physicists*, North-Holland Physics Publ. Amsterdam-Oxford-New York-Tokyo, 7th edn, 1985.
40. R. C. Newman, Defects in silicon, *Rep. Prog. Phys.*, 1982, **45**, 1163.
41. H. J. Queisser and E. E. Haller, Defects in Semiconductors: Some Fatal, Some Vital, *Science*, 1988, **28114**, 945.
42. S. M. Sze, *Physics of devices*, Wiley & Sons, New York, 1981.
43. A. L. Saraiva, A. Baena, M. J. Calderón and B. Koiller, Theory of one and two donors in silicon, *J. Phys.: Condens. Matter*, 2015, **27**(15), 154208.
44. J. H. Nevin and H. T. Henderson, Thallium-doped silicon ionization and excitation levels by infrared absorption, *J. Appl. Phys.*, 1975, **46**, 2130.
45. W. R. Thurber, R. L. Mattis, Y. M. Liu and J. J. Filliben, Resistivity-dopant density relationship for boron-doped silicon, *J. Electrochem. Soc.*, 1980, **127**, 2291.
46. W. R. Thurber, R. L. Mattis, Y. M. Liu and J. J. Filliben, Resistivity-dopant density relationship for phosphous-doped silicon, *J. Electrochem. Soc.*, 1980, **127**, 1807.
47. A. Mostafa and M. Medrai, Binary phase diagrams and thermodynamic properties of silicon and essential doping elements (Al, As, B, Bi, Ga,In, N, P, Sb and Tl), *Materials*, 2017, **10**, 676.
48. R. W. Olesinski and G. J. Abbaschian, The B-Si (Boron-Silicon) System, *Bull. Alloy Phase Diagrams*, 1984, **5**(5), 478.
49. R. W. Olesinski, K. Kanani and G. J. Abbaschian, The P-Si system, *Bull. Alloy Phase Diagrams*, 1985, **6**(2), 130.
50. R. Rockett, D. D. Johnson, S. V. Kahre and B. R. Tuttle, Prediction of dopant ionization energies in silicon: the importance of strain, *Phys. Rev. B: Condens. Matter Mater. Phys.*, 2003, **68**, 233208.
51. U. V. Desnica, Wide band-gap II–VI compounds—can efficient doping be achieved?, *Vacuum*, 1998, **50**(3–4), 463.
52. I. Teramoto, Calculation of the equilibrium distribution of amphoteric silicon in gallium arsenide, *J. Phys. Chem. Sol.*, 1972, **33**, 2089.
53. I. Teramoto, Solid solubility of amphoteric silicon in gallium arsenide, *Jpn. J. Appl. Phys.*, 1974, **13**, 1817.
54. T. Ahlgren, J. Likonen, J. Slotte, J. Räisänen, M. Rajatora and J. Keinonen, Concentration dependent and in dependent Si diffusion in ion-implanted GaAs, *Phys. Rev. B: Condens. Matter Mater. Phys.*, 1977, **56**, 4597.
55. D. Vazquez-Cortas, S. Shimomura, M. Lopez-Lopez, E. Cruz-Hernandez, J. E. Northrup and S. B. Zhang, Dopant and defect energetics: Si in GaAs, *Phys. Rev. B: Condens. Matter Mater. Phys.*, 1993, **47**(11), 6791.
56. S. Gallardo-Hernandez, Y. Kudriavtsev and V. H. Mendez-Garcia, Electrical and optical properties of Si doped GaAs layers studied as a function of the growth temperature, *J. Cryst. Growth*, 2012, **347**, 77.
57. J. E. Northrup and S. B. Zhang, Dopant and defect energetics: Si in GaAs, *Phys. Rev. B: Condens. Matter Mater. Phys.*, 1993, **47**(11), 6791.
58. J. T. Torvik, Dopants in GaN, in *III-Nitride Semiconductors: Electrical, Structural and Defects Properties*, ed. M. O. Masnasreh, Elsevier, 2000.
59. S. Nakamura, N. Iwasa, M. Senoh and T. Mukai, Hole compensation mechanism of P-type GaN films, *Jpn. J. Appl. Phys.*, 1992, **31**, 1258.
60. C. J. Eiting, P. A. Grudowski and R. D. Dupuis, P- and N-type doping of GaN and AlGaN epitaxial layers grown by metal-organic chemical vapor deposition, *J. Electron. Mater.*, 1998, **27**, 206.
61. J. Neugebauer and C. G. Van de Walle, Role of hydrogen in doping of GaN, *Appl. Phys. Lett.*, 1996, **68**(13), 1829.
62. G. Miceli and A. Pasquarello, Self-compensation due to point defects in Mg-doped GaN, *Phys. Rev. B*, 2016, **93**, 165207.

63. U. Kaufmann, P. Schlotter, H. Obloh, K. Köhler and M. Maier, Hole conductivity and compensation in epitaxial GaN:Mg layers, *Phys. Rev. B: Condens. Matter Mater. Phys.*, 2000, **62**, 10867.
64. P. Kozodoy, H. Xing, S. P. DenBaars, U. K. Mishra, A. Saxler, R. Perrin, S. Elhamri and W. Mitchel, Heavy doping effects in Mg-doped GaN, *J. Appl. Phys.*, 2000, **87**, 1832.
65. K. Ellmer and A. Bikowski, Intrinsic and extrinsic doping of ZnO and ZnO alloys, *J. Phys. D: Appl. Phys.*, 2016, **49**, 413002.
66. R. O. Jones, Density functional Theory; Its origin, rise to prominence, and future, *Rev. Mod. Phys.*, 2015, **87**, 897.
67. A. Boonchun and W. R. L. Lambrecht, Electronic structure of defects and doping in ZnO: Oxygen vacancy and nitrogen doping, *Phys. Status Solidi B*, 2013, **250**(10), 2091.
68. C. A. Janotti and C. G. Van de Walle, Fundamentals of zinc oxide as a semiconductor, *Rep. Prog. Phys.*, 2009, **72**, 126501.
69. C. Kumari, A. Pandey and A. Dixit, Zn interstitial defects and their contribution as efficient light blue emitters in Zn rich ZnO thin films, *J. Alloys Comp.*, 2018, **735**, 231.
70. C. G. Van de Walle, Hydrogen as a Cause of Doping in Zinc Oxide, *Phys. Rev. Lett.*, 2000, **85**(5), 1012.
71. C. G. Van de Walle, Defect analysis and engineering in ZnO, *Phys. B*, 2001, **308–310**, 899.
72. J. Koßmann and C. Hättig, Investigation of interstitial hydrogen and related defects in ZnO, *Phys. Chem. Chem. Phys.*, 2012, **14**, 16392.
73. Y. Yan, M. M. Al-Assim and S.-H. Wei, Doping of ZnO by group -IB elements, *Appl. Phys. Lett.*, 2006, **89**, 181912.
74. Z. Ng, K. Chan and S. Muslimin, P-Type Characteristic of Nitrogen-Doped ZnO Films, *J. Electron. Mater.*, 2018, **47**, 5607.
75. H. T. Chang and G. J. Chen, Influence of nitrogen doping on the properties of ZnO films prepared by radio-frequency magnetron sputtering, *Thin Solid Films*, 2016, **618**, 84.
76. B. Yao, *et al.*, Effects of nitrogen doping and illumination on lattice constants and conductivity behaviour of zinc oxide grown by magnetron sputtering, *J. Appl. Phys.*, 2006, **99**, 123510.
77. J. L. Lyons, A. Janotti and C. G. Van de Walle, Why nitrogen cannot lead to pp-type conductivity in ZnO, *Appl. Phys. Lett.*, 2009, **95**, 252105.
78. M. C. Tarun, M. Z. Iqbal and M. D. McCluskey, Nitrogen is a deep acceptor in ZnO, *AIP Adv.*, 2011, **1**(2), 022105.
79. W. Shockley and W. T. Read, Statistics of the Recombination of Holes and Electrons, *Phys. Rev.*, 1952, **87**, 835.
80. R. N. Hall, Electron-Hole Recombination in Germanium, *Phys. Rev.*, 1952, **87**, 387.
81. T. Goudon, V. Miljianovic and C. Schmeiser, On the Schokley-Read-Hall Model: Generation–recombination in semiconductors, *SIAM J. Math.*, 2007, **67**, 1183.
82. G. L. Miller, D. V. Lang and L. C. Kimerling, Capacitance transient spectroscopy, *Annu. Rev. Mater. Sci.*, 1997, **7**, 377.
83. A. Das, V. A. Singh and D. V. Lang, Deep-level transient spectroscopy (DLTS) analysis of defect levels in semiconductor alloys, *Semicond. Sci. Technol.*, 1988, **3**(12), 1177.
84. J. W. Chen and A. G. Milnes, Energy levels in silicon, *Annu. Rev. Mater. Sci.*, 1980, **10**, 157.
85. A. A. Istratov, H. Hieslmair and E. R. Weber, Iron and its complexes in silicon, *Appl. Phys. A: Mater. Sci. Process*, 1999, **69**, 13.
86. G. W. Ludwig and H. H. Woodbury, Electronic structure of transition metal ions in a tetrahedral lattice, *Phys. Rev. Lett.*, 1960, **5**, 98.

87. S. Zhao, J. F. Justo and L. V. C. Assali, Structure and Bonding of Iron-Acceptor Pairs in Silicon, *Braz. J. Phys.*, 2002, **32**, 418.

88. M. Itsumi, Method of determining metal contamination by combining p-type Si and n-type Si recombination lifetime measurements, *Appl. Phys. Lett.*, 1993, **63**, 1095.

89. G. L. Miller, D. V. Lang and L. C. Kimerling, Capacitance transient spectroscopy, *Annu. Rev. Mater. Sci.*, 1997, **7**, 377.

90. A. Das, V. A. Singh and D. V. Lang, Deep-level transient spectroscopy (DLTS) analysis of defect levels in semiconductor alloys, *Semicond. Sci. Technol.*, 1988, **3**(12), 1177.

91. D. Macdonald and L. J. Geerligs, Recombination activity of interstitial iron and other transition metal point defects in p- and n-type crystalline silicon, *Appl. Phys. Lett.*, 2004, **85**, 4061.

92. S. Pizzini, Chemistry and physics of defect interaction in semiconductors, in *Defect Interaction and clustering in semiconductors*, Trans Tech Publications, Switzerland, 2002.

93. R. van Kemp, M. Sprenger, E. G. Sieverts and C. A. Amerlaan, Oxygen-vacancy complex in silicon I. Si- electron-nuclear double resonance, *Phys. Rev. B: Condens. Matter Mater. Phys.*, 1989, **40**, 4037.

94. R. van Kemp, M. Sprenger, E. G. Sieverts and C. A. Amerlaan, Oxygen-vacancy complex in silicon II. Si-electron-nuclear double resonance, *Phys. Rev. B: Condens. Matter Mater. Phys.*, 1989, **40**, 4054.

95. T. Niewelt, J. Schön, W. Warta, S. W. Glunz and M.-C. Schubert, Degradation of Crystalline Silicon Due to Boron–Oxygen Defects, *J. Photovoltaics*, 2017, **7**(1), 383.

96. M. Vaqueiro-Contreras, V. P. Markevich, J. Coutinho, P. Santos, I. F. Crowe, M. P. Halsall, I. Hawkins, S. B. Lastovskii, L. I. Murin and A. R. Peaker, Identification of the mechanism responsible for the boron oxygen light induced degradation in silicon photovoltaic cells, *J. Appl. Phys.*, 2019, **125**, 185704.

97. C. Park, G. Shim, N. Balaji, J. Park and J. Yi, Correlation between Boron–Silicon Bonding Coordination, Oxygen Complexes and Electrical Properties for n-Type c-Si Solar Cell Applications, *Energies*, 2020, **13**(12), 3057.

98. M. Vaqueiro-Contreras, V. P. Markevich, J. Coutinho, P. Santos, I. F. Crowe, M. P. Halsall, I. Hawkins, S. B. Lastovskii, L. I. Murin and A. R. Peaker, Identification of the mechanism responsible for the boron oxygen light induced degradation in silicon photovoltaic cells, *J. Appl. Phys.*, 2019, **125**, 185704.

99. Bond dissociation energies of simple molecules. National Bureau of standards, NSRDS-NBS 31, 1970.

100. M. Reiche and M. Kittler, Electronic and optical properties of dislocations in Silicon, *Crystals*, 2016, **6**, 74.

101. W. T. Read Jr., Theory of dislocations in germanium, *London, Edinburgh Dublin Philos. Mag. J. Sci.*, 1954, **45**(367), 775.

102. H. Veth and M. Lannoo, The electronic properties of charged dislocations in semiconductors, *Philos. Mag. B*, 1984, **50**, 93.

103. V. V. Kveder, M. Kittler and W. Schröter, Recombination activity of contaminated dislocations in silicon: A model describing electron-beam induced current behaviour, *Phys. Rev. B: Condens. Matter Mater. Phys.*, 2001, **63**, 115208.

104. V. V. Kveder and M. Kittler, Dislocations in Silicon and D-Band Luminescence for Infrared Light Emitters, *Mater. Sci. Forum*, 2008, **590**, 29.

105. D. B. Holt, The role of defects in semiconductor materials ad devices, *Scanning Microsc.*, 1996, **10**, 1047.

106. S. Pizzini, E. Leoni, S. Binetti, M. Acciarri, A. Le Donne and B. Pichaud, Luminescence of dislocations and oxide precipitates in Si, *Solid State Phenom.*, 2004, **95–96**, 273.

107. E. C. Lightowlers and V. Higgs, Luminescence associated with the presence of dislocations in silicon, *Phys. Status Solidi A*, 1993, **138**(2), 665.
108. J. Singh, *Semiconductor Devices, Basic Principles*, John Wiley& Sons, 2001.
109. E. Schibli and A. G. Milnes, Lifetime and capture cross-section studies of deep impurities in silicon, *Mater. Sci. Eng.*, 1968, **2**(5), 229.
110. L. J. Geerligs and D. Macdonald, Base Doping and Recombination Activity of Impurities in Crystalline Silicon Solar Cells, *Prog. Photovoltaics: Res. Appl.*, 2004, **12**, 309.
111. A. Ali, M. Shafi and A. Majid, Electron capture cross-section of Au–Fe complex in silicon, *Phys. Scr.*, 2006, **74**(4), 450.
112. D. Macdonald and L. J. Geerligs, Recombination activity of interstitial iron and other transition metal point defects in pp- and nn-type crystalline silicon, *Appl. Phys. Lett.*, 2004, **85**, 4061.
113. M. Seibt, M. Griess, A. A. Istratov, H. Hedemann, A. Sattler and W. Schröter, Formation and Properties of Copper Silicide Precipitates in Silicon, *Phys. Status Solidi A*, 1988, **166**(1), 171.
114. T. Buonassisi, M. A. Marcus, A. A. Istratov, M. Heuer, T. F. Ciszek, B. Lai, Z. Cai and E. R. Weber, Analysis of copper-rich precipitates in silicon: chemical state, gettering, and impact on multicrystalline silicon solar cell material, *J. Appl. Phys.*, 2005, **97**, 063503.
115. A. A. Istratov, C. Flink, H. Hieslmmair, S. A. McHugo and E. R. Weber, Diffusion, solubility and gettering of copper in silicon, *Mater. Sci. Eng. B*, 2000, **72**(2), 99.
116. R. N. Hall and J. H. Racette, Diffusion and Solubility of Copper in Extrinsic and Intrinsic Germanium, Silicon, and Gallium Arsenide, *J. Appl. Phys.*, 1964, **35**(2), 379.
117. D. Shin, J. E. Saal and Z.-K. Liu, Thermodynamic modeling of the Cu–Si system, *Calphad.*, 2008, **32**, 520.
118. R. W. Olesinski and G. J. Abbashian, The Cu–Si phase diagram, *Bull. Alloy Phase Diagrams*, 1986, **7**(2), 170.
119. C. Mikkelsen Jr, An Overview of oxygen in silicon in Oxygen, Carbon, Hydrogen and Nitrogen in crystalline silicon, *MRS. Symp. Proc.*, 1987, **59**, 3.
120. R. Falster, G. R. Fisher and G. Ferrero, Gettering thresholds for transition metals by oxygen-related defects in silicon, *Appl. Phys. Lett.*, 1991, **59**, 809.
121. F. Shimura, Introduction to oxygen in silicon, *Semicond. Semimetals*, 1994, **42**, 1.
122. A. Borghesi, B. Pivac, A. Sassella and A. Stella, Oxygen precipitation in silicon, *J. Appl. Phys.*, 1995, **77**(9), 4169.
123. R. Falster, D. Gambaro, M. Olmo, M. Cornara and H. Korb, Defect and Impurity Engineered Semiconductors and Devices II, *MRS Proc.*, 1998, **510**, 27.
124. A. Sassella, A. Borghesi, P. Geranzani and G. Borionetti, Infrared response of oxygen precipitates in silicon: experimental and simulated spectra, *Appl. Phys. Lett.*, 1999, **75**, 1131.
125. R. Falster and V. V. Voronkov, Intrinsic Point Defects and Their Control in Silicon Crystal Growth and Wafer Processing, *MRS Bull.*, 2000, **25**, 28.
126. R. Falster and V. V. Voronkov, The Engineering of Intrinsic Point Defects in Silicon Wafers and Crystals, *Mater. Sci. Eng. B*, 2000, **73**, 69.
127. A. Sassella, Defects involving oxygen in crystalline silicon, *Solid State Phenom.*, 2002, **85–86**, 285.
128. S. Binetti, S. Pizzini, E. Leoni, R. Somaschini, A. Castaldini and A. Cavallini, Optical properties of oxygen precipitates and dislocations in silicon, *J. Appl. Phys.*, 2002, **92**(5), 2437.
129. S. Pizzini, Chemistry and physics of defect interaction in semiconductors, *Solid State Phenom.*, 2002, **85–86**, 1.
130. S. Pizzini, E. Leoni, S. Binetti, M. Acciarri, A. Le Donne and B. Pichaud, Luminescence of dislocations and oxide precipitates in Si, *Solid State Phenom.*, 2004, **95–96**, 273.

131. R. Falster and V. V. Voronkov, Rapid Thermal Processing and the Control of Oxygen Precipitation Behaviour in Silicon Wafers, *Mater. Sci. Forum*, 2008, **573–574**, 45.

132. M. Di Sabatino, S. Binetti, J. Libal, M. Acciarri, H. Nordmark and E. J. Øvrelid, Oxygen distribution on a multicrystalline silicon ingot grown from upgraded metallurgical silicon, *Sol. Energy Mater. Sol. Cells*, 2011, **95**, 529.

133. R. C. Newman, *Infrared studies of crystal defects*, Tayor & Francis, 1973.

134. J. Coutiño, R. Jones, P. R. Briddon and S. Öberg, Oxygen and dioxygen centers in Si and Ge: Density-functional calculations, *Phys. Rev. B: Condens. Matter Mater. Phys.*, 2000, **62**, 10824.

135. Y. Itoh and T. Nozaki, Solubility, and diffusion coefficient of oxygen in silicon, *Jpn. J. Appl. Phys.*, 1985, **24**, 279.

136. H. Okamoto, O-Si (Oxygen-Silicon), *J. Phase Equilib. Diffus.*, 2007, **28**, 309.

137. B. Hallstedt, Thermodynamic assessment of the Silicon–Oxygen system, *Calphad*, 1992, **16**(1), 53.

138. W. Götz, Thermal donor formation and annihilation at temperatures above 500°C in CZ silicon, *J. Appl. Phys.*, 1961, **84**, 3561.

139. K. F. Kelton, Diffusion-influenced nucleation: a case study of oxygen precipitation in silicon, *Philos. Trans. R. Soc., A*, 2003, **361**, 429.

140. Y. Yang, A. Sattler and T. Sinno, Data-assisted physical modeling of oxygen precipitation in silicon wafers, *J. Appl. Phys.*, 2019, **125**, 165705.

141. Y. Yang, Quantitative Modelling of Oxygen Precipitation in Silicon, *Publicly Accessible Penn Dissertations*, 2017, p. 2816.

142. R. Falster, V. V. Voronkov, V. Y. Resnik and M. G. Milvidskii, Thresholds for Effective Internal Gettering in Silicon Wafers, in *High Purity Silicon VIII: Proceedings of the International Symposium*, 2004, vol. 05, p. 188.

143. A. Ourmazd and W. Schröter, Phosphorus gettering and intrinsic gettering of nickel in silicon, *Appl. Phys. Lett.*, 1984, **45**, 78.

144. J. S. Kang and D. K. Schroder, Gettering in silicon, *J. Appl. Phys.*, 1989, **65**, 2974.

145. D. Gilles, E. R. Weber and S. Hahn, Mechanism of internal gettering of interstitial impurities in Czochralski-grown silicon, *Phys. Rev. Lett.*, 1990, **64**, 196.

146. I. F. Shimura, Intrinsic/Internal Gettering, *Semicond. Semimetals*, 1994, **42**, 577.

147. W. Wijaranakula, P. M. Burke and L. Forbes, Internal gettering heat treatments and oxygen precipitation in epitaxial silicon wafers, *J. Mater. Res.*, 1986, **1**(5), 693.

148. V. Raineri, F. Priolo, M. Kittler and H. Richter, Gettering and Defect Engineering in Semiconductor Technology, *Solid State Phenom.*, 2001, **82–84**, 1.

149. K. Sumino and I. Yonenaga, Interaction of impurities with dislocations: Mechanical effects, *Solid State Phenom.*, 2002, **85–86**, 145.

150. E. R. Weber, Impurity Precipitation, Dissolution, Gettering, and Passivation in PV Silicon. Final Technical Report 30 January 1998–29 August 2001, NREL/SR-520-31528, February 2002.

151. Y. H. Kim, K. S. Lee, H. Y. Chung, D. H. Hwang, H. S. Kim, H. Y. Cho and B.-Y. Lee, Internal Gettering of Fe, Ni and Cu in Silicon Wafers, *J. Korean Phys. Soc.*, 2001, **39**, S348.

152. P. Geranzani, M. Pagani, C. Pello and G. Borionetti, Internal Gettering in Silicon: Experimental and Theoretical Studies Based On Fast And Slow Diffusing Metals, *Solid State Phenom.*, 2002, **82–84**, 381.

153. P. Zhang, H. Väinölä, A. A. Istratov and E. R. Weber, The thermal stability of iron precipitates in silicon after internal gettering, *Phys. B*, 2003, **340–342**, 1051.

154. K. Sumino, Basic aspects of impurity gettering, *Microelectron. Eng.*, 2003, **66**(1–4), 268.

155. M. Seibt, A. Sattler, C. Rudolf, O. Voß, V. Kveder and W. Schröter, Gettering in silicon photovoltaics: Current state and future perspectives, *Phys. Status Solidi A*, 2006, **203**, 696.

156. R. Krain, S. Herlufsen and J. Schmidt, Internal gettering of iron in multi-crystalline silicon at low temperature, *Appl. Phys. Lett.*, 2008, **93**, 152108.

157. M. Seibt and V. Kveder, Gettering Processes, and the role of extended defects, in *Advanced silicon materials for photovoltaic applications*, ed. S. Pizzini, J. Wiley & Sons, 2012.

158. M. Seibt and K. Graff, Characterization of haze-forming precipitates in silicon, *J. Appl. Phys.*, 1988, **63**, 4444.

159. K. Sueoka, S. Sadamitsu, Y. Yasuo Koike and T. Kihara, Internal Gettering for Ni Contamination in Czochralski Silicon Wafers, *J. Electrochem. Soc.*, 2000, **147**(8), 3074.

160. K. Nakamura and J. Tomioka, Effect of oxygen precipitates on the surface precipitation of nickel on CZ Silicon wafers, *Solid State Phenom.*, 2005, **108–109**, 103.

161. M. S. Chandrasekharaiah and J. L. Margrave, The Disilicides of Tungsten, Molybdenum, Tantalum, Titanium, Cobalt, and Nickel, and Platinum Monosilicide: A Survey of Their Thermodynamic Properties, *J. Phys. Chem. Ref. Data*, 1993, **22**, 1459.

162. J. S. Won, K. Sato and Y. Hirotsu, Synthesis of Iron Silicides by Electron-Beam Evaporation: Effects of Substrate Prebaking Temperature and Fe Deposition Thickness, *Jpn. J. Appl. Phys.*, 2007, **46**, 732.

163. P. V. Sushko, S. Mukhopadhyay, A. S. Mysovsky, V. B. Sulimov, A. Taga and A. L. Shluger, Structure and properties of defects in amorphous silica: new insights from embedded cluster calculations, *J. Phys.: Condens. Matter.*, 2005, **17**, S2115.

164. P. Li, Y. Song and X. Zuo, Computational Study on Interfaces and Interface Defects of Amorphous Silica and Silicon, *Phys. Status Solidi RRL*, 2019, **13**, 1800547.

165. K. Kajhara, T. Miura, H. Kamioka, A. Aiba, M. Urimoto, Y. Morimoto, M. Hirano, L. Skuja and H. Hosono, Diffusion and reactions of interstitial oxygen species in amorphous $SiO_2$: A review, *J. Non-Cryst. Solids*, 2008, **354**, 224.

166. M. V. Gomoyunova, D. E. Malygin, I. I. Pronin, A. S. Voronchikhin, D. V. Vyalikh and S. L. Molodtsov, Initial stages of iron silicide formation on the Si(1 0 0)2×1 surface, *Surf. Sci.*, 2007, **601**, 5069.

167. O. Sjödén, S. Seetharaman and L.-I. Staffansson, On the Gibbs energy of formation of wustite, *Metall. Mater. Trans. B*, 1986, **17**, 179.

168. G. Kissinger, D. Kot, M. Klingsporn, M. A. Schubert, A. Sattler and T. Müller, Investigation of the Copper Gettering Mechanism of Oxide Precipitates in Silicon, *J. Solid State Sci. Technol.*, 2015, **4**, N124.

169. J. Vanhellemont, Diffusion limited oxygen precipitation in silicon: Precipitate growth kinetics and phase formation, *J. Appl. Phys.*, 1995, **78**, 4297.

170. O. De Gryse, P. Clauws, J. Van Landuyt, O. Lebedev, C. Claeys, E. Simoen and J. Vanhellemont, Oxide phase determination in silicon using infrared spectroscopy and transmission electron microscopy techniques, *J. Appl. Phys.*, 2002, **91**, 2493.

171. J. Nicolai, J. Burle and B. Pichaud, Determination of silicon oxide precipitate stoichiometry using global and local techniques, *J. Cryst. Growth*, 2013, **363**, 93.

172. A. Konczakowska and B. M. Willamowski, Noise in semiconductor devices, in *Fundamentals of Industrial Electronics*, ed. B.-M. Wilamowski and J. D. Irwin, CRC Press, 2018.

173. J. L. Pankove, *Optical processes in semiconductors*, Dover Publications, New York, 1971.

# 3 Physico-chemical Aspects of Growth Processes of Elemental and Compound Semiconductors

## 3.1 Introduction

Different from metallic and ceramic materials for industrial or ordinary uses that are manufactured with polycrystalline or amorphous morphologies, semiconductors are mostly grown and used as single crystal ingots, apart from multi-crystalline silicon for solar applications and nanocrystalline semiconductors for advanced photonic applications. Only in single crystal substrates, in fact, is the number of impurities and of crystallographic defects brought down to sufficiently low levels to make negligible their role in the electronic properties of the material.

Therefore, the development of microelectronic and photonic technologies has been associated, from the very beginning of the semiconductor era, with the development of dedicated crystal growth processes, which started with germanium and silicon, and continued with compound semiconductors (see Table 3.1).

How chemical and thermodynamic constraints, together with defects and impurities, limit, or even rule, the performance of semiconductor devices has been already discussed in the previous chapters and is amply discussed in the literature.[1–7]

Chemistry of Semiconductors
By Sergio Pizzini
© Sergio Pizzini 2024
Published by the Royal Society of Chemistry, www.rsc.org

**Table 3.1** Thermodynamic properties of selected semiconductors of technological interest.

| Material | Melting point (K) | Heat of fusion (kJ mol$^{-1}$) | Vapour pressure (atm) | Density at 298 K (g cm$^{-3}$) | Ref. |
|---|---|---|---|---|---|
| Ge | 1210 | 34.7 | $10^{-6}$ at 760 °C, 1 at 1330 °C | 5.3267 | 12 |
| Si | 1690 | 50.21 | $10^{-6}$ at 900 °C, 1 at 1650 °C | 2.328 | 12 |
| 3C-SiC | 3100 | n.a. | $\cong 10^{-5}$ | 3.166 | 12 |
| InSb | 800 | 25.5 | $4 \times 10^{-8}$ | 5.775 | 12 |
| GaSb | 985 | 25.1 | $10^{-6}$ | 5.61 | 12 |
| GaAs | 1510 | 87 | 1 | 5.32 | 12 |
| InP | 1333 | 62.7 | 27.5 | 4.81 | 12 |
| GaP | 1730 | 117.6 | 32 | 4.14 | 12 |
| GaN | 2790 | n.a. | $10^4$ at 2000 K | 6.15 | 12 |
| CdSe | 1623 | 44.8 | 0.3 | 5.81 | 10 and 11 |
| CdTe | 1314 | 43.5, 64.852 | 0.65 | 5.86 | 9–11 |
| ZnSe | 1373 | 56 | 0.5 | 5.26 | 10 and 11 |
| ZnTe | 1513 | 51.00 | 0.6 | 5.65 | 10 and 11 |

The aim of this chapter is to discuss the problems of physico-chemical nature involved in the growth of elemental and compound semiconductor single crystal ingots to be utilized, after wafering, as the substrates of microelectronic or photonic devices. This process could be carried-out by the crystallization of a liquid phase on a crystalline seed, for semiconductors presenting thermodynamically stable liquid phases at and above the melting temperature. For semiconductors presenting problems of thermodynamic stability of their solid phases at the melting temperature, vapour phase or thermochemical processes might be best suited as crystal growth processes.

Since sapphire ($\alpha$-Al$_2$O$_3$) is, with silicon and silicon carbide, the substrate used for the heteroepitaxial fabrication of microelectronic and optical devices of semiconductors for which single-crystal wafers are not available, the growth of single crystal ingots of sapphire will also be considered and discussed.[8]

The single crystal ingot growth process starts with the identification and production of the precursors, which consist, in the case of melt growth, of polycrystalline or granular materials[†] whose purity should be compatible with the requested purity of the final product, or with the identification of solid, liquid, or gaseous precursors, whose

---

[†] The number of melt components depends on the semiconductor composition.

thermal sublimation or decomposition allows the seeded growth of a semiconductor ingot with a vapour phase process. Thermochemical processes, as is the case in GaN crystals, are, instead, based on molecular species and activators as precursors.

Ingot growth by liquid crystallization processes is, in principle, the most favourable route, in terms of cost and technological ease, for the semiconductor industry, and also a practical route for the growth of large single crystal ingots of silicon, germanium, gallium arsenide, gallium phosphide and indium phosphide, which represent key segments of the semiconductor market.

Single crystal ingot growth by liquid crystallization processes represents the best, when technically feasible, solution for the preparation of a material with a negligible amount of impurities and structural defects (dislocations, point defect clusters and precipitates).

It can be operated in all circumstances for which the liquid phase is thermodynamically stable at (and slightly above) the melting temperature, the vapour pressure of the melt component(s) is sufficiently low, and the physicochemical properties of the melt are compatible with the mechanical, thermal and corrosion properties of materials of possible use as crucible materials. In fact, the interaction of the melt with the crucible, if the melt wets the crucible walls, almost inevitably causes at least its partial dissolution, with unwanted contamination of the melt and thus with the subsequent contamination of the ingot by impurity segregation phenomena, which will be discussed in Section 3.2.

It is therefore a fortunate circumstance that several refractory materials are available for the fabrication of crucibles, and that the vapour pressure of elemental and of several compound semiconductors of industrial interest is sufficiently low at their melting temperatures (see Table 3.1)[9-12] to allow the growth directly from a stable melt under normal pressure conditions or under technologically acceptable pressurization. In the case of melts spontaneously decomposing at or below the melting temperature, the applied hydrostatic pressure should be sufficiently high to stabilize the melt.

The Czochralski process (CZ),[‡] together with the vertical Bridgman (VB)[‡] method, are typical examples of crystal growth from a melt, both using a refractory crucible as the melt container, while the float zone (FZ) process is a typical example of a crucible-less process.

---

[‡] Also called the directional solidification process.

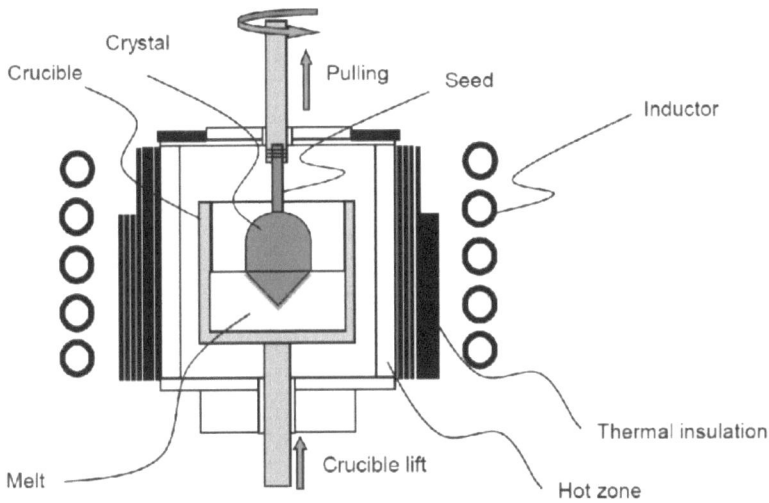

**Figure 3.1** Schematic view of a Czochralski furnace. Reproduced from ref. 8 with permission from Elsevier, Copyright 2012.

With the CZ process,[13] a cylindrical single crystal is grown by liquid crystallization from a melt, using a dislocation-free single crystal seed, which enables the crystallization to start epitaxially on a single crystal substrate. The growth is carried out in a dedicated furnace, see Figure 3.1, electrically heated and maintained under a protective atmosphere to avoid the environmental degradation or contamination of the melt and of the ingot during the growth. Full details about this process will be given in Section 3.4.3.

The vertical Bridgman (VB) process, see Figure 3.2,[14,15] is also a seeded process, but the growth occurs directly into a crucible, which also defines the final ingot diameter. The growth is carried-out under a vertical temperature gradient, which allows the control of the upward shift of the solid/liquid interface, starting from the bottom, in correspondence with the position of the seed.[16] Also this process is carried out under a protective atmosphere.

The float zone (FZ) process, see Figure 3.3,[17–19] is a crucible-less, seeded process carried-out by scanning a molten zone along a cylindrical polycrystalline ingot, starting from a single crystal seed on the bottom. Like the CZ one, the FZ process is carried out under a protective atmosphere to avoid the oxidation of the material. FZ heating is conventionally operated with an RF source, but today optical heating can also be used for special purposes, as is the case in oxide growth.[20]

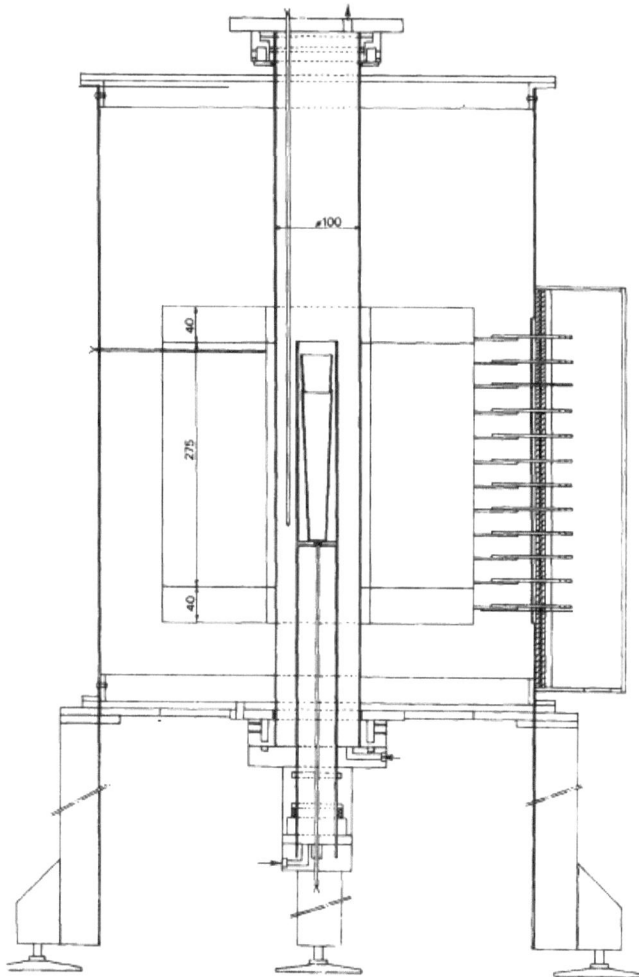

**Figure 3.2**   Schematic view of a vertical Bridgman furnace. Reproduced from
ref. 14 with permission from Elsevier, Copyright 1989.

Both processes will be discussed in great detail in the following sections dedicated to the growth of specific semiconductors.

Crystal growth from the vapour phase should be carried out when the material is unstable at the melting temperature, or the melting temperature is too high, and crucible materials of tolerable properties are not available or do not exist at all, as preliminarily discussed in Section 1.1, but the vapour pressure of the precursor material is sufficiently high to allow its sublimation and its deposition as a single crystal on a seed. We will see that this is the case in the industrial process of the growth of silicon carbide ingots.

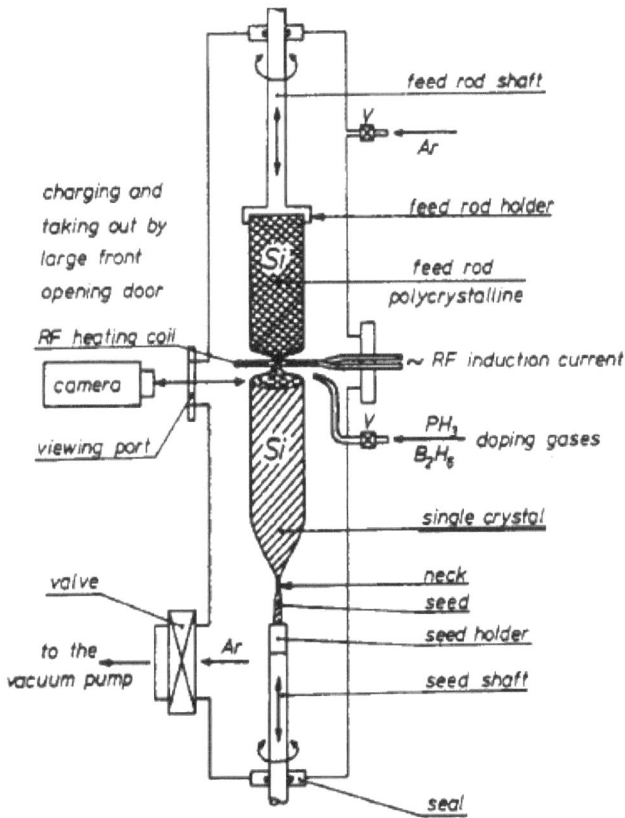

**Figure 3.3** Schematic drawing of an FZ growth process, with the melt zone in the centre of the ingot, in correspondence with the RF heating coil, and input point for p-type and n-type doping. Reproduced from ref. 19 with permission from IOP Publishing, Copyright 1994.

## 3.2 Impurity Segregation Phenomena in a Crystallization Process

We already know that impurities in semiconductors may work as dopants and as minority carrier recombination centres,[21] but could also be used to compensate growth-related ionized defects to bring a semiconductor to semi-insulating properties,[22,23] or to improve the mechanical properties of the material,[24] as is the case of nitrogen doped silicon.

Therefore, the quantitative knowledge of the impurity distribution process among the molten phase and the ingot during a crystal growth process is a prerequisite both for the design of an optimized growth of a doped semiconductor crystal or for the set-up of a purification process of impure feedstocks in view of obtaining crystalline ingots of desired properties.

As we expect from thermodynamic arguments, the equilibrium distribution of an impurity i between a liquid and a solid phase, at the melting temperature of the solid, is ruled by the condition of equal values of the chemical potential $\mu_i$ of the impurity i in the two phases

$$\mu_i^s = \mu_i^l \tag{3.1}$$

where

$$\mu_i = \mu_i^\circ + RT \ln a_i = \mu_i^\circ + RT \ln \gamma(x)x_i \tag{3.2}$$

$\mu_i^\circ$ is the chemical potential of the impurity i in the condensed phase (liquid or solid) containing the impurity i at its solubility concentration, $a_i$ is the activity of the impurity, $\gamma(x)$ is the activity coefficient and $x_i$ is the atomic fraction of i.

On that basis, the ratio of the equilibrium concentration of the impurity i in the two phases $\dfrac{x_i^s}{x_i^l}$, and, therefore, the equilibrium segregation coefficient $k_i^{eq}$, is given by the following equation

$$k_i^{eq}(x, T) = \frac{x_i^s}{x_i^l} = \frac{\gamma_i^l}{\gamma_i^s} \exp - \frac{\mu_i^{\circ,s} - \mu_i^{\circ,l}}{RT} \tag{3.3}$$

where $x_i^s$ is the atomic fraction of the impurity i in the solid phase, $x_i^l$ is the atomic fraction of the impurity i in the liquid, $\gamma_i^s$ and $\gamma_i^l$ are the activity coefficients of the impurity i in the solid and liquid phase, and $\mu_i^{\circ,s} - \mu_i^{\circ,l}$ is a measure of the difference of the enthalpies of solution of the impurity i in the liquid and solid phases, which drives the equilibrium repartition.

If the solubility of the impurities in both the solid and the liquid phase is very small, which is often the case, one could also assume the activities to be equal to the concentrations and ignore in eqn (3.3) the term $\dfrac{\gamma_i^l}{\gamma_i^s}$, which is close to unity.

We expect, therefore, that during a crystal growth process from an impurity contaminated melt, the equilibrium distribution of impurities among the liquid and solid phase would result in a solid phase containing a minor content of impurities than the liquid phase if $k_i^{eq} \ll 1$.

This is the case of the segregation coefficient of TM impurities in silicon, at the melting temperature of silicon (Ti $2 \times 10^{-6}$, Cr $1.1 \times 10^{-5}$, Mn $1.3 \times 10^{-5}$, Fe $5.4 \times 10^{-6}$, Co $2.2 \times 10^{-6}$, Ni $1.3 \times 10^{-4}$, Zn $1 \times 10^{-5}$, Au $2.5 \times 10^{-5}$),[25] but common for all semiconductors, as we will see later in this section.

Therefore, a crystal growth process provides intrinsic conditions of preferential impurity segregation in the liquid phase, working as a purification process of impurity contaminated feedstocks. As an example, the crystallization processes employed at the beginning of this century as the final technological step of the production of solar grade silicon[26,27] from metallurgical silicon feedstocks heavily contaminated by TM impurities are still of interest[§] thanks to the excellent segregation coefficients of TM impurities, despite the poor purification yield of dopant impurities and the chemical interactions occurring among metallic and non-metallic impurities, which influence the absolute values of the segregation coefficients and their segregation patterns.

In the practical circumstances of a crystal growth process, the repartition of impurities might be, however, far from true equilibrium conditions, given the experimental evidence that the absolute values of the segregation coefficients strongly depend on the growth rate,[28,29] as is shown in Figure 3.4, for the case of selected impurities in silicon.[29]

True equilibrium conditions might be, in fact, effectively achieved at growth rates approaching zero and, therefore, the impurity segregation yield in the case of industrial growth processes is ruled by effective segregation coefficients $k_{eff}$, whose actual values would significantly depend on the type of growth system used.

A relationship between the effective segregation coefficients $k_{eff}$, which are experimentally available by simple chemical measurements of the impurity concentrations in the solid and liquid phase, and the equilibrium coefficients $k_{eq}$ was proposed by Burton[28,29]

$$k_{eff} = \frac{k_{eq}}{k^{eq} + (1 - k^{eq})\exp\left(-\dfrac{v\delta}{D_i}\right)} \qquad (3.4)$$

where $v$ is the growth rate, $D_i$ is diffusion coefficient of the impurity in the melt and $\delta$ is the thickness of the diffusion layer. The assumption

---

[§] Ferrosolar communication at the 2022 WCPEC Conference in Milano (Italy).

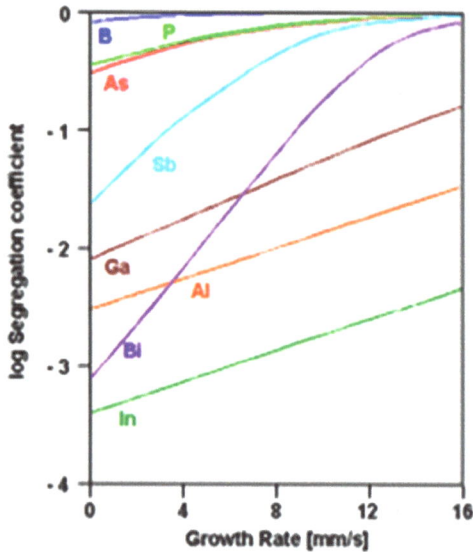

**Figure 3.4**  Dependence on the growth rate of the absolute values of the segregation coefficients of selected impurities in silicon.

made here is that the segregation rate depends on the diffusion of the impurities in a static (diffusion) layer of thickness $\delta$ at the liquid/solid interface, where the fluid motion is laminar and non-mixing, and that impurity diffusivity is negligible in the solid.

Eqn (3.4) shows that the effective segregation coefficients should tend to a value corresponding to $k_{eq}$, at very low growth rates and under good stirring conditions, for which $\delta \to 0$. Conversely, $k_{eff}$ should tend to 1 for large values of the growth rate and bad stirring conditions. The problem with the practical use of this equation is, however, that the diffusion coefficients of impurities in the liquid and solid phases are known, or could be measured with relative ease, while the evaluation of $\delta$ is complicated because it depends on the convection fluxes in the liquid charge.

The literature values of the effective segregation coefficients of impurities of technological importance in Si, Ge, GaAs and InP are reported in Table 3.2.[21,30–48]

Limiting our interest to the segregation coefficients of impurities in silicon, it is apparent from this table that the FZ growth is more efficient than the CZ growth as a purification process, given that the segregation coefficients are systematically lower than those for a CZ process, displayed in the first two columns of Table 3.2. These results show that the FZ process has the best potential to grow extremely pure

**Table 3.2**  Segregation coefficients of donor, acceptor and metallic impurities in Si, Ge, GaAs and InP. The first two columns display the literature values of the segregation coefficients of impurities in Czochralski grown silicon. (The numbers in parentheses are the references.)

| Impurity | CZ-Si (30) | CZ-Si | FZ-Si | Ge | GaAs | InP |
|---|---|---|---|---|---|---|
| C | $5.8 \times 10^{-2}$ | $5 \times 10^{-2}$ (21) | | $\geq 1.85$ (31) | $1.44 \pm 0.08$ (30) | |
| O | $0.5$–$1.3$ | $0.25$ (21) | | $0.11$ (33) | $0.3$ (34); $0.1$ (35) | |
| N | $8 \times 10^{-4}$ | $7 \times 10^{-4}$ (43) | | | | |
| B | $0.8$ | $0.73$ (36) | | $\approx 6$ (32) | | |
| Al | $2 \times 10^{-3}$ | $3 \times 10^{-2}$ (21) | | $7.3 \times 10^{-2}$ (38) | | |
| Ga | $8 \times 10^{-3}$ | $0.008$ (39) | | $8.7 \times 10^{-2}$ (38) | $\approx 10^{-6}$ (40) | $4$ (41) |
| In | $4 \times 10^{-4}$ | | | $10^{-3}$ (38) | $0.1$–$0.07$ (34) | $\approx 10^{-6}$ (40) |
| As | $0.30$ | | | $2 \times 10^{-2}$ (38) | $\approx 10^{-6}$ (40) | $1$ (41) |
| Sb | $0.0020$ | | | $3 \times 10^{-3}$ (38) | | $1$ (41) |
| Ge | | $0.6$ (21) | | | | |
| P | $0.35$ | $0.35$ (36) | | $0.08$ (38) | $2$–$3$ (34) | $\approx 10^{-6}$ (40) |
| Cu | $4 \times 10^{-4}$ | $8 \times 10^{-4}$ (21) | | $1.3 \times 10^{-6}$ (45) | | |
| Ni | $1.5 \times 10^{-5}$ | $3.2 \times 10^{-5}$ (21) | $1.5 \times 10^{-7}$ (21) | $3 \times 10^{-6}$ (45) | $6 \times 10^{-4}$ (5) | |
| Au | $2.5 \times 10^{-5}$ | | $5 \times 10^{-5}$ (21) | $3 \times 10^{-5}$ (45) | | |
| Fe | $8 \times 10^{-6}$ | $6.4 \times 10^{-6}$ (21) | $5 \times 10^{-6}$ (21) | $3 \times 10^{-5}$ (45) | $\geq 10^{-3}$ (34) | |
| Co | $8 \times 10^{-6}$ | $10^{-5}$ (21) | $10^{-7}$ (30) | $10^{-6}$ (45) | | |
| Ti | | $2 \times 10^{-6}$ (21) | $1.8 \times 10^{-6}$ (21) | | | |
| Cr | | $1.1 \times 10^{-5}$ (21) | $2.5 \times 10^{-6}$ (21) | | $10^{-3}$–$5.7 \times 10^{-4}$ (34) | $>4$ (46) |
| Zn | | | | | $\approx 1$ (47) | |
| Si | | | | | | $30$ (48) |

materials in comparison with CZ, not only because it is a crucible-less technique, but because it can work close to true equilibrium conditions, and the melt zone might be multiply scanned along the ingot.

Eventually, it is also apparent that the segregation coefficients of dopant impurities in silicon are close to one, with a predicted modest influence on their solute profile along a grown ingot.[¶]

Concerning the basic physical and chemical factors that determine the absolute values of the equilibrium segregation coefficients, a relationship between solid solubilities and segregation coefficients is expected to hold on the basis of the thermodynamic arguments discussed at the beginning of this section. In fact, an empirical relationship between the segregation coefficients and the solubility of a number of impurities in silicon and germanium was found by Fischler,[49] and is graphically displayed in Figure 3.5, which also allows the estimation of the segregation coefficients once the solubility of a specific impurity is known.

This correlation finds its qualitative rationale in the dependence of the solubility and of the segregation coefficient of impurities on their covalent radii in the solid at the melting point of silicon or germanium,[39] as is shown in Figure 3.6.

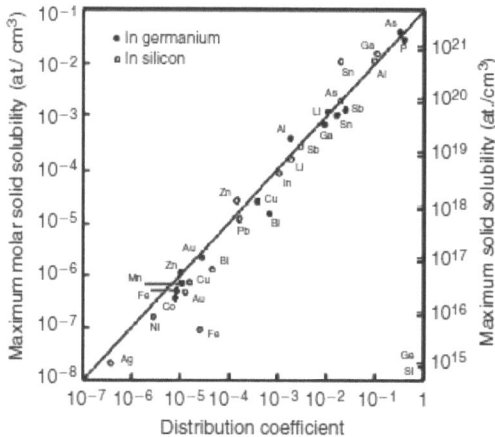

**Figure 3.5** Correlation between solid state solubility and segregation coefficient of impurities at the melting point of Ge and Si. Reproduced from ref. 49 with permission from AIP Publishing, Copyright 1962.

---

[¶] This item will be fully discussed in the next section.

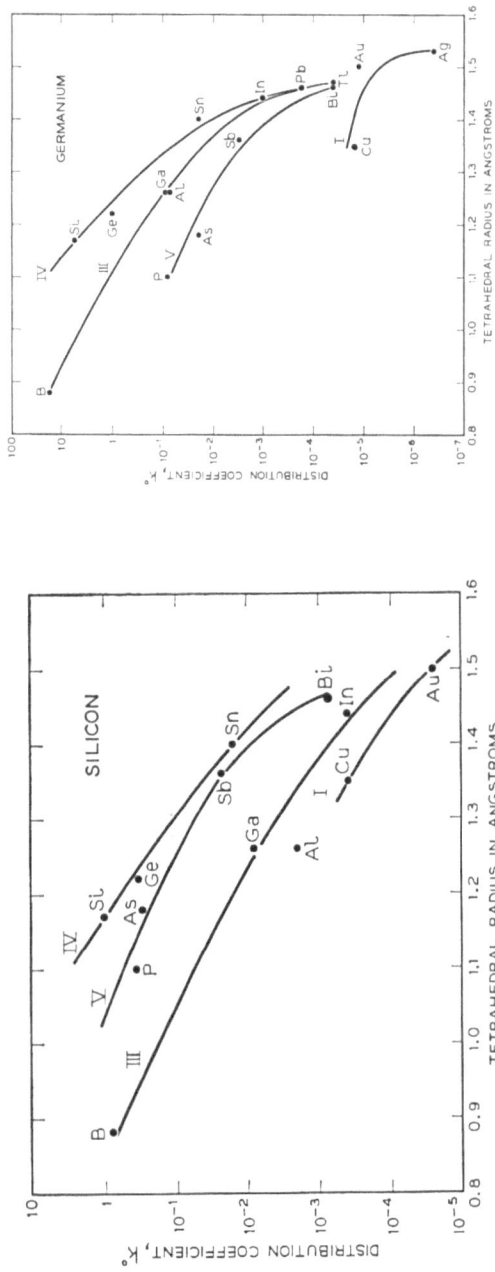

**Figure 3.6** Dependence of the segregation coefficient of selected impurities at the melting point of germanium and silicon on their covalent radii in the solid. Reproduced from ref. 39 with permission from Nokia Bell Labs, Copyright 1960.

## 3.3   Growth Dependence of the Impurity Pick-up and Concentration Profiling

Once the values of the effective segregation coefficients for a specific solute (dopant or impurity) in a specific solvent are known, it is important to foresee and optimize the solute profile along the ingot, which should depend on the solute concentration depletion or increase in the molten charge during the growth. This issue is of tremendous commercial value for silicon producers because it makes it possible for them, as an example, to know the sections of a B-doped or a P-doped ingot presenting specific resistivity values without the need of a manual electrical inspection.

The same approach might be used in a purification process to know the length of the ingot depleted of impurities.

To do this, we shall at first consider that during the crystallization the total amount of impurities in the crystal and in the melt should remain constant, and this condition can be written as

$$x_i^\circ = \int_0^{f} x_i^s dx + (1-f)x_i^l \tag{3.5}$$

where $x_i^\circ$ is the initial concentration (in molar fraction) of the impurity i in the molten charge, $x_i^s$ is the atomic fraction of the impurity i in the solid phase, $x_i^l$ is the atomic fraction of the impurity i in the liquid, and $f$ is the solidified fraction. After differentiation of eqn (3.5) with respect to $f$ one obtains

$$x_i^s + (1-f)\frac{dx_i^l}{df} - x_i^l = 0. \tag{3.6}$$

Since $x_i^s = k_{eff}x_i^l$ we have also

$$\frac{dx_i^l}{x_i^l} = \frac{(1-k_{eff})}{(1-f)}df. \tag{3.7}$$

By applying the boundary condition $x_i^l = x_i^\circ$ for $f = 0$, the solution of eqn (3.7) turns out to be the normal freezing equation[17]

$$x_i^s = k_{eff}x_i^\circ(1-f)^{k_{eff}-1} \tag{3.8}$$

which gives the distribution of impurity concentration in the solid along the growth direction[19] and compares it to the classical Scheil equation,[50] valid for complete mixing processes

$$x_i^s = k° x_i (1-f)^{k°-1} \tag{3.9}$$

where $k°$ is the equilibrium segregation coefficient.

Figure 3.7 shows the distribution of impurities in the solid as a function of the crystallized fraction $f$, calculated for different values of the segregation coefficients.[51]

For $k < 1$, the solid phase contains fewer impurities than the liquid, so that the impurity concentration in the liquid increases with the increase in $f$, as does the concentration of impurities in the solid. For $k > 1$, the solid phase contains more impurities than the liquid, and the impurity concentration in the solid reduces with the increase in $f$ and goes to zero when $f$ approaches 1.

It is immediately apparent that for values of the distribution co-efficient $k_{eff}$ close to one, as is the case of dopants in silicon, an almost flat impurity profile is obtained, which implies a constant value of the resistivity along the ingot.

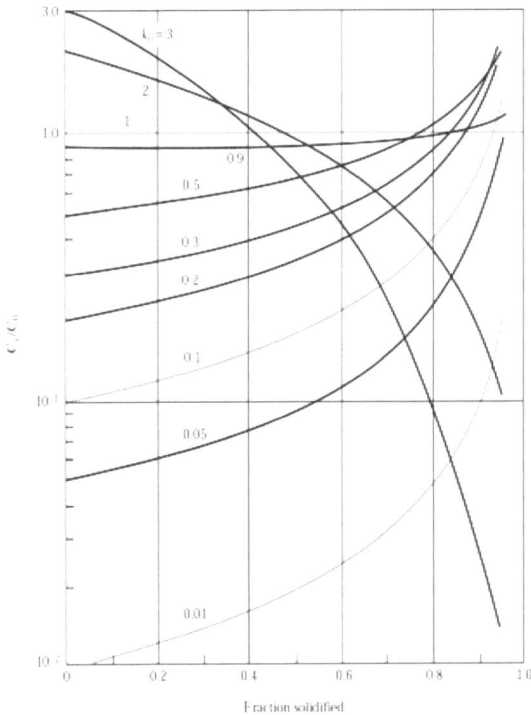

**Figure 3.7** Distribution of impurities as a function of the solidified fraction for different values of the segregation coefficients. Reproduced from ref. 51 with permission from Springer Nature, Copyright 1952.

This behaviour is common to all crystals grown from the liquid phase, quantitative differences depending only on the values of the segregation coefficients and on the growth rate.

The growth of compound semiconductors of defined composition is influenced by additional constraints, such as the reciprocal solubility of the elemental precursors in the liquid phase, the possible non-ideality of both the liquid and solid solution, the absence of stable stoichiometric or quasi-stoichiometric solid phases and the massive interplay of extended defects in the case of heteroepitaxial growth when structural compatibility between the two phases is lacking.

The discussion of these topics, with relevant examples concerning the segregation of impurities in elemental or compound semiconductors and the control of the defectivity of the grown solid and of lattice strain in heteroepitaxial layers will be carried out in the next sections.

## 3.4 Growth of Elemental Semiconductor Crystals from the Melt

This section deals with the growth of silicon as the prototype of elemental semiconductors using for its single crystal preparation the growth from a melt, or a liquid to solid (LS) process.

Germanium synthesis and growth look very similar to that of silicon in terms of chemical and thermal constraints, while diamond presents complex relationships with its parent phases (graphite in this case), see Figure 1.3, and extremely critical temperature/pressure conditions that make its growth with a LS process technically impossible.

In fact, the (high pressure) synthesis of small diamond crystals is industrially carried out using a liquid Fe–Co alloy as the solvent of high purity graphite and synthetic diamond crystals as seeds, on which small diamond crystals grow by applying a pressure of 5.5 GPa at a temperature of 1350 °C.[52]

The production process of silicon ingots can be partialized in four main steps:

a. The production of metallurgical silicon (MG-Si).
b. The synthesis of ultra-pure gaseous precursors starting from MG-Si.
c. The production of polysilicon rods from the gaseous precursors, to be used as high purity feedstock for the final growth process.
d. The growth of silicon ingots.

## 3.4.1 The Production of Metallurgical Si (MG-Si)

All the industrial silicon processes have as a common initial feed-stock metallurgical silicon, which is produced in millions of tonnes annually using the carbothermic reduction of quartz ores or quartzites, available in large amounts on the earth, in a submerged arc furnace electrically heated with graphite electrodes, using a blend of coke, charcoal and woodchips as the reductants, where woodchips contribute to the cooling of the very top of the bed thanks to the evaporation of water that might be sprayed on them, and favour the porosity of the reaction mass and thus the reduction yield.

As known from the literature,[27,53,54] the carbon-assisted thermal decomposition of quartzites, formally depicted by the following reaction

$$SiO_2(s) + 2C(s) \rightarrow Si(l) + 2CO(g) \tag{3.10}$$

occurs as a two-stage process, with the first step occurring at 1512 °C, in correspondence with the first invariance point of the Si–O–C system (see Figure 3.8),[53] with the production of SiC and SiO

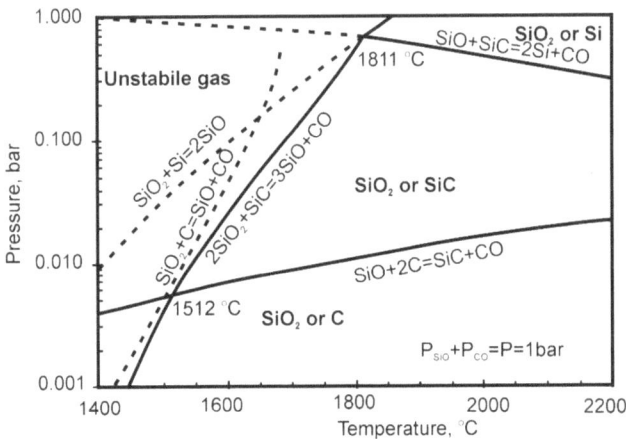

$$SiO_2(s) + C(s) \rightleftharpoons SiO(g) + CO(g) \tag{3.11}$$

$$SiO(g) + 2C(s) \rightleftharpoons SiC(s) + CO(g) \tag{3.12}$$

**Figure 3.8** Phase diagram of the Si–SiO–SiC–SiO$_2$ system. On the pressure scale is the partial pressure of SiO. Reproduced from ref. 53 with permission from H. Tveit.

and the second at 1810 °C

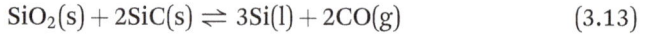

$$SiO_2(s) + 2SiC(s) \rightleftharpoons 3Si(l) + 2CO(g) \qquad (3.13)$$

in correspondence with the second invariance point, where SiC reacts with the unreacted $SiO_2$ with the formation of liquid silicon. Since reaction (3.13) is the stoichiometric coupling of the following reactions (see again Figure 3.8)

$$SiO(g) + SiC(s) \rightleftharpoons 2Si(l) + CO \qquad (3.14)$$

$$SiO_2(s) + Si(l) \rightleftharpoons 2SiO(g) \qquad (3.15)$$

occurring at the invariance point, a source of SiO (eqn (3.11)) should be present in the bottom zone of the reactor to allow the silicon production to occur.

Reactions involving SiO and CO in the region of thermodynamic SiO instability (dotted lines in Figure 3.8) result in the condensation of SiO if the system runs under thermodynamic equilibrium conditions.

The process is carried out in a carbothermic furnace schematically displayed in Figure 3.9, which operates under a continuous replenishment of the charge. The energy input is delivered by the graphite electrode dipped in the top region of the furnace, filled by a mixture of quartz and of electrically conductive coke- and charcoal-chips,

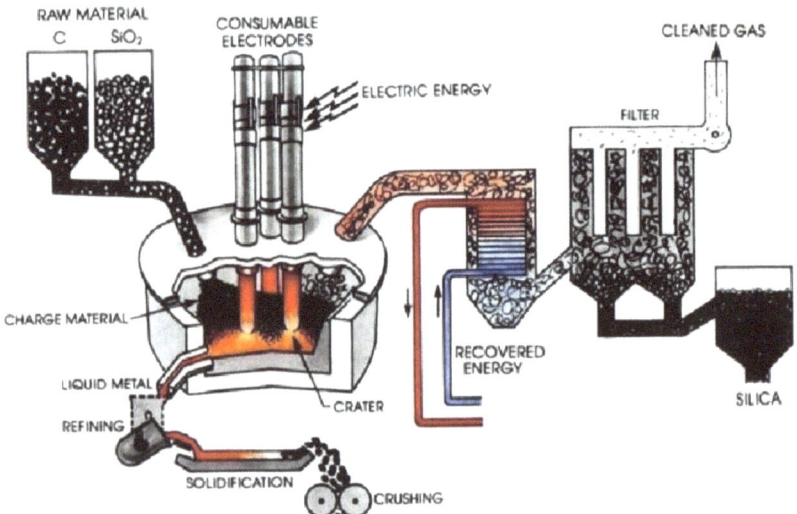

**Figure 3.9**  Schematic view of a carbothermic furnace used to produce MG silicon, including ancillary equipment for the recovery of energy and of production dusts. Reproduced from ref. 55 with permission from Elsevier, Copyright 2017.

which allows the establishment of arc-conditions between the electrodes with the heating of the reactant mass and the production of SiC. In turn, SiC reacts with the residual quartz in the bottom region of the furnace with the production of liquid silicon, which is drawn outside from a graphite orifice shown on the bottom left of the figure.

The optimized silicon output is the result of a careful compromise between the energy input, the thermal and electrical conductivity of the charge and its porosity, which is needed to allow the establishment of thermodynamic equilibrium conditions between two solid and a gaseous phase on the top of the furnace.

The carbothermic process leads to a very impure material, with a $\sim 1\%$ of total impurities, see Table 3.3, which depends on the impurity content of quartzites and carbon, with iron and dopant impurities (Al, B and P) as the main impurities.

To be used for electronic or solar applications, therefore, MG-Si should be submitted to a purification process, which involves, in the present industrial practice, the intermediate production of volatile chlorides according to the Siemens Process discussed in the next section.

Given the high cost of MG-Si purification using the Siemens process, worldwide R&D attempts were carried out at the beginning of this century to develop a low-cost process for the production of silicon of electronic quality at least appropriate for photovoltaic applications, the so-called solar grade silicon.[57–59]

Most of these processes were based on the operation of the carbothermic process in dedicated furnaces, heated with pure graphite electrodes and lined with pure graphite bricks at least in the high temperature section of the furnace, using pure quartz sand pellets and purified charcoal or carbon black pellets as the reactants. The upgraded MG-Si was then used to grow multi-crystalline silicon ingots in a directional solidification (DS) furnace[58] of the type shown in Figure 3.10, where the final purification was carried out.

As mentioned by Øvrelid *et al.*,[59] and as personal experience of the author of this book, the solar silicon research activities failed in getting satisfactory results, both in terms of cost and of material purity. This was mostly due to the improper working of the carbothermic furnace when operated with pelletized reactants, and to the unsatisfactory purification yield of the subsequent ingot growth operated with DS furnaces.

In fact, the impurity concentration was systematically found far from the electronic level, with few exceptions.[60] Renewed interest in

**Table 3.3**  Typical concentration of impurities (in ppm) in quartzites, carbon, metallurgical silicon (MG silicon), and electronic grade silicon (EG silicon).

| Impurity | Quartzite | Carbon | MG silicon | EG silicon (ppb) |
|---|---|---|---|---|
| Al | 620 | 5500 | 1570 | — |
| B | 8 | 40 | 44 | <1 |
| Cu | <5 | 14 | — | 0.4 |
| Fe | 75 | 1700 | 2070 | 4 |
| P | 10 | 140 | 28 | <2 |
| Cr | | | 137 | 1 |
| Mn | | | 70 | 0.7 |
| Ni | | | 4 | 6 |
| Ti | | | 163 | — |
| V | | | 100 | — |
| C | | | 80 | 0.6 |

this process by Ferrosolar and the Instituto de Energia Solar, ETSI Telecomunicacion, Universitad Politecnica de Madrid, Spain, after more than ten years of attempts, resulted in PERC solar cells with 22.39% efficiency,[||] showing that the solar silicon route is far from being obsolete.

## 3.4.2 The Siemens Process for the Purification of Metallurgical Si (MG-Si)

In present industrial practice, MG-Si purification is carried-out using technical improvements of the original Siemens process, based on its reaction with dry hydrochloric acid to produce trichlorosilane $SiHCl_3$ (TCS) and then silane ($SiH_4$), which are the main precursors used for the industrial production of electronic grade (EG) silicon.

TCS is industrially produced by reacting with dry HCl the MG-Si powder in a fluidized bed reactor at a temperature around 250–300 °C and at atmospheric pressure[27]

$$Si + 3HCl \rightarrow SiHCl_3 + H_2. \tag{3.16}$$

The reaction is autocatalytic due to the impurities present in MG-Si and occur with the simultaneous side production of tetrachlorosilane

[||] Presented at the WCPEC Conference in Milano (Italy), September 2022, del Canizo *et al.* Progress in the development of upgraded metallurgical silicon solar cells for industrial production. Proc. WCPEC-8 solar Energy Conference, Milano 2022.

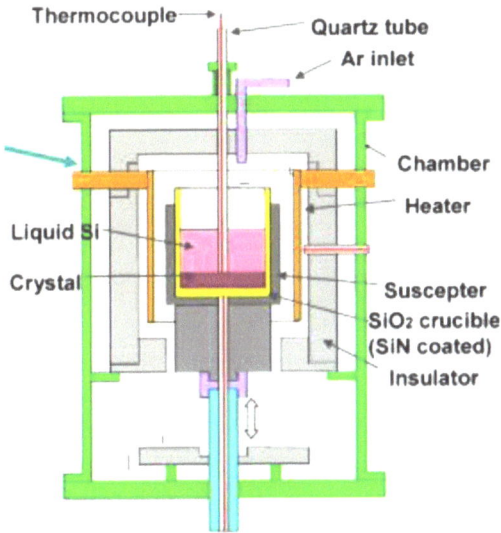

**Figure 3.10** Directional solidification (DS) furnace for the growth of multi-crystalline silicon. Reproduced from ref. 58 with permission from Elsevier, Copyright 2009.

$SiCl_4$ (STC)

$$Si + 4HCl \rightarrow SiCl_4 + 2H_2 \qquad (3.17)$$

which must be recycled.

$SiCl_4$ can, however, be also produced by direct chlorination of MG-Si with chlorine

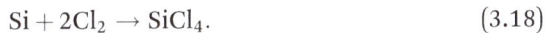

$$Si + 2Cl_2 \rightarrow SiCl_4. \qquad (3.18)$$

A second process for the production of TCS uses the equilibrium reaction of STC with MG-Si in the presence of hydrogen

$$Si + 2H_2 + 3SiCl_4 \rightleftharpoons 4SiHCl_3. \qquad (3.19)$$

The main advantage of this last process, although endothermic and operated at 500–600 °C under high pressure (25–35 bar), is the process yield in terms of TCS (close to 30%) and the absence of a side reaction typical of the former process (eqn (3.17)).

The trichlorosilane is brought to high purity conditions (1 ppb or less of volatile B- and P-halides) by fractional distillation directly in the industrial plant where the polysilicon is produced with several distillation columns whose number depends on the desired purification grade.

One can remark on the excellent purification yield of this process, which starts from MG-Si with more than 1000 ppmw of impurities to arrive in the ppb range.

Although silane might be produced by different methods,[61–63] the most widely applied process, developed by Union Carbide,[64,65] is the catalytic conversion of trichlorosilane to silane, following the following main equilibrium reactions

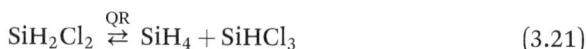

$$2SiHCl_3 \overset{QR}{\rightleftarrows} SiH_2Cl_2 + SiCl_4 \tag{3.20}$$

$$SiH_2Cl_2 \overset{QR}{\rightleftarrows} SiH_4 + SiHCl_3 \tag{3.21}$$

using quaternary ammonium ion exchange resins (QR) as catalysts.

### 3.4.3 Production of Polysilicon Rods or Granular Silicon

The production of polysilicon rods or granular silicon, used as the feedstock for the final ingot growth processes, is carried out with two main competing processes, the Siemens and the fluidized bed reactor (FBR) ones.

The Siemens process provides cylindrical polycrystalline rods prepared by the high temperature gas-phase deposition of silicon on cylindrical seeds of polycrystalline Si, using trichlorosilane as the precursor.[66]

The process is industrially carried out in stainless steel reactors at ~1100 °C, see Figure 3.11, using a mixture of TCS and $H_2$, which converts to silicon under a self-catalysed, equilibrium reaction

$$SiHCl_3 + H_2 \overset{Si}{\rightleftarrows} Si + 3HCl \tag{3.22}$$

occurring at the surface of silicon.

The deposition process occurs with a Si yield around 30%, since reaction (3.22) occurs in equilibrium with the following reactions

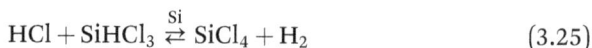

$$2SiHCl_3 \overset{Si}{\rightleftarrows} SiH_2Cl_2 + SiCl_4 \tag{3.23}$$

$$2SiHCl_3 \overset{Si}{\rightleftarrows} SiH_2Cl_2 + SiCl_4 \tag{3.24}$$

$$HCl + SiHCl_3 \overset{Si}{\rightleftarrows} SiCl_4 + H_2 \tag{3.25}$$

leading to the side-production of dichloro-$SiH_2Cl_2$ and tetrachloro-$SiCl_4$ silane.

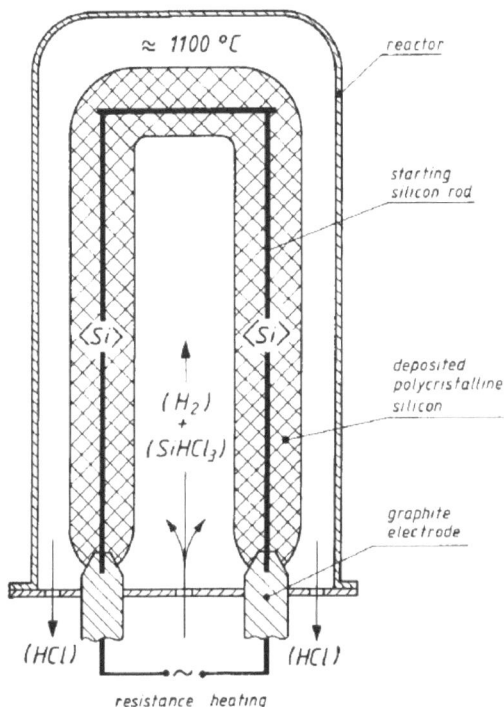

**Figure 3.11** Schematic drawing of a Siemens reactor. Reproduced from ref. 66 with permission from Elsevier, Copyright 1983.

The recycling of the unreacted TCS and a reconversion of the side products is therefore necessary, which is carried out in dedicated ancillary plants.

The purity of polycrystalline silicon produced with the Siemens process (99.9999999%) fully meets the requirements of both the electronic and photovoltaic market.

The FBR process, patented by SUN Edison[67] is, instead, addressed at the direct production of granulated silicon with the thermal decomposition of silane, silicon tetrachloride or trichlorosilane in a fluidized bed (see Figure 3.12)[68] of silicon powder that works as a catalyst and a growth substrate, whose average size increases with the progress of the process. The advantage of $SiH_4$ over $SiHCl_3$ is that its thermal decomposition occurs at temperatures from 600 to 900 °C with the formation of silicon and hydrogen

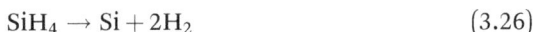

$$SiH_4 \rightarrow Si + 2H_2 \qquad (3.26)$$

without the side-products typical of the TCS route.

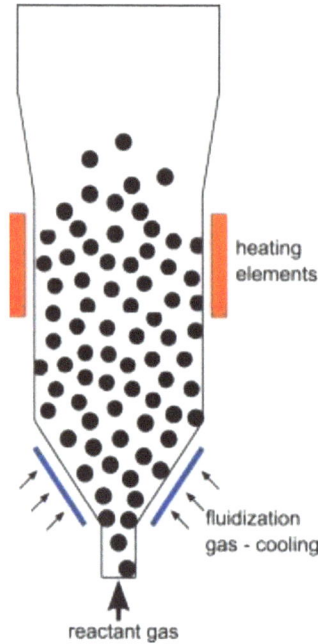

**Figure 3.12**   Schematic drawing of a fluidized bed reactor for the production of granulated silicon. Reproduced from ref. 68 with permission from Elsevier, Copyright 2010.

Despite the lower cost, and the advantage of working as a continuous process, the FBR process was charged for years with having a lower product purity with respect to the Siemens process, mostly due the repeated contacts of the fluidized powder on the FBR walls, leading to metal contamination of the silicon grains, and to the co-production of fine powder.

These problems have been brought to solution recently by two Chinese companies, who started the production of 300 000 Mt of granulated silicon in Inner Mongolia for the PV market,** and claim that their process matches the Siemens process on purity.[69] This polysilicon plant will be the largest in China, since the other polysilicon plants are in the range of 80 000 Mt.

## 3.4.4   Growth of CZ and FZ Silicon Ingots

The cylindrical rods of the Siemens process could be directly used, after polishing to a constant diameter, to produce cylindrical ingots of single

---

** GCL has purchased from SUN Edison all the rights dedicated to the FBR production of Si.

crystal silicon with the float zone process, or, after mechanical breakage to fragments of convenient size, with the CZ process. The granulated FBR silicon can be, instead, directly used to feed the CZ process.

Square section, multi-crystalline silicon ingots could be, instead, grown with the directional solidification (DS) technique[56,70] directly into a silica crucible (see Figure 3.10) that works as a container and as a mold[58] using polycrystalline silicon as the feedstock. The crucible is coated with a silicon nitride layer, using proprietary treatments,[56] to avoid the sticking of silicon on the crucible walls.

Since production costs of DS multi-crystalline silicon wafers are lower than those of CZ silicon wafers, multi-crystalline silicon has been systematically used for the fabrication of solar cells for several decades, despite lower conversion efficiencies attainable, due to grain boundary recombination losses. The use of multi-crystalline silicon has, however, been dismissed when low-cost CZ wafers became commercially available.

At the time of writing this book, almost all silicon wafers available on the market arise from ingots grown with the CZ or FZ process, while the DS technique is only used for the ingot growth of compound semiconductors (see Section 3.6 on GaAs).

### 3.4.4.1  The Czochralski Process

The Czochralski process, carried out in the process configuration illustrated in Figure 3.13, enables the growth of ingots up 2 m in length, and with diameters ranging from 20 to 300 mm, while the growth of ingots 450–500 mm in diameter is under progress.

The growth system consists of a rotating, fused silica crucible, which works as the container of the molten silicon charge, of a susceptor,[††] generally manufactured with graphite, which holds the crucible and favours constant temperature conditions, and of the heaters, generally consisting of shaped graphite resistors. A thermal shield arranged around the crucible and the heaters favours conditions of constant temperature of the molten silicon mass in the crucible. A mechanical system that holds the seed and the ingot allows the rotation and the vertical extraction of the ingot from the molten silicon charge.

The whole growth system is protected by a water-cooled, double-walled stainless-steel frame, which allows the growth to be carried out in a deoxygenated argon atmosphere.

---

[††] To hold the fused silica crucible which softens at temperatures around 1100 °C.

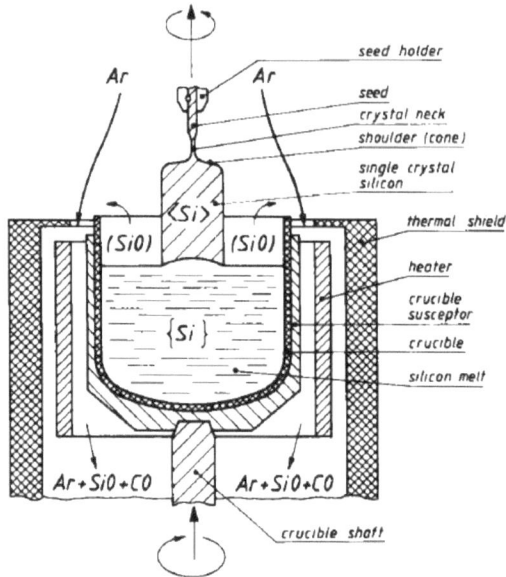

**Figure 3.13**  A Czochralski furnace for the growth of silicon ingots. Reproduced from ref. 66 with permission from Elsevier, Copyright 1983.

The silicon ingot, which rotates counter-clockwise, is slowly extracted from the melt in the vertical direction, starting the growth on an oriented single crystal seed, which determines the orientation of the single crystal ingot. The seed orientation is, case by case, selected according to the function of the final application of the grown material.

Necking[71,72] is a specific and necessary practice of the CZ process that provides the condition to grow dislocation-free crystals. It is carried out at the very beginning of the growth process when the single crystal seed is put in contact with the molten silicon charge, and consists in reducing the diameter of the originally dislocation-free seed crystal (see Figure 3.14)[73] to eliminate the dislocations generated in the seed when the seed encounters the molten silicon charge. The seed orientation is arranged perpendicular or oblique with respect to the (111) slip plane to favour the dislocations glide-out of the crystal. Spontaneous dislocation gliding is, in fact, possible since the silicon crystal is within its plastic regime at the growth temperature.

A necking-free growth practice that allows the growth of dislocation-free ingots is however possible with heavily B-doped ($10^{18}$ at cm$^{-3}$) Si melts.[74]

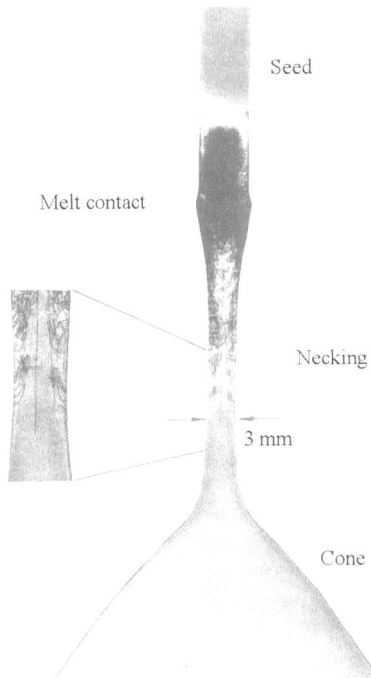

**Figure 3.14** Picture of a section of the seed and of the initial ingot cone after the necking step. The array of gliding dislocations in the necking sector is displayed in the enlarged section of the seed on the left. Reproduced from ref. 73 with permission from Springer Nature, Copyright 2017.

Counter-clockwise rotation of the crucible and of the ingot ensures a good stirring of the melt and allows the thermal and compositional profile of the liquid charge and the quality of the growth interface to be kept constant. Unwanted heat losses by irradiation from the top liquid surface are limited using a properly shaped metallic cover.

Thanks to the experimental and modelling work concerning the heat inputs and the heat flows within the system carried out in the last 40 years in modern CZ furnaces, the entire growth process is fully computer-controlled, allowing growth rate conditions to be optimized and the best structural and chemical properties of the as-grown material to be gained, including the concentration of dopants and oxygen that arises from the partial dissolution of the silica crucible, as will be discussed in the next section.

The growth rate, in turn, depends on the extraction dynamics of the heat of fusion $\Delta H_f$, which should be, possibly, homogeneous on the whole solid/liquid interface interested to solidification.

According to Von Ammon *et al.*,[75] the crystallization rate $u_n$ normal to the crystal-melt interface is given by the following equation

$$u_n = \frac{1}{\Delta H_f \rho} \left[ \left( \lambda \frac{dT}{dn} \right)_{melt} - \left( \lambda \frac{dT}{dn} \right)_{cryst} \right] \tag{3.27}$$

where $\Delta H_f$ is the heat of fusion of silicon, $\rho$ is the density of the crystal, $\lambda$ is the thermal conductivity of liquid and solid silicon, and the $\frac{dT}{dn}$ terms are the temperature gradients in the solid and in the melt, normal to the crystal melt interface.

Heat extraction is carried-out while removing from the melt a crystallized portion of the ingot at a pulling rate of several $cm\,h^{-1}$, depending on the crystal diameter. The solid portion removed from the melt cools spontaneously, since the extraction of the heat of fusion occurs by heat transfer along the solid ingot and *via* surface radiation against the water-cooled stainless-steel walls of the growth chamber, which works as a heat sink. The cooling rate depends, therefore, on the thermal conduction coefficient of the solid silicon, on its emissivity and on the crystal diameter. As expected, heat losses are rather inhomogeneous along the length of the ingot extracted from the melt, with important consequences on the distribution of defects and microprecipitates, which will be dealt with in Section 3.4.7.

### 3.4.4.2 The Float-zone Process

The float zone (FZ) process is a seeded, crucible-less process carried out in the growth configuration schematically illustrated in Figure 3.15[76] that shows the main process steps of the FZ growth. The first step is the zone-melting of the bottom of a polished cylindrical polycrystalline feed rod and the top surface of the seed crystal. The second step is the melting and recrystallization of the feed rod at the surface of the seed crystal. The third and intermediate step is the fusion and recrystallization of the feed rod, by scanning the molten zone from the bottom to the top of the feed rod, with the generation of a single crystal ingot. By this process impurities of the poly rod are transferred into the molten zone simultaneously with the recrystallization process.

Zone melting is carried out by induction heating, although optical heating is also used for the growth of oxide crystals. The whole growth system is enclosed in a cylindrical quartz tube that allows working in

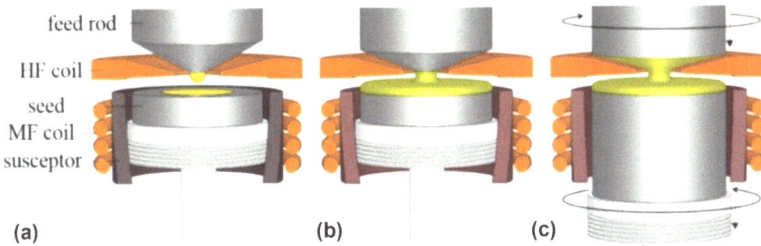

**Figure 3.15** Schematic illustration of the main process steps of float zone growth. (a) Melting of the seed crystal surface and of the tip of the feed rod, (b) generation of a molten zone, (c) crystallization by downward movement and counter-rotation of feed rod and seed. The necking process is not illustrated in the figure. Reproduced from ref. 76 with permission from Elsevier, Copyright 2019.

a protective argon atmosphere. Like in the case of the CZ process, the extraction of the heat of fusion occurs by heat conduction along the solid ingot and by surface heat radiation.

As in the case of the CZ process, necking is needed to grow a dislocation-free crystal. It is carried out on the seed in contact with a drop of melt formed at the bottom of the poly rod. The neck is then allowed to increase in diameter up to the desired diameter of the ingot for steady-state growth.

Different from the CZ process, where oxygen contamination of the liquid charge and then of the solid ingot occurs because of the partial dissolution of the crucible, the FZ process leads to an oxygen-free material, with typical oxygen concentration below $5 \times 10^{15}$ cm$^{-3}$, against values of $10^{18}$ cm$^{-3}$ for the case of CZ silicon.

Another advantage of the FZ process over the CZ one is that repetitive melting scans along the ingot drastically improve the quality of the crystal, due to sequential impurity segregation processes. Compared to CZ silicon the impurity concentration is at last a factor of $10^{2}$ lower.

As an example, the typical concentration of shallow (dopant) impurities after four FZ passes was [As] $10^{11}$, [P] $2 \times 10^{12}$, [B] $\sim 5 \times 10^{12}$ cm$^{-3}$, much below their typical concentrations in CZ silicon, which makes FZ silicon a high resistivity material.

### 3.4.5 Factors Affecting the Chemical Quality of the As-grown CZ Silicon

The chemical quality of the CZ silicon depends:

- On the original impurity content of the polycrystalline silicon charge.

- On the partial dissolution of the crucible with its impurities in the liquid charge.
- On the transfer of volatile impurities to the melt from the graphite heaters, from the thermal shields of the furnace chamber and from the current feedthroughs.
- On the effective values of the segregation coefficients of the impurities at the process temperature (the melting temperature of silicon), which will determine the impurity distribution in the grown crystal.

Impurity contamination is a thermodynamically spontaneous (irreversible) process, as is causes a decrease in the free energy of the system by an increase in its configurational entropy

$$\Delta S = R \sum_{1}^{n} x_i \ln x_i. \tag{3.28}$$

One of the main impurity inputs comes from the interaction of the silicon melt with the crucible. Because of mechanical and chemical interactions with the liquid charge, a matter flux from the crucible walls to the liquid charge occurs until thermodynamic equilibrium conditions are achieved. Fortunately, this equilibration process requires, generally, much longer times than those of the entire growth process, but still the impurity release could be important.

To prevent massive contamination, the crucible material should be, therefore, suitably selected to ensure the best thermodynamic and chemical (and mechanical) compatibility with the liquid phase.

This issue is particularly required for the CZ growth of semiconductor grade silicon from a charge of polycrystalline silicon, which is today the purest synthetic material ever produced, with a typical metallic impurity content in the ppt or sub-ppb range (As, Sb, Co, Ag, Au $\leq 1$ ppt, B, Cu, Fe, Zn, P $\leq 0.1$ ppb), and will be discussed in detail in the next section.

## 3.4.6 Crucible Compatibility Problems and Furnace Pick-up of Impurities

Due to the high temperatures involved in the silicon growth, refractory material crucibles should be used as containers for liquid silicon melts, among which only silica $SiO_2$ ($\alpha$-quartz) and high purity graphite are suitable candidates for industrial mass production.

Graphite is thermally very stable up to 4000 K but in contact with fused silicon, carbon dissolves until thermodynamic equilibrium conditions are reached ($\mu_C^C = \mu_C^{Si}$) with the formation of a C-saturated liquid solution of silicon of activity $a_C = \exp - \dfrac{\mu_C^{Si}}{RT}$, where $a_C$ is the activity of C in liquid silicon, which coincides with its concentration since the solution is very diluted. During silicon growth, in turn, carbon segregates in the solid, its equilibrium concentration being determined by the segregation coefficient of carbon $k_C^{Si} = \dfrac{x_C^{Si,s}}{x_C^{s,l}} = 0.07$.

The presence of carbon in solid silicon might lead to the nucleation and growth of micrometric precipitates of β-SiC at the liquid/solid interface that might be incorporated in the silicon ingot during solidification.

Carbon segregation in (EG) silicon causes severe detrimental effects, since SiC precipitates behave as recombination centres of minority carriers.[77] Therefore, the use of graphite crucibles is excluded in the electronic grade (EG) silicon metallurgy, favouring the use of silica glass crucibles in the industrial practice worldwide.

Silica crucibles could be made with α-quartz powder or with amorphous silica powder, but silica glass crucibles made with amorphous silica powder are used to prevent the problems associated with the α-quartz → cristobalite phase transformation that occurs around 1400 °C[78] and causes the rupture or the embrittlement of the crucible and the loss of the molten silicon charge.

Silica glass melts at 2023 K, well above the melting point of silicon (1687 K), is mechanically and thermodynamically stable in contact with a silicon melt at the melting temperature of silicon but softens at temperatures above 1300 K. Therefore, fused silica crucibles are embedded in a graphite susceptor (see Figure 3.13) to avoid their mechanical collapse during the growth process. Thermodynamic equilibrium conditions of a SiO$_2$ phase with a molten silicon phase are accomplished by the dissolution of silicon dioxide in molten silicon

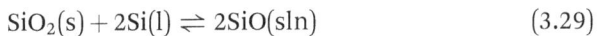

$$SiO_2(s) + 2Si(l) \rightleftharpoons 2SiO(sln) \tag{3.29}$$

with the formation of a solution of oxygen, formally of SiO molecules, in liquid silicon with an oxygen concentration of 39 ppm ($3.4 \times 10^{18}$ atoms per cm$^3$)[79] at the melting temperature of silicon.

In turn, oxygen segregates in solid silicon during the growth process, where it dissolves as interstitial species, with a segregation coefficient $k_{Si}^{O} = \dfrac{x_{O}^{Si,s}}{x_{O}^{s,l}} = 0.8 \pm 0.1$ at the melting temperature of silicon.[80]

Since the presence of oxygen in silicon provides conditions of internal gettering (IG) of metallic impurities, which we discussed in Section 2.5, the use of silica glass crucibles fits with the need to have a known (and potentially controllable) oxygen content in the silicon ingots.

The dissolution of silica is, however, not the unique source of oxygen in silicon. In fact, also traces of oxygen in the covering atmosphere of the growth furnace lead to the oxidation of both the liquid silicon and the surface of solid silicon ingots

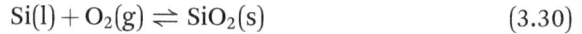

$$Si(l) + O_2(g) \rightleftharpoons SiO_2(s) \tag{3.30}$$

with a process driven by the high Gibbs formation energy of a $SiO_2$ phase $\left(\Delta G_f^{SiO_2} = -610 \text{ kJ mol}^{-1}\right)$ at the melting temperature of silicon. To minimize the high temperature oxidation of the silicon melt and of the silicon ingot, leading to an unwanted and slightly controllable oxygen input, an inert gas (argon) covering atmosphere is systematically employed in the CZ growth process, with the use of oxygen getters to bring down the oxygen content to negligible amounts.

In addition to these thermodynamic constraints, which influence the compatibility of a ceramic material with liquid silicon, its compatibility depends also on its wettability and its chemical inertness.

Wetting occurs in the presence of a work of adhesion $W > 0$. In the case of pure physical wetting, the interfacial tension terms $\sigma_{ij}$, determine the amount of work of adhesion $W$

$$W = \sigma_{SV} + \sigma_{LV} - \sigma_{SL} = \sigma_{SV}(1 + \cos\theta) \tag{3.31}$$

(with S for solid, L for liquid and V as vapour) and the contact angle

$$\cos\theta = \frac{(\sigma_{SV} - \sigma_{SL})}{\sigma_{LV}}. \tag{3.32}$$

When physical wetting is associated with a chemical reaction[81,82] occurring at the liquid–solid $Si/SiO_2$ interface, with the formation of a

surface layer of SiO

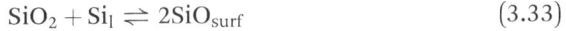

$$SiO_2 + Si_l \rightleftharpoons 2SiO_{surf} \tag{3.33}$$

the work of adhesion $W$ includes also an additional chemical reaction term $W_{react}$

$$W = W_{equil} + W_{react} \tag{3.34}$$

which represents the contribution to the work of adhesion of the chemical reaction that occurs at the liquid/solid interface.

In this case, a decrease in the interfacial tension $\sigma_{SL}$ is expected, given by the following equation

$$\sigma_{SL}(t) = \sigma_{SL}^{\circ} + \frac{d(\Delta G_R)}{d\Omega_{SL} \cdot dt} - \Delta\sigma_{SL} \tag{3.35}$$

where $\sigma_{SL}(t)$ is the interfacial tension at the time $t$, $\sigma_{SL}^{\circ}$ is the interfacial tension in the absence of reaction contributions, the term $\frac{d(\Delta G_R)}{d\Omega_{SL} \cdot dt}$ is the contribution of the free energy of reaction $\Delta G_R$ per unit of interfacial surface $\Omega_{SL}$ and per unit of time, and $\Delta\sigma_{SL}$ is a measure of the difference in the interfacial tension at the liquid/solid interface when the pristine clean interface has been substituted by the reacted interface.[81]

The actual values of $\Delta G_R$ and $\Delta\sigma_{SL}$ depend on the nature of the system considered and is experimentally or theoretically determined.[82]

In the case of Si/SiO$_2$ interfaces, $\Delta G_R$ for the formation of a SiO surface phase, according to eqn (3.33), is $-375.125$ kJ mol$^{-1}$ at the melting temperature of silicon. Thus, the formation of the interfacial SiO phase is thermodynamically favoured, enabling crucible wetting conditions.

Eventually, the reaction of silica with liquid silicon with the formation of SiO is followed by its sublimation as amorphous SiO on the cold walls of the furnace,[83] while a fraction of SiO dissolves in the liquid charge and dissociates to silicon and oxygen[83]

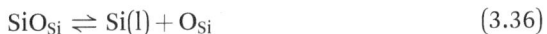

$$SiO_{Si} \rightleftharpoons Si(l) + O_{Si} \tag{3.36}$$

leading a solution oxygen in liquid silicon.

With a rate that depends on temperature, the main oxygen route in a CZ process is, therefore, its transport from the crucible walls to the bulk of the silicon ingot, where it segregates as silicon oxide

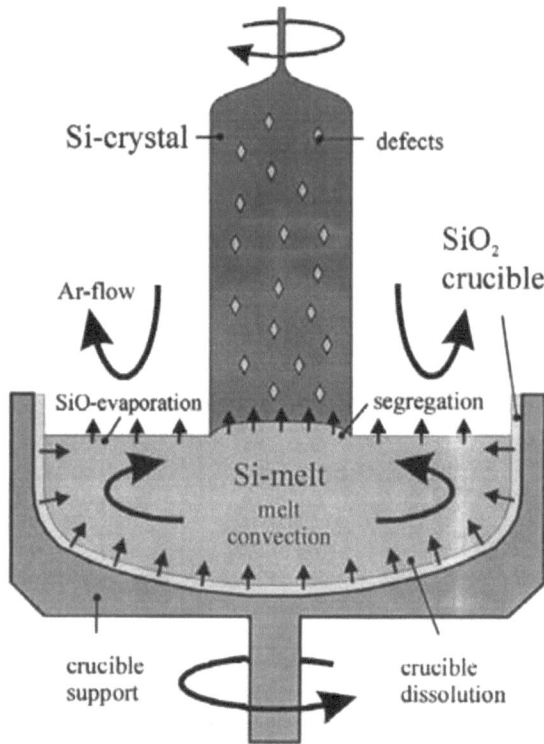

**Figure 3.16** Schematic representation of the oxygen transport processes in a CZ furnace. Reproduced from ref. 84 with permission from Elsevier, Copyright 1999.

precipitates, as shown schematically in Figure 3.16,[84] where its precipitation is controlled by defect engineering procedures illustrated in the next section.

Despite its superior performance with respect to other ceramic materials, fused silica might contribute also to the metallic contamination of the liquid silicon charge. In fact, if the quartz sand from which the silica crucibles are manufactured contains even minute amounts of transition metal oxides (see Table 3.4), all thermodynamically less stable than $SiO_2$, their reduction to metals by reaction with liquid silicon would occur spontaneously at the melting temperature of silicon for most of them. Considering, as an example, the case of iron, the reduction reaction of FeO at the surface of the crucible occurs

$$Si + 2FeO \rightleftharpoons 2Fe + SiO_2 \tag{3.37}$$

**Table 3.4** Typical metallic impurities content of a quartz powder and of rock samples of a natural quartz (concentrations in ppmw). Metals are present in the quartz rock as metal oxides.

| Impurity | Quartz powder | Quartz rock |
|---|---|---|
| Fe | 0.08 | 4.80 |
| Cr | <0.01 | n.d. |
| Ti | 0.70 | 0.50 |
| Cu | 0.01 | n.d. |
| Al | 11.0 | 20.50 |
| Ca | 0.09 | 0.40 |
| Mg | 0.02 | n.d. |
| B | 0.10 | n.d. |
| K | 0.41 | 1.60 |
| Na | 1.60 | 4.50 |
| Li | 0.54 | 2.00 |
| Mn | <0.01 | n.d. |
| Zr | <0.01 | n.d. |

given the value of the Gibbs energy of reaction, which is $\Delta G_r = -322.354$ kJ mol$^{-1}$, with the consequent iron contamination of the melt and then of the solid ingot.

To add complexity to the system, carbon is also involved in the chemistry of oxygen.

In fact, SiO vapours react with the hot graphite parts of the furnace (susceptor and heaters) to give CO

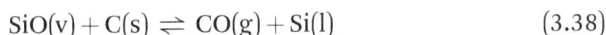

$$SiO(v) + C(s) \rightleftharpoons CO(g) + Si(l) \tag{3.38}$$

as do traces of oxygen in the inert covering atmosphere of the growth furnace

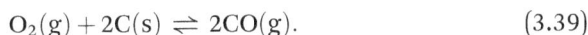

$$O_2(g) + 2C(s) \rightleftharpoons 2CO(g). \tag{3.39}$$

Eventually, CO dissolves in the molten silicon charge with the formation of carbon in solution, in equilibrium amounts that depend on the oxygen content of liquid silicon,[85] according to the following equilibrium reaction

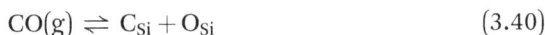

$$CO(g) \rightleftharpoons C_{Si} + O_{Si} \tag{3.40}$$

with experimental values of O and C concentrations corresponding to their saturation (C, 90 ppma; O, 44 ppma) for a partial pressure of CO of $3.8 \times 10^{-4}$ atm at the melting temperature of silicon.

A complete thermodynamic assessment of the Si–O system, which includes the oxygen segregation features, we will deal with in detail in the next section and can also be found in the papers of Schnurre,[86] Okamoto[87] and Borghesi.[88]

### 3.4.7 Incorporation of Oxygen and Oxygen-induced Defects in Silicon

The physics of the oxygen transport and segregation processes occurring during the silicon growth have been the subject of systematic experimental and theoretical studies carried out in the last 20 years,[84,86,89–93] the period of the most intense development of silicon growth technologies under the urgence of the EG and PV markets.

Although an electrochemical method for the direct measurement of the (total) oxygen distribution[‡‡] in the silicon melt of a CZ furnace was made available by Müller *et al.*,[84] the modelling of oxygen transport during the CZ growth is a more convenient strategy to foresee oxygen incorporation in the silicon ingot.

The kinetics of oxygen incorporation in silicon during the growth process is determined by the oxygen concentration in liquid silicon at the growth interface, which, in turn, depends on the dynamic equilibrium between the oxygen fluxes into the melt originated by the silica crucible dissolution, and by the SiO sublimation fluxes.

This process has been simulated, as an example, by Smirnov and Kalaev,[91] who developed a mathematical model, with the following boundary conditions for the oxygen incorporation:

- At the crystallization interface, oxygen is incorporated with an equilibrium segregation coefficient $k_O^{Si} = \frac{x_O^s}{x_O^l} \approx 1.$[§§] Since the growth rate of the CZ process is relatively slow, the oxygen incorporation into the crystal would be negligibly small, thus providing zero flux as the boundary condition

$$\frac{\delta c_O}{\delta n} = 0 \tag{3.41}$$

where $c_O$ is the equilibrium oxygen concentration in the melt.

---

[‡‡] Oxygen concentration measurements by FTIR (Fourier transform infrared) spectroscopy on silicon samples extracted from the melt and solidified give the total oxygen content in the melt measured as interstitial oxygen.

[§§] The literature values of the segregation coefficient of oxygen in silicon range between 0.25 and 2.5, strongly depending on the experimental methods used for its determination. For the CZ process a value $k_O \approx 1$ is generally used.[86]

- On the crucible walls the silica dissolution reaction (3.30) is supposed to occur, and the boundary condition is given by the (experimentally determined) concentration of oxygen.
- At the melt/argon cover gas interface, the SiO sublimation reaction

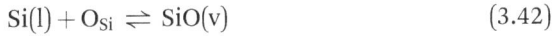

$$Si(l) + O_{Si} \rightleftharpoons SiO(v) \tag{3.42}$$

occurs, and the boundary condition is the partial pressure of SiO at the equilibrium temperature, given by the equation

$$p_{SiO} = K_p^{SiO}(T)\, \gamma_O x_O \tag{3.43}$$

which depends on the equilibrium constant $K_p^{SiO}$ of eqn (3.42) and on the activity $a_O^1 = \gamma_O x_O$ of the dissolved oxygen.

The oxygen transport in the melt is eventually given by the following equation

$$\frac{\delta \rho C_O}{\delta t} + \nabla\left(\rho \vec{V} C_O\right) = \nabla(D_{eff} \nabla C_O) \tag{3.44}$$

where $\rho$ is the melt density, $C_O$ is the oxygen concentration, $\vec{V}$ is a velocity vector and $D_{eff}$ is the dynamic diffusivity.

The results of the simulation are given in Figure 3.17, which displays the oxygen concentration at the crystallization front (the planar liquid–solid interface of a growing silicon ingot) at different process times and shows the uneven distribution of oxygen in liquid silicon at the growth interface, which would limit its uniform distribution in the crystal.

The main features of the results of the model are consistent with the experimental values of oxygen concentration obtained by the authors in realistic CZ growth.

To obviate, at least partially, the problem of uneven distribution of oxygen at the crystallization interface, which substantially depends on the convective features of the oxygen transport in silicon melts and on fluctuating growth rates,[94,95] the use of transverse magnetic fields has been attempted and successfully applied for the optimization of the oxygen distribution in the liquid,[96] once it was demonstrated that magnetic fields stabilize the melt convection.[97,98]

As we already discussed in Section 2.4.2, the final fate of the oxygen segregation in silicon during the crystal growth is the formation of oxygen precipitates and of thermal donors (TDs) that occurs in the

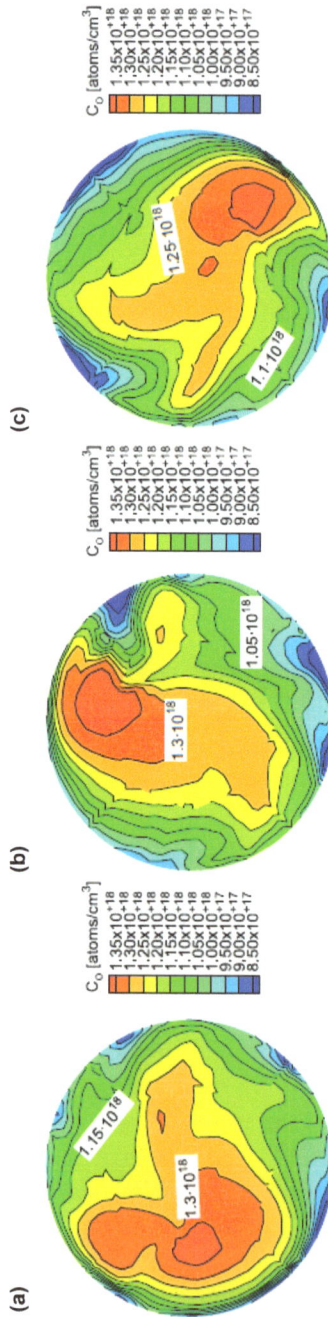

**Figure 3.17** Oxygen concentration distribution at the crystallization front (a) at $t = 0$, (b) at $t = 900$ s, (c) at $t = 1800$ s. Reproduced from ref. 91 with permission from Elsevier, Copyright 2008.

ingot sections remaining at temperatures between 800–300 °C during the cooling down of a silicon ingot extracted from the melt.

Nucleation of oxygen precipitates occurs in the 600–800 °C range, and its thermodynamics and formation kinetics depends, according to Falster *et al.*[99] and Kulkarni,[100] on the contribution of intrinsic point defects (silicon vacancies $V_{Si}$), which are generated in equilibrium conditions at the growth temperature and assumed to be the unique point-defect responsible of the stress-relief mechanism capable of delivering the requested excess volume (see Section 2.4.2) and then reduce the thermodynamic cost of forming the silicon dioxide phase.

Native defects, in turn, might coalesce under form of microdefects as vacancy clusters (voids) and self-interstitial dislocation loops (swirls),[101,102] when a silicon ingot is cooled down from the melting temperature of some hundred degrees K.[99]

Vacancies interact, eventually, with oxygen to form stable $V–O_n$ complexes[103]

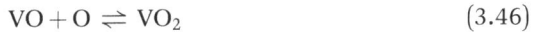

$$V + O \rightleftharpoons VO \tag{3.45}$$

$$VO + O \rightleftharpoons VO_2 \tag{3.46}$$

whose temperature-dependent concentration is reported in Table 3.5.

This makes growing a silicon ingot a unique example of a complex, non-isothermal thermodynamic system, where the interplay between vacancies and self-interstitial generation processes, local vacancies and self-interstitials supersaturation, oxygen and self-interstitials diffusion and clustering is at the origin of uneven distribution of chemical and structural inhomogeneities in the silicon ingot.

These last are of serious prejudice for the reliable and reproducible performance of the technological processes to which silicon wafers are subjected during the device manufacturing processes, and,

**Table 3.5** Temperature dependence of the concentration of vacancies, self-interstitials, oxygen and oxygen-vacancy complexes in silicon.[100,101] Data from ref. 100 and 101.

| |
|---|
| $C_V = 7.52 \times 10^{26} \exp - 4.0(\text{eV})/kT$ |
| $C_i = 6.17 \times 10^{26} \exp - 4.0(\text{eV})/kT$ |
| $C_O = 9 \times 10^{22} \exp - 1.52(\text{eV})/kT$ |
| $C_{VO_2} = 0.2c_O^\circ \times 10^{-22} \exp - 0.5(\text{eV})/kT$ |

therefore, the subject of dedicated studies aimed at the growth of a "perfect silicon", which would imply the virtual absence of point defect aggregates and the uniform oxygen profile in axial and radial direction of silicon ingots, suitable for optimized internal gettering processes at oxygen microprecipitates.[104–108]

A fundamental step toward the achievement of this objective was a study of Voronkov and Falster[99,101] addressed at the control of defect agglomeration phenomena, who succeeded in demonstrating a direct dependence of the defect aggregation process on the $V/G$ ratio, where $V$ is the growth rate and $G$ is the temperature gradient at the surface/melt interface.

They could, in fact, show that at large $V/G$ values, the crystal will be grown vacancy-type, while at low $V/G$ values the crystal is grown interstitial-type,[101,109] and that a critical $V/G$ ratio, the Voronkov ratio $\xi$, could be defined

$$\xi = \frac{E}{kT^2} \frac{(D_I C_I - D_V C_V)}{C_V - C_I} \tag{3.47}$$

which defines the critical condition at which a change from vacancy-type to interstitial type of growth occurs, where $E = \dfrac{(E_V + E_I)}{2}$ is the average free energy of formation of vacancies and interstitials, $D$ is their diffusivity and $C$ is their concentration at the growth temperature.[99]

A value of 0.12 mm² min⁻¹ K⁻¹ is given as the critical value of $\xi$, using for the free energy of formation of vacancies and self-interstitials the values of 4 eV and 4.6–4.8 eV, respectively.[109]

A value of $0.12 \text{ mm}^2 \text{ min}^{-1} \text{ K}^{-1}$ is given as the critical value of $\xi$, using for the free energy of formation of vacancies and self-interstitials the values of 4 eV and 4.6–4.8 eV, respectively.[109]

Because in conventional CZ growth processes the temperature at the growth interface is rather non-uniform, due to the non-uniform fusion-heat disposal, the axial temperature gradient $G$ increases from the centre to the periphery, as does[¶¶] the critical growth rate $V_{cr} = \xi G$.

It is therefore expected that the $G/V$ values should vary radially across the surface, as was demonstrated by visualizing the contours of the domains containing vacancy or self-interstitial excesses, using a copper decoration technique[110] for the defect delineation.

The growth of a virtually defect-free silicon ingot might be, therefore, conceived under the hypothesis of being able to improve the temperature uniformity at the growth interface and to grow the crystal

---

[¶¶] Details of optimized internal gettering conditions were discussed in Section 2.4.2.

at growth rates close to the Voronkov ratio $\xi$. In these circumstances the concentration of vacancies and of interstitials is very close and they will recombine before having the chance to interact with oxygen and segregate the oxide or to start a clustering process as the crystal cools.

As for the case of the magic denuded zone (MDZ) process, discussed in Section 2.4.2, the technological details of the process are in general proprietary know-how of each silicon producer.

Since the formation of thermal donors (TDs)[111,112] could not be avoided with the perfect silicon process, because their formation is a clusterization process involving single oxygen atoms, it is an industrial practice to anneal at 500 °C CZ the silicon wafer to annihilate TDs and grant desired resistivity properties.

## 3.5　Melt Growth of Compound Semiconductors

The availability of compound semiconductor ingots and wafers is mandatory to provide conditions of device fabrication with homo-epitaxial processes, which would favour optimized device working conditions, as we have already mentioned. Melt growth of compound semiconductors presents, however, several constraints that are not present in the case of elemental semiconductors, of which the most critical are the thermodynamic instability of the liquid phase at the melting temperature, the non-stoichiometry of their solid phases, *i.e.* the extended homogeneity region around the stoichiometric composition, as could be seen for the case of GaAs schematically illustrated in Figure 3.18,[113] and the solid phase composition at the congruent melting point, not necessarily corresponding to the stoichiometric composition.

The width and the shape of the homogeneity region, when experimentally determined with precision, are the quantitative signatures of the temperature-dependent deviations from the stoichiometry of the solid phase and of the corresponding amount of point defects that compensate the deviations from the stoichiometry, as we saw in Section 1.5.

The actual shape and width of the homogeneity region does not only rule the defectivity of the compound semiconductor solid phase, but also influences the practical growth procedures, as can be seen in Figure 3.19 for the case of a hypothetical compound semiconductor AB, with an almost symmetrical homogeneity region but with the equilibrium melt composition at the congruent melting temperature little shifted toward a B-rich melt.

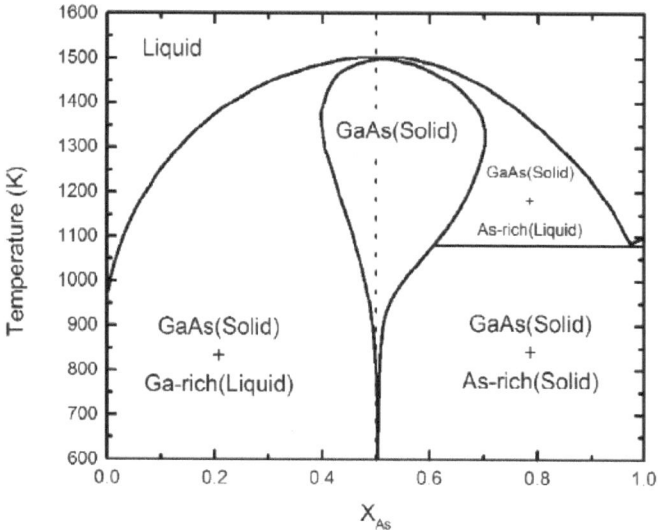

**Figure 3.18**   Schematic phase diagram of GaAs. Reproduced from ref. 113 with permission from Elsevier, Copyright 2016.

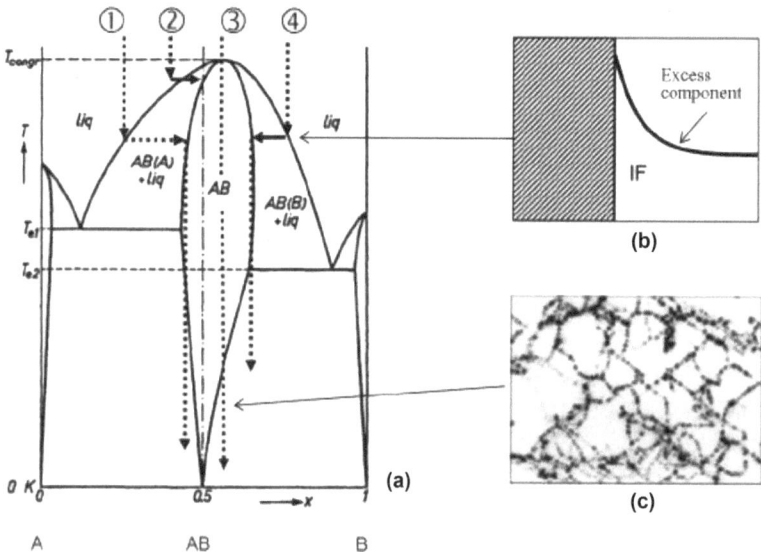

**Figure 3.19**   (a) Non-stoichiometry related growth effects. Possible crystallization path (b) enrichment in the B component by diffusion across the interface IF, (c) AB precipitates at temperatures below the eutectic. Reproduced from ref. 114 with permission from John Wiley & Sons, Copyright © 2003 Wiley-VCH Verlag GmbH & Co. KGaA, Weinheim.

It is apparent that if the growth is carried out along track 3, the phase that grows is the B-rich homogeneous $AB_{1+x}$ phase, which decomposes under cooling in a two-phase system consisting of the non-stoichiometric **$AB_{1+x}$** phase and of the solid solution of A in B (see Figure 3.19c). If the growth is carried-out from a B-rich melt along track 4, a non-stoichiometric phase **$AB_{1+x}$** is initially grown that decomposes in a two-phase system after cooling below the eutectic temperature. Eventually, only if the growth is carried out from a A-rich melt along track 2 can one get the AB phase of stoichiometric composition, which remains such also under cooling.

A perfect knowledge of its phase diagram is, therefore, a mandatory issue for successful growth of a compound semiconductor.

## 3.5.1 Seeded Growth of SiC from the Liquid Phase

Direct seeded melt growth of SiC is not applicable in view of the absence of a stoichiometric liquid phase at atmospheric pressure at the melting temperature of SiC (2830 °C),[115,116] see Figure 3.20.

For this reason, SiC is bulk-grown by sublimation, as will be discussed in Section 3.9, although alternative solutions have been attempted such as the top seeded solution growth (TSSG) from saturated SiC solutions in liquid silicon, a practice followed for other

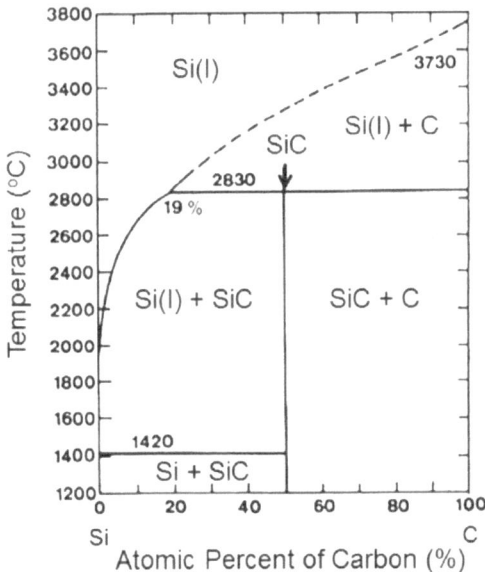

**Figure 3.20** Phase diagram of SiC. Reproduced from ref. 115 with permission from Elsevier, Copyright 2016.

**Table 3.6** Total vapour pressure of Si species ($\sum P_{Si}$) and pressure of monomeric Si ($P_{Si}$) above liquid silicon. The monomeric species is prevalent at every temperature. Data from ref. 117.

| $T$ (K) | $\sum P_{Si}$ (MPa) | $P_{Si}$ (MPa) |
|---|---|---|
| 1700 | $6.9 \times 10^{-8}$ | $6.78 \times 10^{-8}$ |
| 1800 | $3.2 \times 10^{-7}$ | $3.17 \times 10^{-7}$ |
| 1900 | $1.3 \times 10^{-6}$ | $1.22 \times 10^{-6}$ |
| 2000 | $4.6 \times 10^{-6}$ | $4.53 \times 10^{-6}$ |
| 2100 | $1.4 \times 10^{-5}$ | $1.33 \times 10^{-5}$ |
| 2200 | $3.9 \times 10^{-5}$ | $3.83 \times 10^{-5}$ |
| 2300 | $1.00 \times 10^{-4}$ | $9.30 \times 10^{-5}$ |
| 2400 | $2.36 \times 10^{-4}$ | $2.17 \times 10^{-4}$ |
| 2500 | $5.20 \times 10^{-4}$ | $4.74 \times 10^{-4}$ |
| 2600 | $1.08 \times 10^{-3}$ | $9.73 \times 10^{-4}$ |
| 2700 | $2.11 \times 10^{-3}$ | $1.89 \times 10^{-3}$ |
| 2800 | $3.95 \times 10^{-3}$ | $3.35 \times 10^{-3}$ |
| 2900 | $7.08 \times 10^{-3}$ | $6.22 \times 10^{-3}$ |
| 3000 | $1.22 \times 10^{-2}$ | $1.06 \times 10^{-2}$ |
| 3100 | $2.3 \times 10^{-2}$ | $1.75 \times 10^{-2}$ |

compound semiconductors. The drawbacks of this method applied to SiC are the high temperatures needed to get acceptable carbon solubilities in silicon (the solubility of C in liquid Si ranges from 0.01% at the melting point of Si to 19% at 2830 °C), the high vapour pressure of silicon (see Table 3.6)[117] at these temperatures[III] and the catastrophic degradation of the graphite crucibles used as containers, due to the permeation of silicon in the graphite pores with the formation of a surface layer of SiC.

The low carbon content of the saturated liquid solution implies large composition changes of the liquid solution occurring with the segregation of SiC, and a carbon-diffusion limited growth rate, while the high silicon pressure leads to important silicon losses by sublimation, unless operating under applied pressures.

Despite these difficulties, Epelbaum *et al.*[118] succeeded in growing SiC crystals of 20–25 mm in diameter and 20 mm long at a pull rate of 5–15 mm h$^{-1}$ with TSSG. The process was carried out in a graphite crucible, operated in a vertical gradient temperature arrangement, with a bottom temperature of the crucible of 2400 °C and a top temperature of 1900 °C, where the growth occurs at a SiC seed. It is supposed that at the crucible bottom a solution of SiC is formed by reaction of carbon with silicon, which works as the feed, and turns

---

[III] The vapour phase in equilibrium with liquid silicon in the temperature range 1762–1998 K consists predominantly of Si and Si$_2$ dimers.[119]

out to be supersaturated conditions on the crucible top where SiC growth occurs.

The authors, however, mention that the grown SiC sample is monocrystalline, with the structure of the 6H polytype, for the first 3–5 mm from the seed, while the rest is polycrystalline, with inclusions and crystallographic defects, leading to overall poor characteristics.

Given the severe difficulties encountered, this technology has been dismissed in favour of the physical vapor transport (PVT) methods, which will be discussed in Section 3.8.

Saturated solutions of SiC in silicon have been, instead, successfully used by Yakimova *et al.*[120] for the liquid phase epitaxy (LPE) of defective SiC wafers decorated by micropipe defects, consisting of cylindrical holes that penetrate deep in the crystal and that are present in high density (1–100 cm$^{-2}$) in sublimation grown ingots.[121]

The system, schematically illustrated in Figure 3.21, is operated in a temperature gradient of 3 °C mm$^{-1}$, with a top temperature of 1750 °C in correspondence with the substrate, with a solvent consisting of a saturated solution of SiC in silicon and could be successfully operated, healing successfully the micropipe defects.

## 3.5.2 Liquid Phase Epitaxy (LPE) Growth of GaN from the Liquid Phase

It is known that the problems encountered in the fabrication of the first blue and white Ga-nitride LEDs and lasers, as well as the difficulty in fabricating highly efficient light-emitting devices using GaN active layers and the difficulty of p-doping GaN (see Section 2.4.2)[122–124] were overcome using doped InGaN quantum well structures deposited with CVD or MO–CVD techniques on sapphire, SiC or on silicon substrates, which present the best structural compatibility with GaN, see

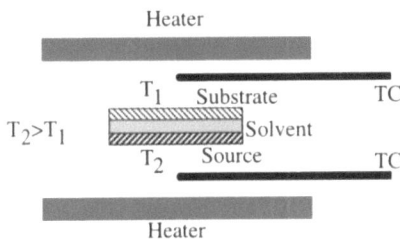

Figure 3.21 Schematic drawing of the experimental set-up used for the liquid phase epitaxy of SiC substrates. Reproduced from ref. 120 with permission from Springer Nature, Copyright 2011.

**Table 3.7**  Lattice constant of GaN, sapphire, SiC and Si. Data from ref. 12.

| Material | Lattice constant ($A$) |
|----------|------------------------|
| GaN      | $a$ 3.189              |
|          | $c$ 5.185              |
| Sapphire | $a$ 4.785              |
|          | $c$ 12.991             |
| SiC      | $a$ 3.08               |
|          | $c$ 15.12              |
| Si       | 5.4307                 |

**Table 3.8**  Activation energies for the formation and dissociation of CN bonds in selected local configurations. Data from ref. 137.

| Local conditions | Formation energy (eV) | Dissociation energy (eV) |
|------------------|-----------------------|--------------------------|
| Na–Ga melt | 0.90 | 3.0 |
| Na–GaN interface | 0.57 | 2.8 |
| Na–GaN interface with kinks and excess Ga in the melt | 1.1 | 1.3 |

Table 3.7.[12] The remaining problem, the light-emission degradation associated with the use of heteroepitaxial deposition processes due to lattice-mismatch-induced dislocations (see Table 3.8), was mitigated by the use of substrates on which a GaN buffer layer was overgrown epitaxially, although the best solution would be the availability of GaN ingots and wafers, which would allow homoepitaxial deposition processes.

Among the few, possible routes to grow large, high quality GaN crystals, based on the crystallization of GaN from a liquid solution of GaN in a low melting metal, free of undesired electrical effects if segregated in the crystal, was considered a potential solution, since the direct growth of GaN from a liquid GaN phase must be excluded due the critical temperature and pressure conditions involved.[125,126]

A thermodynamically feasible, and practically experimented solution was the growth of GaN from a saturated solution of GaN in liquid Ga, Al or In by Porowski *et al.*,[126–130] operated at convenient temperature/pressure conditions.

The thermodynamics of the Ga–GaN system was extensively studied by Karpinski *et al.*[131,132] who determined the experimental temperature dependence of the equilibrium pressure of nitrogen gas over GaN and Ga (see Figure 3.22a).

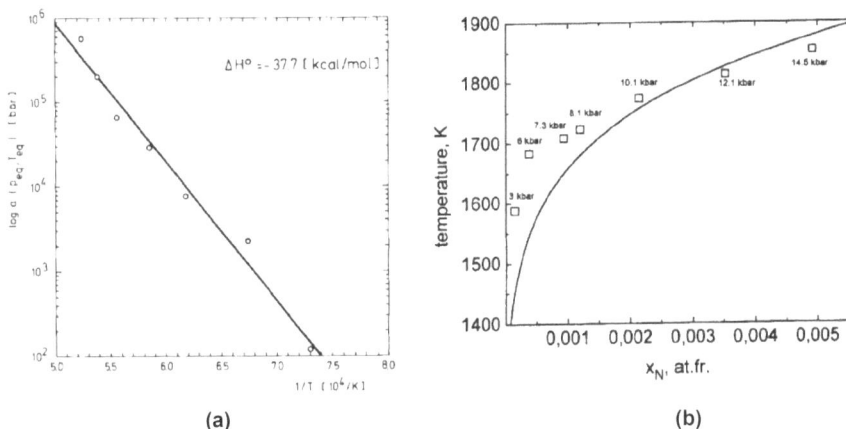

**Figure 3.22** (a) Equilibrium nitrogen pressures over GaN + Ga(l). Reproduced from ref. 132 with permission from Elsevier, Copyright 1984. (b) Solubility of nitrogen in liquid Ga (in molar fraction). Reproduced from ref. 126 with permission from Elsevier, Copyright 1997.

$$\ln p_{N_2} = -\frac{\Delta H_{GaN}^\circ}{RT} \tag{3.48}$$

(with $\Delta H_{GaN}^\circ = -157.74$ kJ mol$^{-1}$) and the standard free energy of formation of GaN

$$\Delta G_{GaN}^\circ = -157.74 + 135.69 \times 10^{-3}T \ (kJ \ mol^{-1}) \tag{3.49}$$

taking for the entropy contribution*** a value of 135.69 J mol$^{-1}$ K$^{-1}$, close to the value obtained from EMF measurements reported in Section 1.2.

It should be noted that the dissolution of diatomic nitrogen in liquid Ga is hindered by its dissociation energy, which amounts to $-941.7$ kJ mol$^{-1}$, making molecular nitrogen the strongest bounded molecule in nature and its dissociation from atomic nitrogen difficult. Theoretical studies of the nitrogen dissociation kinetics at the surface of liquid Ga and of other metals have, however, shown that the metal surfaces have a strong catalytic effect, with energy barriers toward the formation of Me–N bonds ($-320.33$ kJ mol$^{-1}$ for Al, $-328$ kJ mol$^{-1}$ eV for Ga and $-540.3$ kJ mol$^{-1}$ for In) lower than the dissociation energy.[130]

---

*** Second term of eqn (3.49).

The solubility of nitrogen in liquid Ga at temperatures around 1800 K and 1200 MPa was determined by Porowski and Grzegory,[126] (see Figure 3.22b), showing that in this range of temperatures and pressures the concentration of nitrogen in liquid gallium is significant, around 1% atomic, making a GaN growth process from the liquid phase at least potentially feasible.

On that basis, a growth process called high pressure solution growth (HPSG), carried out in vertical or horizontal furnaces under a nitrogen pressure up to of 20 kbar, was developed.[129,133]

The process was operated in the temperature range 1300–1600 °C, under a temperature gradient of 30–100 °C cm$^{-1}$, in a solution of nitrogen in liquid gallium that allows the upset of supersaturation conditions in the cooler zone of the furnace, where the nucleation of GaN crystals occurs from the liquid solution. Though not explicitly mentioned, boron nitride (BN) crucibles were used as containers.

In the absence of intentional seeding, GaN crystallites segregate at the surface of the liquid, in the upper and cooler part of the furnace, and act as nucleation sites for larger crystals, which normally take the shape of small (4–6 mm) hexagonal platelets and crystallize with the wurtzite structure.

When a needle-shaped seed was instead used, small GaN single crystals were also grown after 50 h of growth. Both GaN platelets and crystals were successfully used as substrates for homoepitaxial growth of InGaN layers.

### 3.5.3  GaN Growth from Alkali Metal–Gallium Melts

Recent progress in this field has been obtained by the Kawamura group[134–137] operating the growth of GaN in Na–Ga or Na–Ca–Ga melts, which allow the growth of 2″ in diameter, 3 mm thick GaN crystals at temperatures of the order of 800 °C.

The growth is carried out in an alumina crucible (see Figure 3.23), on the bottom of which a sapphire substrate coated by a thin GaN film MOCVD grown serves as the seed. The crucible filled with the starting materials is transferred to a stainless-steel container which is heated at 800 °C under a nitrogen pressure of 50 atm and left at this temperature for several hours.

Under these conditions, nitrogen dissolution in the liquid melt takes place, and the epitaxial growth of GaN on the GaN film-coated sapphire substrate simultaneously occurs with the formation of polycrystalline GaN grains at the surface of the melt as soon as GaN supersaturation conditions occur. It was found that the formation of

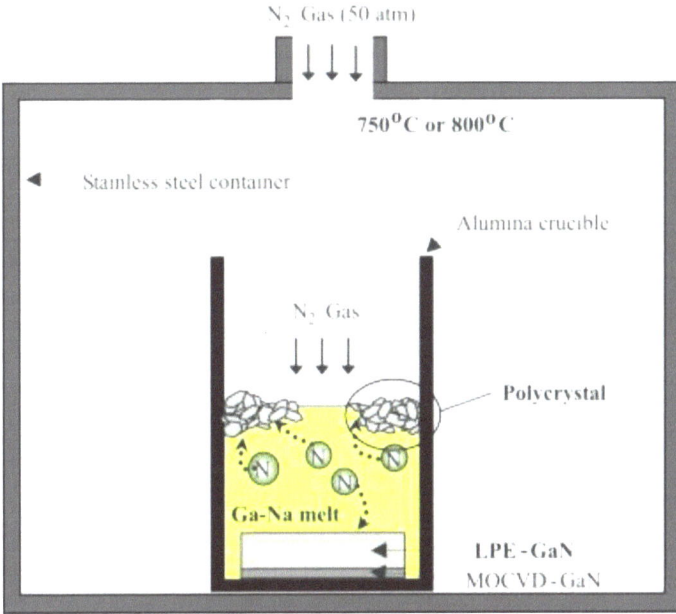

**Figure 3.23** Schematic illustration of the growth system used to grow GaN crystals in Na–Ga melts. Reproduced from ref. 135 with permission from Elsevier, Copyright 2008.

GaN polycrystals reduces the epitaxial growth rate of GaN on the GaN film-coated sapphire substrate, but that the addition of 1 at% of C to the melt suppresses the formation of the polycrystals and allows the growth of a 3 mm thick GaN overlayer on the sapphire substrate.

The reason why C-additions have a critical influence on the growth rate and inhibit the formation of surface polycrystals has been tentatively proposed by the same authors[136,137] using first principles molecular dynamics (MD) simulations, showing that C in Na–Ga melts is strongly bonded to N atoms, with the formation of stable C≡N species, whose formation energies are 6.3 eV per bond, higher than Ga–N bonds (2.26 eV per bond), in agreement with literature data relative to the Ga–N bond energies (−212.3 kJ per bond) and of C≡N bond energies (−873.2 kJ per bond).[138] It is further demonstrated that the nitrogen transport and the growth rate of GaN in Ga- and Na–Ga melts is enhanced by carbon due to the formation of C≡N species.

The calculated activation energies for the formation and dissociation rates of the CN species[†††] in three different system configurations[137]

---

[†††] These energies are not thermodynamic values but parameters of first order kinetics.

are reported in Table 3.8, which allow us to conclude that the suppression of GaN polycrystal growth is due to the negligible concentration of nitrogen in the melt, given that nitrogen is strongly bonded with C, and that the activation energy for the CN dissociation is so high (3 eV) to prevent its dissociation.

In the case of a GaN/melt interface with a Ga-rich melt, the activation energy for CN formation at kink sites (1.1 eV) is found to be comparable with its dissociation (1.3 eV), allowing CN dissociation and GaN epitaxial growth to occur.

A different procedure was selected by Wu *et al.*[139] for the growth of thicker crystals, using Na, Li, Ca and Ga melts, which allows the reduction of the loss of Na by sublimation at a growth temperature of 750 °C under a $N_2$/Ar (5:1) atmosphere at a pressure kept around 3 GPa.

A $10 \times 10$ mm$^2$ GaN seed was used (see Figure 3.24a) and the growth was carried out for 150 and 300 h, obtaining film thicknesses of 370 and 460 µm, respectively. From Figure 3.24b one can remark that the as-grown GaN layer is transparent but strongly decorated by morphological defects that are better visible in the surface morphologies displayed in Figure 3.25. After 150 h of processing hexagonal hillocks are formed at the surface (a), that disappear after 300 h, but lead to the formation of triangular holes (b).

Different from the samples of Kawamura, which exhibit dislocation densities of around $10^7$ cm$^{-2}$, the dislocation density of Wu's samples after 300 h of growth was only $10^3$ cm$^{-2}$.

Despite these interesting results, this technique cannot compete with the HVPE deposition processes, which we will deal with in Section 3.10, both in terms of process time and process conditions, which implies the dissolution of the reaction system and a chemical polishing of the as- grown samples to get samples ready for device processing.

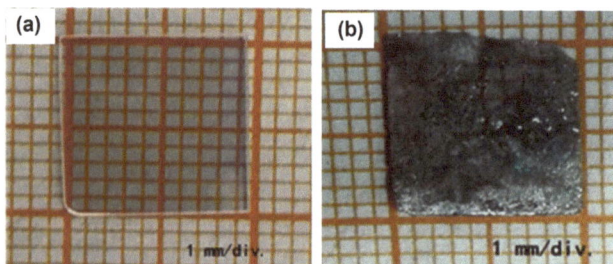

**Figure 3.24**  (a) Optical image of the seed and (b) of the GaN sample after LPE growth. Reproduced from ref. 139 with permission from Elsevier, Copyright 2019.

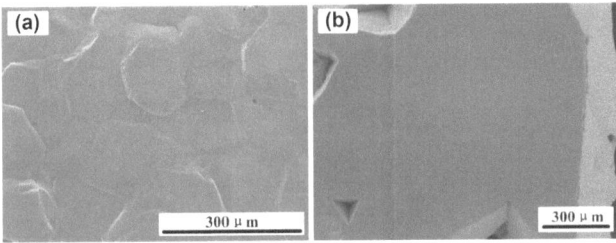

**Figure 3.25** Surface morphology of as-grown GaN crystals after 150 h (a) and 300 h (b). Reproduced from ref. 139 with permission from Elsevier, Copyright 2019.

## 3.6 Growth of GaAs

Together with InP and ZnSe, GaAs is the key member of a family of compound semiconductors still used, despite hard competition from GaN, for red LEDs, high frequency applications and radiation resistant detectors.

GaAs suffers, like all compound semiconductors of the III–V group, from thermal stability problems that lead to its decomposition (see Section 1.2.2) at temperatures below its melting temperature (1238 °C). In addition, the As vapour pressure (as $As_4$ species) is much higher than that of Ga, at the melting point of GaAs ($7 \times 10^{-5}$ atm for Ga and 1 atm for As), leading to large As- non-stoichiometries, if As-losses are not properly controlled.

To overcome these thermodynamic constraints, its growth from the liquid phase is, therefore, carried out either by applying an hydrostatic counter-pressure to balance the decomposition pressure, or by hermetically encapsulating the GaAs melt in a suitable liquid using the liquid encapsulation (LEC) technique to suppresses the As losses, while simultaneously dynamically balancing the dissociation pressure with an equivalent pressure of inert gas[113,140,141] (see again details in Section 1.1.2). Eventually, the connection of the growth chamber to an As-vapour source is a viable alternative solution.

The phase diagram of GaAs displayed in Figure 3.26 shows that the congruent melting point does not correspond with the stoichiometric composition of the solid GaAs phase but it is slightly shifted towards As-rich conditions. Therefore, the GaAs growth from an As-rich melt goes through, by cooling below the eutectic point at 810 °C, see Figure 3.19, a two-phase region consisting of non- stoichiometric GaAs and of solid As. The non-stoichiometric GaAs phase is

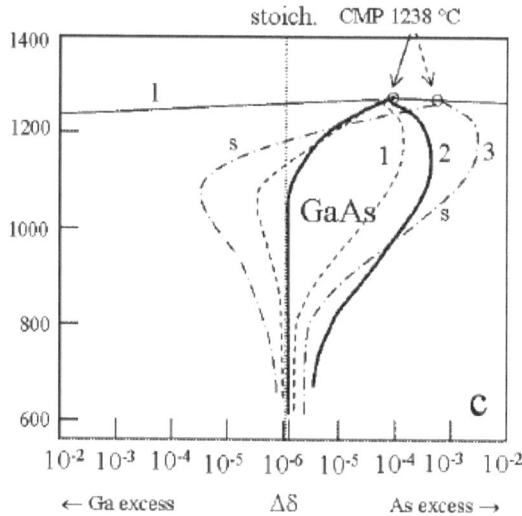

**Figure 3.26** Phase diagram of GaAs: details of the width of the homogeneous region. Reproduced from ref. 114 with permission from John Wiley & Sons, Copyright © 2003 Wiley-VCH Verlag GmbH & Co. KGaA, Weinheim.

characterized by an As-excess, balanced by Ga vacancies

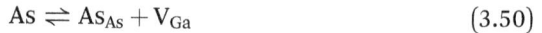

$$As \rightleftharpoons As_{As} + V_{Ga} \tag{3.50}$$

and by the formation of high density $(10^{16}$ cm$^{-3})$ donor defects localized at about 750 meV below the conduction band that have been shown to compromise the charge collection efficiency of GaAs radiation detectors.

These defects, called EL-2 defects (see Section 1.3), are supposed to consist of As$_{Ga}$ antisites[141]

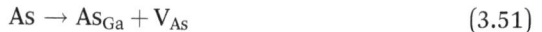

$$As \rightarrow As_{Ga} + V_{As} \tag{3.51}$$

although a different structure has also been proposed that consists of a three-centre complex made by the association of an arsenic vacancy $(V_{As})$, an arsenic antisite $(As_{Ga})$ and a gallium antisite $(Ga_{As})$.[142] Eventually, the presence of an As excess leads also to the formation of As-precipitates/inclusions, which first nucleate and then segregate at dislocations,[‡‡‡] leaving As-free regions around dislocations.[141,143]

The only way to grow a stoichiometric phase is to grow GaAs from a stoichiometric or slightly Ga-rich melt because Ga-rich melts also do not cross an eutectic by cooling toward room temperature, see Figure 3.19.

---

[‡‡‡] The dislocation density typical of melt growth GaAs is around $10^4$–$10^5$ cm$^{-2}$.

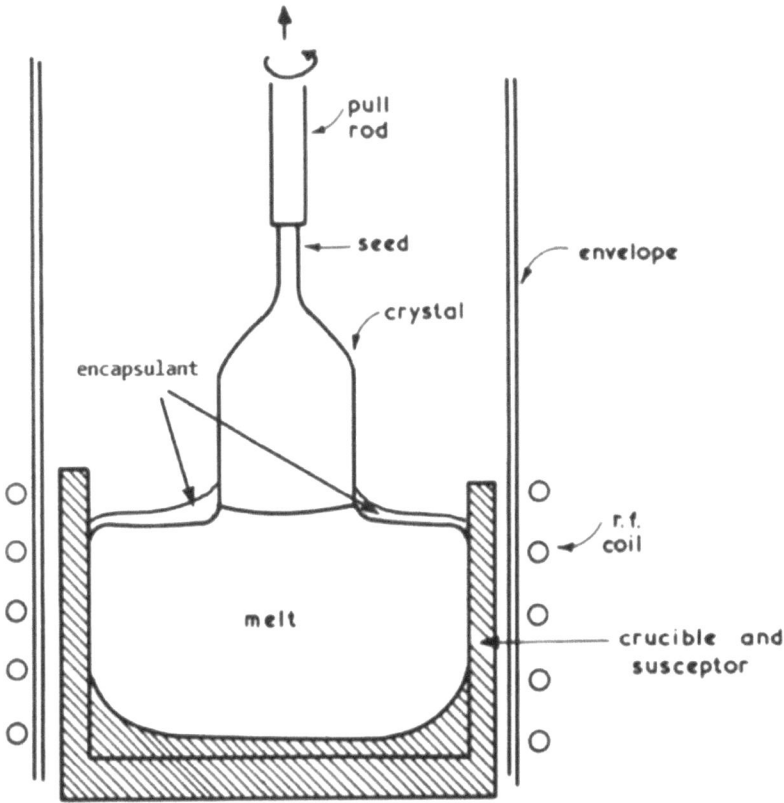

**Figure 3.27** Schematic view of a CZ furnace used for the growth of GaAs single crystal ingots. The layer of the encapsulant on top of the GaAs melt is well visible. Reproduced from ref. 113 with permission from Elsevier, Copyright 2016.

GaAs can be grown using the Czochralski technique (see Figure 3.27[113]) and with the vertical Bridgman (VB) method, see Figure 3.28,[145] but the VB method has shown distinct advantages in terms of dislocations density in the final ingot present at densities three orders of magnitude lower than in CZ grown crystals, such as those industrially used by Sumitomo for the growth of 6–8″ GaAs wafers for laser fabrication.[144]

In both cases, an encapsulant minimises the As losses and the process is carried out under an argon overpressure (2 MPa or 20 bar),[141] although lower pressures are also used with excellent results, considering that the decomposition pressure at the melting temperature (1238 °C) is of the order of 1 MPa (10 bar), see Figure 1.6b.

The system might be optimized in the vapour pressure controlled CZ (VCZ) configuration, where an external source of As vapours at constant pressure is maintained in function during the whole growth process.

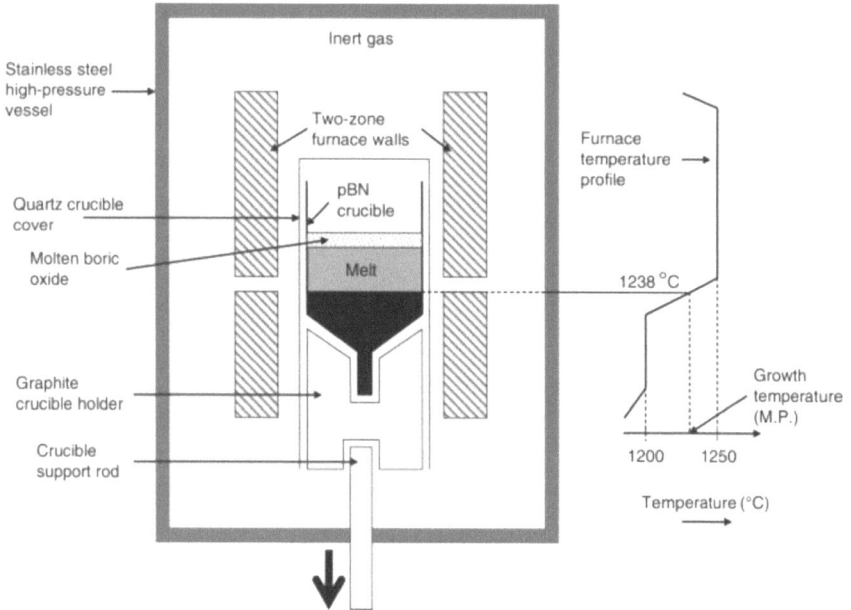

**Figure 3.28** Schematic view of a vertical Bridgman furnace for the growth of GaAs. Reproduced from ref. 145 with permission from Elsevier, Copyright 2011.

The encapsulant should be almost[§§§] insoluble in the semi-conductor melt, it should float over the GaAs melt, and its elemental components should not behave, when segregated in the semi-conductor crystal, as dopants or deep centres. The encapsulant should also be chemically inert toward the crucible used to contain the semiconductor melt to avoid corrosion.

Silica glass or pyrolytic boron nitride (pBN) can be used as crucibles, although only pyrolytic boron nitride (pBN) crucibles are currently used for the growth of GaAs[141] since the use of silica leads to unwanted Si-doping of the GaAs crystals,[146] which behaves as donor or acceptor depending whether it sits in a Ga or As site, respectively, as a $Si_{Ga}$ or a $Si_{As}$ species.

Although a benign effect of silicon can be expected when it behaves as an acceptor,[147] since it could, at least partially, compensate the EL2 defects that have deep donor properties and dominate the electrical properties of As-rich GaA,[148] the benefit is not sufficient to support the use of silica crucibles.

---

[§§§] Full insolubility is a thermodynamical nonsense.

Besides its chemical inertness toward molten GaAs, the premium features of pBN are its high melting temperature (2973 °C), its high decomposition temperature ($>2600$ °C),[12] and its excellent compatibility with GaAs melts and with boron oxide used as encapsulant in LEC growth of GaAs.

Boron oxide is, in fact, used as the encapsulant in GaAs LEC growth, being virtually insoluble in GaAs in dry conditions, with a melting temperature well below (996 °C) that of GaAs and with a density in the molten state (2.460 g cm$^{-3}$) well below the density of molten GaAs (6.096 g cm$^{-3}$), leaving it to float above the GaAs molten charge (see Figure 3.27).

A problem, however, arises with boron oxide encapsulant since the final quality of GaAs crystals is shown to depend on the chemical quality of the encapsulant.[149]

The GaAs crystals grown using the LEC route systematically present, in fact, the unintentional incorporation of C, O, B and H impurities, with an important influence on their electronic properties.

Boron behaves as an amphoteric dopant[150] with isoelectronic properties when it sits in Ga positions ($B_{Ga}$), and acceptor properties when it sits in As positions ($B_{As}$), a condition favoured for GaAs grown in Ga-rich melts.

In turn, oxygen incorporated in As positions $O_{As}$ gives rise to the EL-0 defect, with a mid-gap level at 825 meV below the CBM and deep acceptor properties[151,152] close to that of the EL-2 defect ($As_{Ga}$).

Since the EL-0 concentration depends on the oxygen concentration in the GaAs crystals, which is generally low in CZ- and in Bridgman-grown GaAs samples, the EL-0 defect concentration is generally below that of the EL-2 defects, and only partially compensates the EL-2 defect.

Also carbon dissolves in GaAs[153] at estimated levels of $2 \times 10^{17}$ cm$^{-3}$ with the formation of substitutional carbon in Ga [$C_{Ga}$] and As positions [$C_{As}$].

From *ab initio* calculations[154] a shallow-acceptor behaviour is expected for $C_{As}$, while a deep level behaviour is expected for $C_{Ga}$.

Eventually, hydrogen forms $H_2$ molecules in tetrahedral sites of the GaAs lattice,[155] while isolated atomic hydrogen is a metastable amphoteric species able to passivate¶¶¶ shallow donors, shallow acceptors and deep centres.[156]

According to Korb *et al.*[157] the C, B, O and H incorporation could be understood by considering that the C and O concentrations in GaAs

---

¶¶¶ Electrical passivation of defects is the result of a chemical reaction of hydrogen with defects. See details in Chapter 4.

should depend on the CO presence in the furnace atmosphere that arises from the oxidation of the graphite heaters of the furnace by traces of oxygen in the cover gas (argon or nitrogen), while the B and H concentrations depend on the water content of the encapsulant.

The thermodynamic modelling of the system $B_2O_3$–$H_2O$ shows that the B content in GaAs decreases with the increase of the water content, due to the thermodynamic stabilization of $B_2O_3$ by hydration[158]

$$B_2O_3 + 3H_2O \rightleftharpoons 2H_3BO_3 \tag{3.52}$$

$(\Delta H_{3.52} = -75.94$ kJ mol$^{-1})$, as experimentally demonstrated by Fischer *et al.*[146,159] and shown in Figure 3.29, while the O and H content, instead, increases with the increase in the water content in the encapsulant.

## 3.7 Growth of Sapphire Ingots

Due to its excellent physical properties, such as its high melting temperature (2030 °C) and congruent melting, high mechanical strength, high thermal and chemical stability, high dielectric strength and high optical transmission in the IR portion of the spectrum,

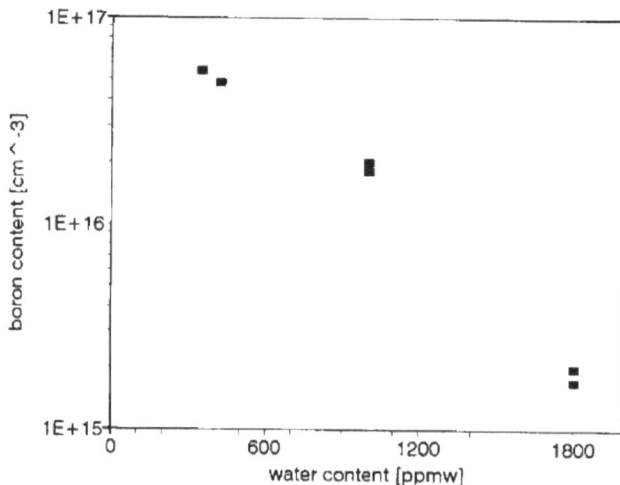

**Figure 3.29** Dependence of the B content in LEC-grown GaAs on the water content of the $B_2O_3$ encapsulant. Reproduced from ref. 146 with permission from Elsevier, Copyright 1995.

sapphire ($\alpha$-Al$_2$O$_3$), pure and doped, was more than a century ago the first material grown in single crystalline form. It is employed in a wide range of ancient and more recent applications, among which its use as substrate for GaN LEDs is the most recent, while its use as an active device for ruby, ($\alpha$-Al$_2$O$_3$–Cr), and Ti-doped lasers is already an historical example.[8,160]

Other, important, applications include its use for IR transmission windows for aerospace applications, windows for watches and for Mg-doped sapphire-based nuclear track detectors.

Sapphire crystallizes with the wurtzite structure, see Figure 3.30, and could be grown from a thermodynamically stable melt with several techniques, among which the CZ, the heat exchanger and the vertical directional solidification ones are the most used.

Its CZ growth is carried-out in a conventional furnace, using iridium, tungsten, or molybdenum crucibles. Heating can be delivered by graphite resistors or by induction. Growth is generally carried out in vacuum to avoid the oxidation of the crucibles.

A typical ingot grown using the CZ process is displayed in Figure 3.31. The process does not present sensible problems, except the difficulty to grow a crystal along the *c*-axis,[||||] the limited size of

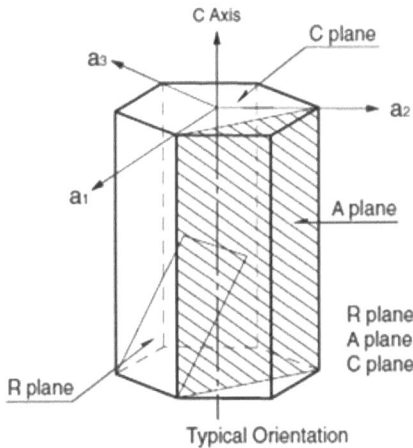

**Figure 3.30** Arrangement of key planes in the sapphire crystal lattice. Reproduced from ref. 160 with permission from John Wiley & Sons, Copyright © 2014 WILEY-VCH Verlag GmbH & Co. KGaA, Weinheim.

---

[||||] The growth is carried-out along the *a*-axis direction.

**Figure 3.31**  Sapphire crystal 4″ in diameter and 8″ long grown with the CZ process. Reproduced from ref. 8 with permission from Elsevier, Copyright 2012.

the crystals and the set-up of considerable stress along the crystal during cooldown, due to the poor thermal conductivity of sapphire and to the large temperature gradients intrinsic to the CZ technique.[161] Since GaN LEDs are conventionally grown on *c*-plane sapphire substrates, these substrates should be drilled laterally from sapphire crystals grown along the *a*-axis with a lot of material loss.

Common defects of sapphire grown with this method are bubbles, inclusions and dislocations,[162] whose density is minimized by its growth using the heat exchanger method (HEM), developed by Chattak and Schmid[161] and further studied by others,[163] which overcomes the size and stress problems typical of CZ grown ingots thanks to some unique capabilities, *i.e.* to conduct the growth of very large boules in the virtual absence of thermal gradients and the annealing of the boule after the growth, directly in the furnace.

These capabilities are due to the innovative design of the heating and cooling structures of the furnace, see Figure 3.32, where a molybdenum crucible, filled with sapphire powder, is homogeneously heated by graphite resistors and the heat of fusion is dissipated by

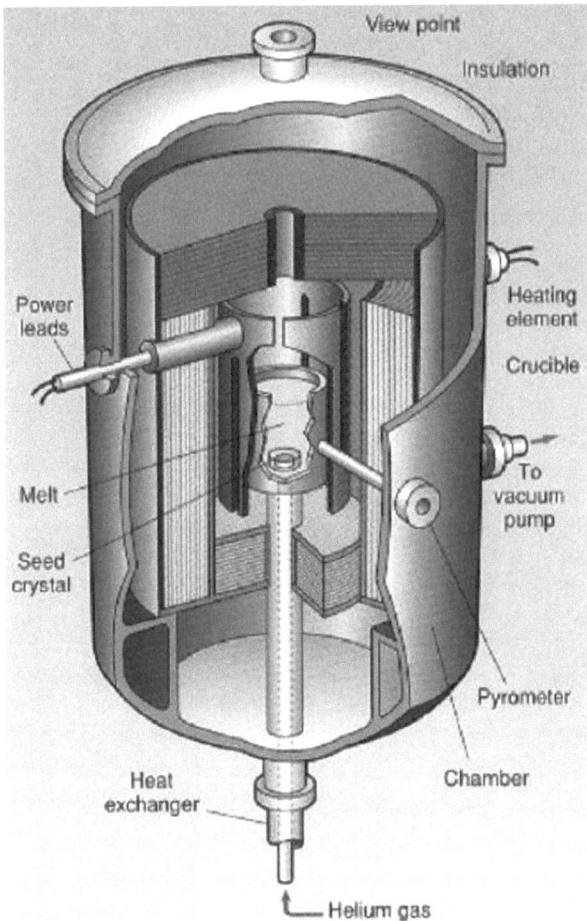

**Figure 3.32** Schematic view of an HEM furnace. Reproduced from ref. 161 with permission from Elsevier, Copyright 2001.

flowing helium in a heat exchanger sitting below the bottom of the crucible.**** The melting of the seed crystal, loaded on the bottom of the crucible, is prevented by flowing helium in the heat exchanger, which keeps its temperature slightly below the melting temperature of sapphire.

With this process the growth rate could be controlled by the flow rate of helium in the heat exchanger, while keeping the heating temperature constant, slightly above the melting temperature of sapphire.

After the growth, the boule is slowly cooled in the furnace to room temperature with a process that minimizes defects density associated with thermal stress. Almost defect-free boules of 34 cm in diameter, 10 cm in height and 65 kg were grown with this process.

Eventually, the horizontal, directional crystallization process (Bagdasarov method) (see Figure 3.33) was also used to grow large sapphire crystals (more than 30 kg in weight), using a shaped molybdenum crucible where a charge of small sapphire chunks is zone melted starting from the seed position on the extreme right, by moving the crucible horizontally across a small hot zone.

The main advantage of this process is that it can be succesfully used to also grow rectangular cross section sapphire substrates with a *c*-plane corresponding to the melt surface, from which *c*-plane wafers could be drilled with a limited loss of material, compared with the CZ or HEM crystals.

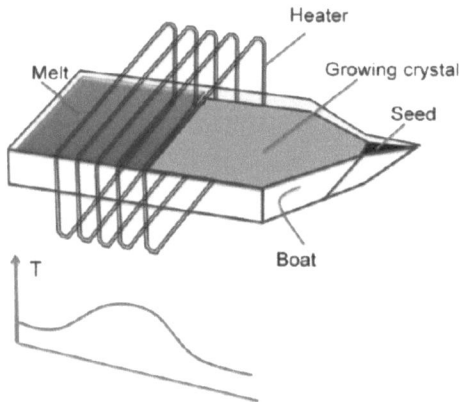

**Figure 3.33**   Schematic illustration of the horizontal, directional crystallization furnace used for the growth of large sapphire boules. Reproduced from ref. 8 with permission from Elsevier, Copyright 2012.

---

**** Helium is used for its excellent thermal conductivity.

# 3.8 Vapour Phase Growth of SiC Ingots

The interest in SiC arises from several, unique properties, such as its wide bandgap, excellent thermal conductivity, chemical and physical stability, and superior mechanical properties, which favour the use of bulk SiC in specific electronic device applications, operating also in extreme conditions or/and harsh environments.

Polytypism is another typical feature of SiC.[164,165] It presents more than 200 different polytypes, which originate from different periodic stacking sequences of the hexagonally packed double layers of Si and C along the cubic [111] or hexagonal [0001] direction. The most common polytypes of SiC are the cubic (3C-SiC or βSiC) ones, and the two hexagonal ones (4H and 6H), which present slightly different band gaps, ranging between 2.416 eV for the cubic material, 3.23 eV for the 4H phase and 3.02 eV for the 6H phase.[12]

The hexagonal polytypes are the most thermodynamically stable, with a Gibbs energy of formation $\Delta G_f = -94\,770 + 24.24T$ J mol$^{-1}$ for the 6H one.

Seeded sublimation growth is the current industrial process to produce single crystal boules of SiC with a modified Lely process[166] developed by CREE, which is still the top world producer.

The Lely process and its subsequent modifications are based on the incongruent sublimation of SiC vapours at temperatures ranging between 1950 and 2450 °C,[167–172] in a graphite or tantalum crucible from a source consisting of SiC powder or lumps, and on their deposition on a substrate consisting of a single crystal seed of SiC.

The induction heated growth furnace consists of an outer single walled[169] quartz tube, see Figure 3.34, and of a thermal insulation frame, which embeds the graphite susceptor and the graphite, or tantalum, a crucible, which contains the SiC charge and is also the reaction chamber. On top of the reaction chamber is suitably arranged the deposition substrate, consisting of a single crystal wafer of a SiC polytype (3C, 4H or 6H).

Argon is used as the covering gas. A vertical temperature gradient of about 20–40 °C cm$^{-1}$ (see Figure 3.34, right panel) is applied to drive the matter flux from the source to the deposition substrate, and to provide local supersaturation conditions and controlled growth at the seed surface.

The process starts with several degassing and backing stages of the growth zone. When the growth temperature is reached (2300–2450 °C in the case of 4H and 6H SiC), the Ar gas pressure is reduced to 5–30 mbar to favour the sublimation.

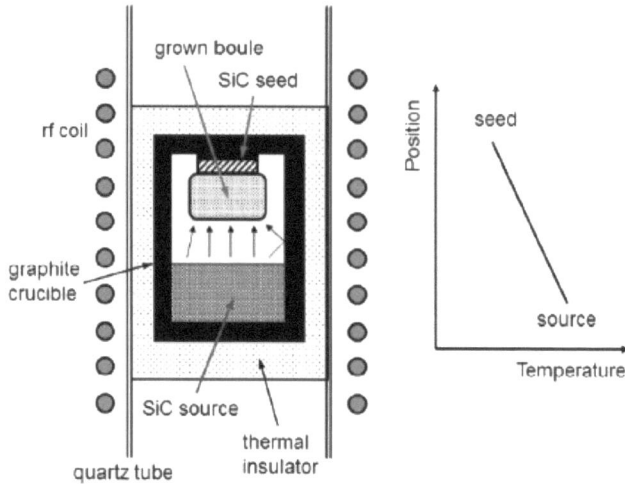

**Figure 3.34** Schematic illustration of the configuration of a Lely furnace for the growth of SiC ingots. The insert on the right displays the temperature gradient inside the furnace. Reproduced from ref. 171 with permission from Elsevier, Copyright 2016.

The main stages of the growth process itself are the sublimation of SiC, the vapour phase transport of the molecular species arising from the dissociative sublimation of SiC, and its deposition onto the top seed.

The process of sublimation growth of SiC bulk crystals is very complex, with many parameters to be controlled[169] to obtain a material with the desired characteristics.

This is in part due to the polytypism of SiC, and to the difficulty of addressing the growth toward one of the three polytypes (6H, 4H, 3C) of most common use in industrial practice, in part to the incongruent sublimation of SiC, which occurs at the surface of the sublimation source

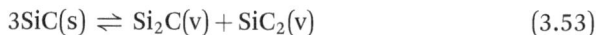

$$3SiC(s) \rightleftharpoons Si_2C(v) + SiC_2(v) \tag{3.53}$$

coupled to the dissociation of the $Si_2C$ molecules in the vapour phase

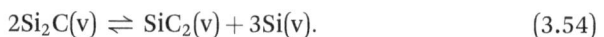

$$2Si_2C(v) \rightleftharpoons SiC_2(v) + 3Si(v). \tag{3.54}$$

Therefore, the composition of the vapour phase above the sublimation source consists of Si, $Si_2C$ and $SiC_2$ molecules (see Table 3.9), whose actual concentration depends, however, on the sublimation environment (crucible nature, temperature, vacuum, or argon as the

Table 3.9 Partial pressure (in atm) of different equilibrium species over SiC. Data from ref. 175.

| $T$ (K) | Si | SiC | SiC$_2$ | Si$_2$ | Si$_2$C | Si$_2$C$_2$ | Si$_3$ |
|---|---|---|---|---|---|---|---|
| 2149 | $2.1 \times 10^{-5}$ | | $1.9 \times 10^{-6}$ | $3.8 \times 10^{-8}$ | $1.4 \times 10^{-6}$ | | |
| 2168 | $2.7 \times 10^{-5}$ | | $2.5 \times 10^{-6}$ | $4.8 \times 10^{-8}$ | $1.9 \times 10^{-6}$ | | |
| 2181 | $3.3 \times 10^{-5}$ | $2.2 \times 10^{-9}$ | $4.2 \times 10^{-6}$ | $6.7 \times 10^{-8}$ | $2.6 \times 10^{-6}$ | | |
| 2196 | $4.1 \times 10^{-5}$ | | $4.4 \times 10^{-6}$ | $1.1 \times 10^{-7}$ | $3.9 \times 10^{-6}$ | $8.5 \times 10^{-9}$ | |
| 2230 | $6.5 \times 10^{-5}$ | | $6.5 \times 10^{-6}$ | $1.6 \times 10^{-7}$ | $5.1 \times 10^{-6}$ | $1.6 \times 10^{-8}$ | $3.2 \times 10^{-8}$ |
| 2247 | $8.3 \times 10^{-5}$ | $6.3 \times 10^{-9}$ | $1.1 \times 10^{-5}$ | $2.1 \times 10^{-7}$ | $8.1 \times 10^{-6}$ | | |
| 2316 | $2.0 \times 10^{-4}$ | $9 \times 10^{-8}$ | $3.1 \times 10^{-5}$ | $7 \times 10^{-7}$ | $2.2 \times 10^{-5}$ | $7.5 \times 10^{-8}$ | $1.6 \times 10^{-8}$ |

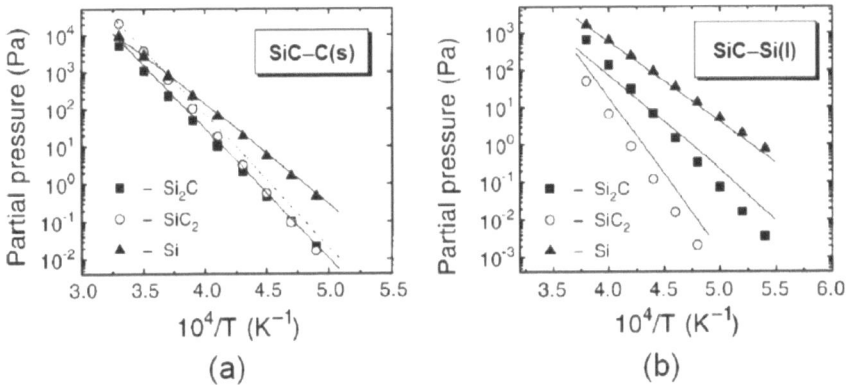

Figure 3.35 Arrhenius plots of the vapour pressures of the most relevant species over SiC saturated with C (a) or with silicon (b). Reproduced from ref. 171 with permission from Elsevier, Copyright 2016.

covering gas) and by the presence of silicon additions to the SiC sublimation source, see Figure 3.35, with relevant influence on the quality of the SiC deposit.[174,175]

How to address the growth to a specific SiC polytype is still matter of concern, although it is qualitatively known that the Si/C ratio in the vapour phase strongly affects the nature of the deposited polytype. For example, high Si–C ratios favour the growth of the 3C-SiC polytype, while carbon excess in the vapour phase favours the growth of 4H-SiC,[171] although also the growth temperature, the temperature gradient in the growth chamber and the substrate structure affect the structure of the deposited SiC phase,[175] also inducing the simultaneous deposition of different polytypes, which would lead to foreign polytypes inclusions in the SiC sublimation-grown crystals.

In turn, polytype inclusions result in the formation of structural defects (dislocations, mosaicity and micropipes, these last looking like tubular voids in the crystal) typical of SiC,[176] which we will discuss at the end of this section.

Yakimova *et al.*[176] were able to show that also the purity of the SiC powder, the seed morphology and the supersaturation conditions at the growth interface have a definite influence on the polytype uniformity and structural quality.

The C–Si ratio in the vapour phase, in turn, has a big role in the properties of the SiC deposits, as shown by Mokhov *et al.*,[174,177] who concluded that the growth of perfect SiC is only possible if thermodynamic equilibrium conditions among the vapour and the SiC are maintained along the entire growth process. In equilibrium conditions the vapour is silicon rich (SiC–Si conditions) but this condition is difficult to preserve when the growth is carried out in a graphite crucible, where C-rich (SiC–C conditions) prevail, leading to the graphitization of the source and of the deposited SiC, with the formation of voids, second phase inclusions, dislocations and other morphological and structural defects.

To reduce or even suppress the C-enrichment in the vapour phase, silicon can be added to the sublimation source, which leads to a 10-fold increase in the silicon partial pressure in the vapour phase, as can be seen in Figure 3.35, although the Si-addition also presents disadvantages, as the formation of 3C-SiC inclusions in the SiC deposit.

The growth rate of SiC can be calculated by means of theoretical modelling of the mass transport of the atomic and molecular species (Si, $Si_2C$, $SiC_2$) involved in the growth process of SiC, in C-rich and Si-rich conditions,[173] assuming that a free molecular transport occurs in conditions of very low background pressures, and that the desorption rate $D_i$ of each atomic or molecular species i from the source depends on their partial pressure $P_i$, according to the Hertz–Knudsen equation

$$D_i = \frac{\alpha_i P_i}{(2\pi m_i kT)^{1/2}} \tag{3.55}$$

where $\alpha_i$ is the evaporation coefficient of the species i and $T$ is the temperature of the source. In turn, the flux $\Phi_i$ of each atomic or molecular species can also be calculated with the Hertz–Knudsen equation

$$\Phi_i = \frac{P_i T}{(2\pi m_i kT)^{1/2}}. \tag{3.56}$$

To avoid mass losses through the gap between the source and the substrate, a mass transfer coefficient $\theta$ is introduced, which represents the fraction of molecules desorbed from the source and arriving at the substrate, and is defined as $\theta = 1 - \omega \dfrac{d}{R_s}$, where $\omega$ is a dimensionless factor taken to be equal to 3.7 for the Maxwellian velocity distribution of the desorbed molecules, $d$ is the source–substrate distance and $R_s$ the radius of the seed substrate.[173]

On that basis, the growth rate $V$ is given by the following equation

$$V = \sigma^{-1}[\Phi_{Si_2C} + 2\Phi_{SiC_2} + \theta R(T_1) - R(T_2)] \tag{3.57}$$

where $\sigma$ is the surface density of lattice sites of the seed substrate, $R$ is a function of the silicon partial pressure, $T_1$ is the temperature of the source material and $T_2$ is the temperature of the seed-substrate.

Figure 3.36 schematically displays the configuration of the experimental set-up used by Karpov *et al.*,[173] consisting of a sandwich of sublimation cells, with a source substrate and a deposition substrate independently heated, suitably enclosed in a container, on the bottom of which a supplementary source (called an environmental source) of silicon vapours or SiC vapours allows addressing the growth to carbon-rich or Si-rich conditions.[173]

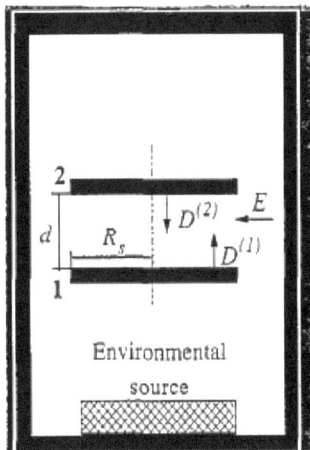

**Figure 3.36** Schematic view of the container and of the sublimation sandwich cells used for the growth of SiC. (1) SiC source wafer. (2) Deposition substrate. The environmental source is a graphite container containing SiC or a liquid silicon container. Reproduced from ref. 173 with permission from Elsevier, Copyright 1997.

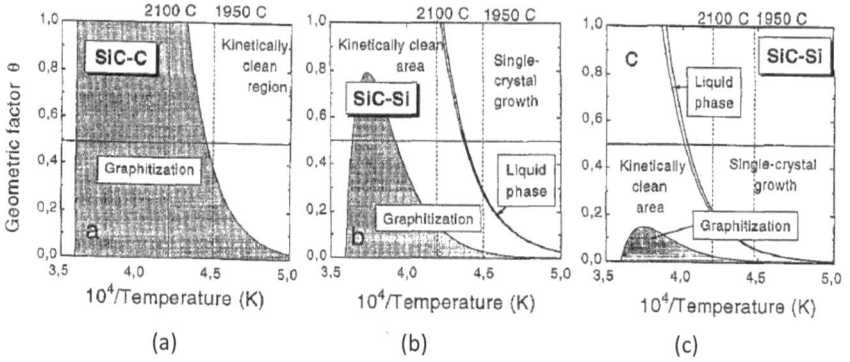

**Figure 3.37** Effect of the growth environment on the formation of excess phases during the sublimation growth of SiC. (a) C-rich type $T_{env} = 2027$ °C. (b) Si-rich type $T_{env} = 2027$ °C. (c) Si-rich $T_{env} = 2327$ °C. Reproduced from ref. 174 with permission from Elsevier, Copyright 1997.

Using this set-up, Mokov *et al.*[174] were also able to calculate the effect of the Si–C ratio in the vapour phase on the segregation of second phases (carbon and silicon) in SiC during the growth process carried out with a SiC sublimation temperature between 1950 and 2100 °C.

The results of this study are displayed in Figure 3.37, where the vertical dotted lines delimitate the temperature range of the growth process. It is possible to observe that for carbon-rich conditions (SiC–C panel (a)), and for a temperature of the environmental source held at 2027 °C, graphitization of the SiC deposits occurs for all values of $\theta$, leaving a margin for carbon-free deposits only at growth temperatures lower than 1950 °C.

The graphitization region is strongly reduced in size for Si-rich conditions (SiC–Si, panel (b)), with the environmental source held at 2027 °C, but in this case segregation of liquid silicon occurs, leaving a region of clean deposition at low values of temperature and high values of $\theta$.

Eventually (panel (c)), when a tantalum crucible is used as the container of liquid silicon, and a higher environmental source temperature (2327 °C), the region of clean SiC deposition is further enlarged, while the region of carbon deposition is confined to a small range of temperatures and $\theta$ values.

The major problem related to the SiC sublimation process is the large number of structural defects (dislocations, stacking faults, inclusions, micropipes) present in the ingot, due to the close heats of formation and structures of the 4H- and 6H-polytypes (see Table 3.10),

Table 3.10    Physical properties of the most common polytypes of silicon carbide (ZB = zinc blende, W = wurtzite).

| SiC polytype | Structure | Lattice constant (nm) 300 K | Thermal conductivity $(W cm^{-1} {}^{\circ}C^{-1})$ | Density $(g cm^{-3})$ 300 K | Heat of formation $(kJ mol^{-1})$ | Energy gap (eV) |
|---|---|---|---|---|---|---|
| 3C | Cubic: ZB | 0.43590 | 3.6 | 3.166 | −64 | 2.36 |
| 6H | Hexagonal: W | $a$: 0.3073 $b$: 1.0053 | 4.9 | 3.211 | −63.5 | 3.23 |
| 4H | Hexagonal: W | $a$: 0.3073 $b$: 1.0053 | 3.7 | 3.21 | −66.6 | 3.0 |

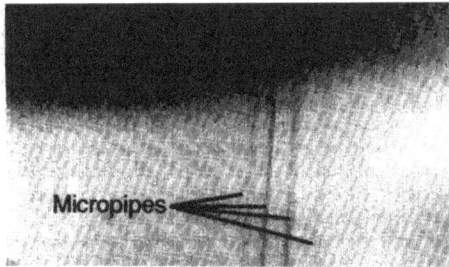

Figure 3.38    Transmission optical micrograph (400×) of a det of four micropipes running across a SiC pn junction. Reproduced from ref. 178.

that favour foreign polytype inclusions or stacking faults formation.[169,170]

Micropipes are a major defect present in 6H- and 4H-SiC wafers, which run across SiC wafers (see Figure 3.38) and cause junction failures on devices fabricated on them.

Their formation mechanism is still speculative, but it is suggested that they consist of open core screw dislocations, possibly nucleated by contaminant particles introduced during the growth process.

Doping of SiC can be done directly during its sublimation growth, or by diffusion and ion implantation on grown samples. We intend to discuss here only the case of sublimation doping, where the maximum level of doping could be achieved at process temperatures of the order of 2400 °C.

Table 3.11 displays the high temperature solubility of impurities of potential use as dopants,[179] showing that for few acceptor (Al, B, Be, Ga) and donor impurities (N and P) satisfactory high concentrations can be achieved,[179] which would favour their use as dopants, although only Al, B and N are systematically used.

**Table 3.11**  Solubility of donor and acceptor impurities in SiC. Data from ref. 179.

| Impurity | Concentration (cm$^{-3}$) | Impurity | Concentration (cm$^{-3}$) | Impurity | Concentration (cm$^{-3}$) |
|---|---|---|---|---|---|
| B | $2.5 \times 10^{20}$ | Ga | $1.8 \times 10^{19}$ | Al | $2 \times 10^{21}$ |
| N | $8 \times 10^{20}$ | P | $2.8 \times 10^{18}$ | As | $5 \times 10^{16}$ |
| Li | $1.8 \times 10^{18}$ | Be | $8 \times 10^{20}$ | Y | $2 \times 10^{16}$ |

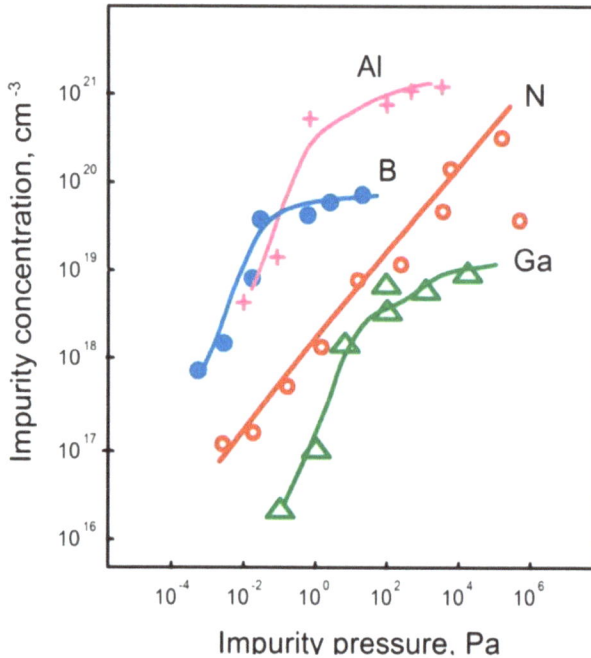

**Figure 3.39**  Dependence of the impurity concentration on the impurity partial pressure for sublimation growth and diffusion at 2000 °C. Reproduced from ref. 179, https://doi.org/10.5772/intechopen.82346, under the terms of the CC BY 3.0 license, https://creativecommons.org/licenses/by/3.0/.

In fact, n-type doping is mainly done using nitrogen as dopant,[††††] by sublimation growth of SiC in a nitrogen–argon atmosphere, where the partial pressure of nitrogen controls almost linearly the dissolved nitrogen concentration in the $10^{17}$ to $10^{21}$ cm$^{-3}$ range, as can be seen in Figure 3.39.[179]

p-type doping of SiC with Al, B and Ga vapours occurs, see Figure 3.40, with important deviation from the linear dependence of

---

[††††] Nitrogen-doped SiC wafers are used as the substrates for the heteroepitaxial growth of gallium nitride-based blue light emitting diodes.

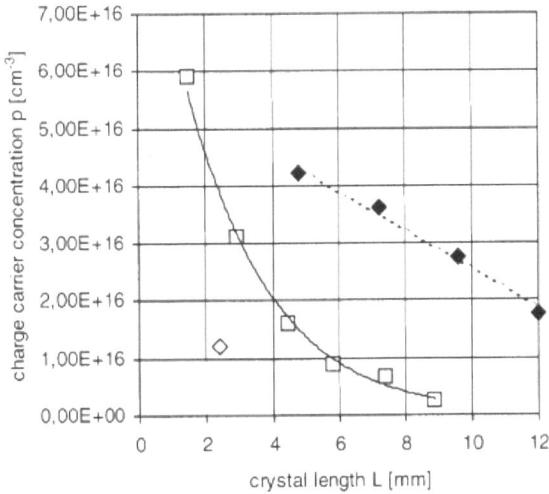

**Figure 3.40** Axial doping gradient in a 4H SiC crystal □ from a Si–Al source ◆ from an outer Al source ◇ sample compensated by residual nitrogen. Reproduced from ref. 182 with permission from Elsevier, Copyright 2002.

the dopant concentration on the dopant vapour pressure at the highest values of the partial pressure, and impurity concentrations comparable with that of nitrogen are only obtained with aluminium, which is the almost standard p-type dopant of SiC.[‡‡‡‡]

The almost ideal behaviour of nitrogen in the vapor phase over a source consisting only of SiC can be understood if we assume the absence of chemical interactions between nitrogen and the molecular species Si, $Si_2C$ and $SiC_2$ in the vapour phase. Different would be the behaviour of Al vapours in equilibrium with an heterogeneous Al–SiC mixture, held at the temperature of SiC sublimation ($\geq 2000\,°C$), since not only the partial pressures of Al is considerably higher than those of the Si, $Si_2C$ and $SiC_2$ species, and would lead to a quick exhaustion of the metal supply source. However, chemical interaction among aluminium and the graphite crucible can also occur, as suggested by Rastagaev *et al.*,[180] with the formation of an $Al_4C_3$ phase, stable up to 2425 K, where its peritectic decomposition occurs[181] such that Al, $Al_4C_3$, SiC and Si are the condensed phases present in the source and Al, Si, $Si_2C$, $SiC_2$ the species present in the vapour phase. Incidentally, no other phase can be present, since a direct interaction of Al with SiC is thermodynamically forbidden, and aluminium silicide $Al_4Si_3$ decomposes at low temperatures.

---

[‡‡‡‡] Used for the production of SiC-based high frequency and high power devices.

The deviation from the linearity of the dopant concentration with the dopant vapour pressure, see Figure 3.39, as well as the axial doping gradients, see Figure 3.40, observed in Al-doped SiC samples,[182] can therefore be explained if the SiC–Al systems in the source deviate from thermodynamic equilibrium conditions, and that the Al sublimation ratio depends on its interaction with the graphite crucible. In the absence of interactions, in fact, the Al sublimation ratio should remain constant as long as Al is present in the source, and the axial doping gradients cannot be observed.

When Al vapours are supplied by a separate, individual Al source (see Figure 3.41), using argon as a carrier gas for the Al vapours[182] sublimated from liquid Al embedded in a graphite crucible held at constant temperature, the axial Al gradient is substantially decreased, see Figure 3.40, but not entirely suppressed. The very reason for the remaining axial gradient is still under question. We believe that also in this case Al interaction occurs with the graphite crucible, leading to a reduced flux of Al vapours from the liquid Al surface contaminated by $Al_4Si_3$ inclusions.

Additional questions arise since the SiC doping success depends on the lattice position that has been substituted by donor or acceptor impurities (see Section 2.3.2).

In the SiC structure, independent of the polytype, each silicon atom is tetrahedrally coordinated with four carbon atoms, with short (0.189 nm) $sp^3$ hybridized, almost fully covalent, bonds. Silicon and carbon atoms are distributed in Si–C bilayers, and the stacking

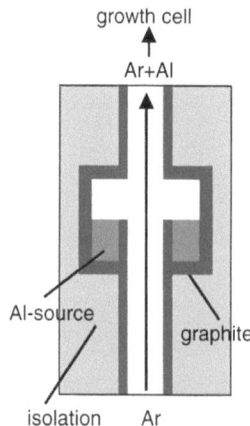

**Figure 3.41**  Schematic illustration of the use of a separate Al source for the growth of Al-doped SiC. Reproduced from ref. 182 with permission from Elsevier, Copyright 2002.

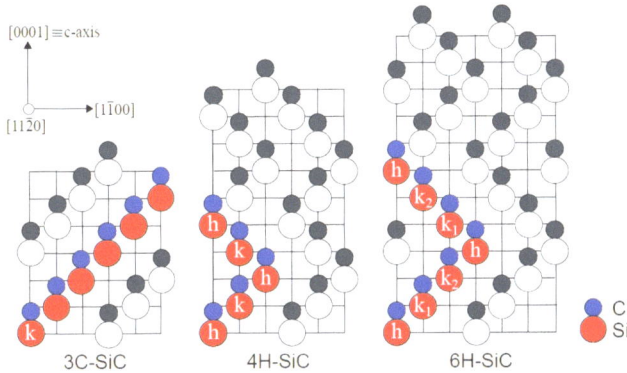

**Figure 3.42** Stacking sequence of Si–C bilayers for the three most common SiC polytypes. Reproduced from ref. 183 with permission from the author and from the correlator.

sequence of bilayers determines the structure of the polytype. An example of stacking differences of Si–C bilayers is given in Figure 3.42 for the case of 3C, 4H and 6H polytypes,[183] which also determine the structural equivalence of lattice sites in the cubic structure of 3C-SiC, where all sites are cubic (k) and the structural inequivalence of lattice sites in the hexagonal polytypes, where the 4H polytype has equal numbers of cubic (k) and hexagonal (h) lattice sites, while there is one hexagonal site h and two kinds of inequivalent quasi-cubic lattice sites, denoted k1 and k2 sites in the 6H-SiC polytype, each of them potentially substituted by donor or acceptor impurities.[184–186]

Under these circumstances any doping strategy is complicated, since the lattice location probability depends on defect formation energies, and the experimental determination of the actual location of impurities in the SiC lattice is very difficult, as recently discussed by Costa,[183] who studied both deep transition metal impurities and In as the dopant impurities. He experimentally demonstrated that In takes as its stable location the substitutional silicon position $In_{Si}$, in good agreement with the theoretical conclusions of Masanori *et al.*,[187] who showed, also, that nitrogen, oxygen, and sulphur atoms have lower formation energies on carbon sites, whereas the favourable sites for boron and selenium depend on the composition. Except for the above elements, impurity atoms always prefer to substitute a silicon site.

The ionization energies of dopant (and TM) impurities in SiC are displayed in Tables 3.12 and 3.13[12,183] which show that acceptor-type impurities behave as deep acceptors, while only nitrogen behaves as a shallow acceptor.

Table 3.12    Ionization energies (in eV) of typical donors and acceptors in 3C-SiC at 300 K.[12]

| Donors | | Acceptors | | |
|--------|------|------|------|------|
| N | V | Al | Ga | B |
| 0.06 | 0.66 | 0.26 | 0.344 | 0.735 |

Table 3.13    Experimental values of the average ionization energy (in meV) level for Al, B, and N in 4H- and 6H-SiC.[183] For nitrogen the lattice positions occupied are indicated by h, k1 and k2.

|        | Al | B | $N^h$ | $N^{k1}$ | $N^{k2}$ |
|--------|------|------|------|------|------|
| 4H-SiC | $220 \pm 20$ | $330 \pm 30$ | $50 \pm 5$ | $90 \pm 5$ | |
| 6H-SiC | $220 \pm 20$ | $220 \pm 20$ | $80 \pm 5$ | $140 \pm 5$ | $145 \pm 5$ |

The case of nitrogen is singular, since despite its lower formation energy in carbon sites, only nitrogen in silicon sites $N_{Si}$ behaves as a shallow donor (see Tables 3.12 and 3.13), while no new states are introduced when nitrogen substitutes a carbon atom.[188]

A significant side-effect of nitrogen doping is to reduce the density of basal plane dislocations, with a tremendous impact on SiC-based devices,[189] while extremely high nitrogen doping induces polytype transformation.[190]

Despite these difficulties, the development of robust growth and doping industrial technologies has succeeded in overcoming these problems, bringing us to today's production of SiC devices by several world producers.

## 3.9   Growth of GaN with the Ammonothermic Process

We discussed in Sections 3.5.2 and 3.5.3 the intrinsic problems of any growth process carried out from metallic melts saturated with GaN. The industrial growth of bulk GaN is ammonothermal growth, operated in supercritical, liquid ammonia solutions.

The basic chemistry of the ammonothermic process of bulk GaN is the same as the process developed for the growth of aluminium nitride AlN.[191]

This last process is based on the reaction of aluminium with supercritical ammonia in the presence of potassium amide ($KNH_2$), which enhances the solubility of AlN, which otherwise would be insoluble in supercritical ammonia,[192] and leads to the formation of the precursor molecule $KAl(NH_2)_4$

$$Al + KNH_2 + 3NH_3 \rightleftharpoons KAl(NH_2)_4 + 1.5H_2 \qquad (3.58)$$

which decomposes to AlN

$$KAl(NH_2)_4 \rightleftharpoons KNH_2 + AlN + 2NH_3. \qquad (3.59)$$

The reaction (3.59) is reversible at 600 °C, and at the pressure of 2 kbar is shifted to the right.

Since using alkali amides as solubility enhancers, or mineralizers, means the solubility of the nitride phase in supercritical ammonia decreases with the increase of the temperature,[193] the process could be carried out in a two zone reactor, leaving reaction (3.58) to occur in the low temperature zone, from which $KAl(NH_2)_4$ is transferred by convection to the high temperature zone, where reaction (3.59) occurs with the spontaneous crystallization of the nitride phase on a monocrystalline seed.

Nucleation and, to a minor extent, the crystal growth of AlN, is basically guided by the ammonia supersaturation ratio and by the quality of the seed.[192]

As the driving force of the transport process is primarily given by the gradient of ammonia density, the hotter part of the autoclave where the AlN segregation occurs sits on the bottom side.

A similar process works to produce GaN, using $NaGa(NH_2)_4$ as the precursor, and sodium amide $NaNH_2$ as the mineralizer agent.[194–200]

The process is typically operated at pressures slightly lower than 400 MPa, with the bottom region held at 570–550 °C, where seed crystals are also placed, and a top region held at 400–450 °C, where the feedstock is housed, see Figure 3.43.[198]

A typical growth process lasts 70–80 days, leading to an average growth rate of about 50 μm per day. Crystals of 5 cm in diameter and 4 mm in thickness might be obtained, often with faceted surfaces, where also hillocks, steps and ridges are visible, see Figure 3.44. According to Dwilinski *et al.*,[196] in the best cases the dislocations densities of GaN crystals grown by the basic ammonothermic process range down to $5 \times 10^3$ cm$^{-2}$.

The ammonothermal process can also be carried out in the presence of acid mineralizers, such as $NH_4Cl$, with the advantage that in acidic conditions the GaN solubility increases with the increase in

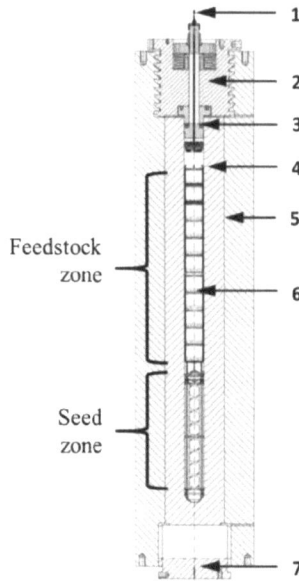

**Figure 3.43** Schematic view of an ammonothermal furnace. (1) Ammonia input capillary. (2) Upper closure. (3) Seal. (4) Growth chamber. (5) Metallic shell. (6) Row of feedstock and seed elements. (7) Bottom closure. Crucibles with feedstock are located in the upper part, while seeds embedded in special containers are in the bottom zone. A baffle separates the feedstock and the growth zone. Reproduced from ref. 198 with permission from Elsevier, Copyright 2020.

temperature,[192] see Figure 3.45, where one can also see the key role of the GaN/NH$_4$Cl ratio on the GaN solubility.

This condition allows performing the process with a more convenient configuration, having the top section of the autoclave at a lower temperature, with the additional advantage that the process can occur at reasonable rates at 500 °C and 120 kbar.

In this case, the precursor is Ga and reaction product is directly GaN

$$Ga(l) + NH_4Cl(g) \rightleftharpoons GaN(s) + HCl(g) + \frac{3}{2}H_2(g) \tag{3.60}$$

which segregates from the solution when supersaturation conditions occur.

The acid ammonothermic process has been successfully tested with the production of mm thick GaN crystals, although with severe corrosion of the autoclave construction materials in the presence of

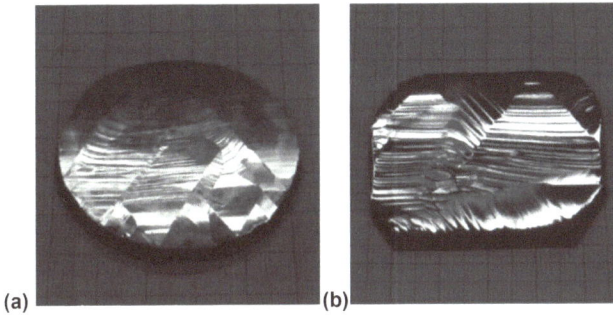

**Figure 3.44** Typical morphology of GaN crystals grown with the basic ammonothermic process (a) top view (b) bottom view, with the irregular morphology of the seed. Reproduced from ref. 198 with permission from Elsevier, Copyright 2020.

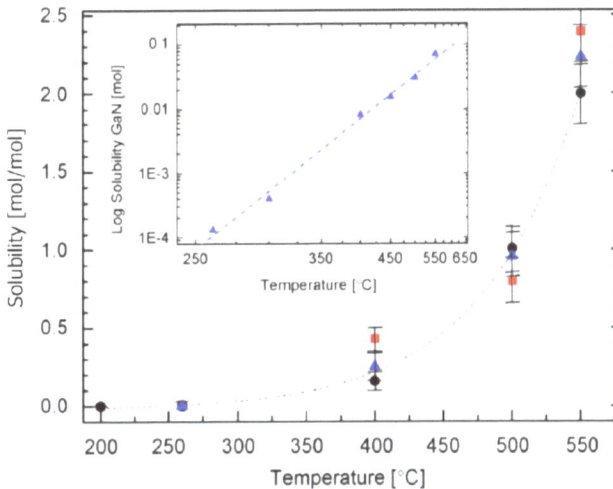

**Figure 3.45** Temperature dependence of the solubility of GaN (main curve) for three molar GaN/NH$_4$Cl ratios (0.0127, 0.032 and 0.127) at 400, 500 and 550 °C. The inset displays the Arrhenius plot of the solubility for the largest value of the mineralizer concentration (0.127). Reproduced from ref. 192 with permission from Elsevier, Copyright 2007.

acidic mineralizers, which could be prevented by Pt liners in the case of small autoclaves.

At the time of writing this chapter, several academic and industrial research groups are involved in the improvement of the ammonothermal technology, which today, together with the Na-flux method, is only used for the growth of seeds needed for the hydride vapour phase epitaxy (HVPE) process[199] (see next section).

## 3.10 Growth of Bulk GaN with the Hydride Vapor Phase Epitaxy (HVPE) Process

The hydride process was originally developed for the heteroepitaxial growth of GaN layers on SiC or sapphire substrates used for the fabrication of GaN LEDs, but today[201,202] it is used also for the growth of bulk crystals with a diameter of 50 mm.

Typically, the HVPE process is carried out in horizontal or vertical quartz reactors operated close to the atmospheric pressure.

The main advantage of this process in comparison with the ammonothermic and the liquid phase ones, discussed in Section 3.5.2, is the high growth rate ($>200$–$300\ \mu m\,h^{-1}$) and the softer operational temperature/pressure conditions (see Table 3.14). Different from the other processes, it is basically an heteroepitaxial deposition process on sapphire substrates.[201–207]

The process is carried-out in a two zone quartz reactor (see Figure 3.46), at pressures slightly above the atmospheric, and is based on the reaction of gallium with gaseous hydrogen chloride at 850 °C in the low temperature zone of the reactor, which occurs with the formation of gallium chloride $GaCl_3$

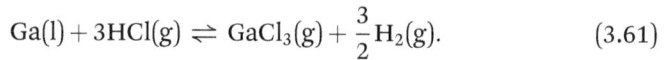

$$Ga(l) + 3HCl(g) \rightleftharpoons GaCl_3(g) + \frac{3}{2}H_2(g). \tag{3.61}$$

$GaCl_3$ is, then, reduced to GaCl in the hot zone of the reactor to avoid parasitic processes associated with the use of $GaCl_3$[202]

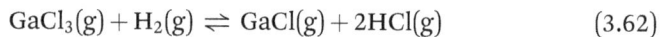

$$GaCl_3(g) + H_2(g) \rightleftharpoons GaCl(g) + 2HCl(g) \tag{3.62}$$

**Table 3.14** Main parameters of the different GaN bulk crystals growth processes. The quality of the crystals is reported in terms of the measured dislocation density $N_D$.

| Process | Temperature range (°C) | Pressure range (Pa) | Growth rate | $N_D$ (cm$^{-2}$) |
|---|---|---|---|---|
| Ammonothermic | 500–700 | 290–210 | 50 μm per day | $10^6$–$10^7$ |
| HVPE | 550–1100 | 100 kPa | $>200$–$300\ \mu m\,h^{-1}$ | $10^6$ |
| Liquid phase | Max 1550 | 1000–2000 MPa | $1\ \mu m\,h^{-1}$ | $10^3$–$10^6$ |
| Na-flux | 700–900 | 3.3 MPa | 38–46 μm h$^{-1}$ | $10^3$ |

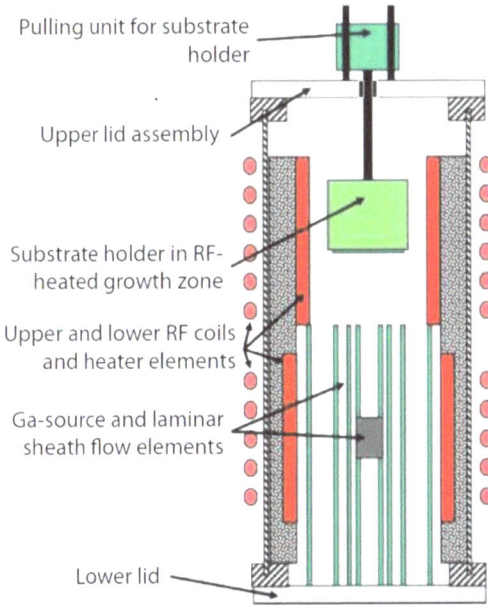

**Figure 3.46**   Schematic view of a vertical Aixtrom HVPE reactor for the growth of GaN. Reproduced from ref. 201 with permission from IOP Publishing, Copyright 2019.

which reacts with ammonia

$$GaCl(g) + NH_3 \rightleftharpoons GaN(s) + HCl(g) + H_2(g) \tag{3.63}$$

leading to the deposition of GaN on a seed suitably housed on the substrate holder, although a halogen-free route was also attempted with success[201]

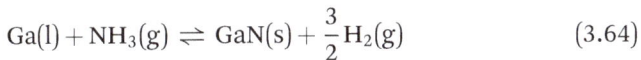

$$Ga(l) + NH_3(g) \rightleftharpoons GaN(s) + \frac{3}{2}H_2(g) \tag{3.64}$$

by reacting ammonia directly with liquid gallium.

# References

1. J. Singh, *Semiconductor devices, Basic Principles*, John Wiley & Sons, New York, 2001.
2. P. Ruterana, M. Albrecht and J. Neugebauer, *Nitride Semiconductors, Handbook on Materials and Devices*, Wiley-VCH, Weinheim, 2003.
3. C. Claeys and E. Simoen, *Germanium based technologies: from materials to devices*, Elsevier, 2007.

4. S. M. Sze and K. K. Ng, *Physics of Semiconductor Devices*, John Wiley & Sons, 2007.
5. P. Friedrichs, T. T. Kimoto, L. Ley and G. Pensl, *Silicon Carbide, Vol. 1 Growth, Defects and Novel Applications*, Wiley-VCH, Weinheim, 2009.
6. K. Nakajima and N. Usami, *Crystal growth of Si for solar cells*, Springer Verlag, 2009.
7. S. M. Sze and M. K. Lee, *Semiconductor Devices: Physics and Technology*, John Wiley & Sons, 2012.
8. M. S. Akselrod and F. J. Bruni, Modern trends in crystal growth and new applications of sapphire, *J. Cryst. Growth*, 2012, **360**, 134.
9. C. Henager and J. R. Morris, Atomistic simulation of CdTe solid-liquid coexistence equilibria, *Phys. Rev. B: Condens. Matter Mater. Phys.*, 2009, **80**, 245309.
10. K. Yamaguki, K. Kameda, Y. Takeda and K. Itagaki, Measurements of high temperature heat content of the II-VI and IV-VI compounds, *Mater. Trans., JIM*, 1994, **35**, 118.
11. J. Wang and M. Isshiki, Wide-Bandgap II-VI semiconductors: Growth and Properties, *Springer Handbook of Electronic and Photonic Materials*, Springer-Verlag, 2007, p. 325, ISBN: 978-0-387-26059-4.
12. Joffe NSM Archive; http://www.ioffe.rssi.ru
13. G. Müller, J. J. Metois and P. Rudolph, *Crystal Growth from the Melt. From fundamentals to technology*, Elsevier Science & Technology, 2004.
14. F. Allegretti, B. Borgia, R. Riva, F. De Notaristefani and S. Pizzini, Growth of BGO single crystals using a directional solidification technique, *J. Cryst. Growth*, 1989, **94**, 373.
15. W. G. Pfann, Temperature Gradient Zone Melting, *JOM*, 1955, 7, 961.
16. P. S. Dutta, Bulk Growth of Crystals of III–V Compound Semiconductors, *Compr. Semicond. Sci. Technol.*, 2011, **3**, 36.
17. W. G. Pfann, Zone melting, *Science*, 1962, **135**, 1101.
18. W. Dietze, E. Doering, P. Glasow, W. Langheinrich, A. Ludsteck, H. Mader, A. Mühlbauer, W. V. Münch, H. Runge, L. Schleicher, E. Sirtl, E. Uden and W. Zulehner, *Technology of Si, Ge, and SiC/Technologie Von Si, Ge und SiC (Landolt-Bornstein)*, Springer Verlag, 1983, vol. 17.
19. W. Zulehner, The growth of highly pure silicon crystals, *Metrologia*, 1994, **31**, 255.
20. D. Souptel, W. Löser and G. Behr, Vertical optical floating zone furnace: Principles of irradiation profile formation, *J. Cryst. Growth*, 2007, **300**, 538.
21. V. Coletti, D. McDonald and D. Yang, Role of impurities in solar silicon, in *Advanced silicon materials for photovoltaic applications*, ed. S. Pizzini, J Wiley & Sons, 2011.
22. V. Murphy, C. R. Alpass, A. Giannattasio, S. Senkader, R. J. Falster and P. R. Wilshaw, Nitrogen in silicon: Transport and mechanical properties, *Nucl. Instrum. Methods Phys. Res., Sect. B*, 2006, **253**, 113.
23. A. Yonenaga, Nitrogen effects on generation and velocity of dislocations in Czochralski-grown silicon, *J. Appl. Phys.*, 2005, **98**, 023517.
24. G. Wang, D. Yang, D. Li, Q. Shui, J. Yang and D. Que, Mechanical strength of nitrogen-doped silicon single crystal investigated by three-point bending method, *Phys. B*, 2001, **308–310**, 450.
25. T. Yoshikawa, K. Morita, S. Kawanishi and T. Tanaka, Thermodynamics of impurity elements in solid silicon, *J. Alloys Compd.*, 2010, **490**(1–2), 31.
26. K. Kakimoto, Crystallization of silicon by directional solidification, in *Crystal growth of Si for solar cells*, ed. K. Nakajima and N. Usami, Springer Verlag, 2009.
27. B. Ceccaroli and S. Pizzini, Processes, in *Advanced silicon materials for photovoltaic applications*, ed. S. Pizzini, Wiley & Sons, 2012.

28. A. Burton, R. C. Prim and W. P. Slichter, The Distribution of Solute in Crystals Grown from the Melt. Part I. Theoretical, *J. Chem. Phys.*, 1953, **21**, 1987.
29. H. Kodera, Diffusion coefficients of impurities in silicon melts, *Jpn. J. Appl. Phys.*, 1963, **2**(4), 212.
30. *Impurities and defects in Group IV elements and IIIV compounds*, ed. M. Schults, Landolt-Börnstein, Springer Verlag, Berlin, 1989, vol. 22b.
31. E. E. Haller, W. H. Hansen, P. Luke, R. Murray and B. Jarret, Carbon in high purity germanium, *IEEE Trans. Nucl. Sci.*, 1981, **29**, 745.
32. T. Kobayashi and J. Osaka, Effective segregation coefficients of carbon in LEC GaAs crystals, *J. Cryst. Growth*, 1985, **71**, 240.
33. J. W. C. O'Hara, R. B. Herring and L. P. Hunt, *Handbook of Semiconductor Silicon Technology*, Noyes Publ., 1990.
34. K. W. Böer, *Semiconductor Physics*, Van Nostrand-Reinhold, 1992, vol. II, p. 155.
35. Y. Itoh, M. Takai, H. Fukushima and H. Kirita, The Study of Contamination of Carbon, Boron, and Oxygen in LEC-GaAs, *MRS Online Proc. Libr.*, 1990, **163**, 1001.
36. A. Giannattasio, A. Gianquinta and M. Porrini, The accuracy of the standard resistivity–concentration conversion practice estimated by measuring the segregation coefficient of boron and phosphorous in Cz-Silicon, *Phys. Status Solidi A*, 2011, **208**, 564.
37. A. G. Ostrowski and G. Muller, A model of effective segregation coefficients accounting for convection in the solute layer at growth interface, *J. Cryst. Growth*, 1992, **121**, 587.
38. K. Lehovec, Thermodynamics of binary semiconductor-metal alloys, *J. Phys. Chem. Solids*, 1962, **23**, 695.
39. F. A. Trumbore, Solid Solubilities of Impurity Elements in Germanium and Silicon, *Bell Syst. Tech. J.*, 1960, **39**, 212.
40. J. B. Mullin, *Compound semiconductor processing*, J. Wiley & Sons, 2013.
41. R. Coquille, Y. Toudic, L. Haji, M. Gouneau, G. Moisan and D. Lecrosnier, Growth of low dislocation semi insulating InP (Fe-Ga), *J. Cryst. Growth*, 1987, **83**, 167.
42. X. Wallart, S. Godey, Y. Douvry and L. Desplanque, Comparative Sb and As segregation at the InP on GaAsSb interface, *Appl. Phys. Lett.*, 2008, **93**, 123117.
43. Y. Yatsurugi, N. Akiyama and Y. Endo, Concentration, Solubility, and Equilibrium Distribution Coefficient of Nitrogen and Oxygen in Semiconductor Silicon, *J. Electrochem. Soc.*, 1973, **120**, 975.
44. S. Pizzini, M. Acciarri and S. Binetti, From electronic grade to solar grade silicon: chances and challenges in photovoltaics, *Phys. Status Solidi A*, 2005, **15**, 2928.
45. C. L. Claeys, R. Falster, M. Watanabe and P. Stallhofer, *High purity silicon 9*, ECS Transactions, 2006, vol. 3, Number 4.
46. B. Labert, Y. Toudic, G. Grandpierre, M. Gauneau and B. Deveaud, Semi-insulating InP co-doped with Ti and Hg, *Semicond. Sci. Technol.*, 1987, **2**, 78.
47. S. Kasap and P. Capper, *Springer Handbook on Electronic and Photonic Materials*, Springer, 2017.
48. V. L. Wrick, K. T. Ip and L. F. Eastman, High purity LPE InP, *J. Electron. Mater.*, 1978, **7**, 253.
49. S. Fischler, Correlation between maximum solid solubility and distribution coefficient for impurities in Ge and Si, *J. Appl. Phys.*, 1962, **33**, 1615.
50. E. Scheil, Bemerkungen zur Schichtkristallbildung, *Z. Metallkd.*, 1942, **34**, 70.
51. W. G. Pfann, Principles of Zone-Melting, *JOM*, 1952, **4**, 747.
52. H. Sumiya and S. Satoh, High-pressure synthesis of high-purity diamond crystal, *Diamond Relat. Mater.*, 1996, **5**(11), 1359.

53. A. Schei, J. Kr. Tuset and H. Tveit, *Production of High silicon alloys*, Tapir Forlag, 1998.
54. B. Andersen, The metallurgical silicon process revisited, in *Silicon for the Chemical and solar industry X*, Alesund (Norwayy), 2010.
55. Z. Chen, W. Ma, K. Wei, J. Wu, S. Li, K. Xie and V. Lv, Artificial neural network modeling for evaluating the power consumption of silicon production in submerged arc furnaces, *Appl. Therm. Eng.*, 2017, **112**, 226.
56. S. Pizzini, *Advanced silicon materials for photovoltaic applications*, Wiley & Sons, 2012.
57. B. Ceccaroli and O. Lohne, in *Solar silicon Feedstock in Handbook of Photovoltaic Science and Engineering*, ed. A. Luque and S. Hegedus, Wiley, 2003.
58. M. Dhamrin, T. Saitoh, K. Kamisako, K. Yamada, V. Araki, I. Yamaga, H. Sugimoto and M. Tajima, Technology development of high-quality n-type multicrystalline silicon for next-generation ultra-thin crystalline silicon solar cells, *Sol. Energy Mater. Sol. Cells*, 2009, **93**, 1139.
59. E. Øvrelid, S. Pizzini and B. Ceccaroli, The metallurgical silicon route, in *Solar Silicon Processes*, Taylor & Francis Group, 2016.
60. E. Øvrelid, S. Pizzini and B. Ceccaroli, Elkem Solar and the Norwegian PV Industry through 40 years (1975–2015), in *Solar Silicon Processes*, Taylor & Francis Group, 2016.
61. J. E. Boone, D. M. Richards and J. A. Bossier, Process for preparation of silane, *US Pat.*, 5075092, 1991.
62. V. Bulanov, V. S. Mikheev, O. Troshin, A. Yu and A. Lashkov, Reaction of Silicon Tetrafluoride with Calcium Hydride as a Propagating Wave, *Russ. J. Inorg. Chem.*, 2008, **53**(1), 6.
63. H. E. Ulmer, D. Pickens, F. J. Rahl and P. A. Lefrancois, Producing silane from silicon tetrafluoride, *US Pat.*, 4407783, 1983.
64. C. J. Bakay, Process for making silane: *US Pat.*, 445667, Assig. Union Carbide Corp., New York, N.Y., 1974.
65. W. C. Breneman, A Process For High Volume Low Cost Production of Silane, Quarterly Report No. 1, JPL Contract No. 954334, 1976.
66. W. Zulehner, Czochralski growth of silicon, *J. Cryst. Growth*, 1983, **65**, 189.
67. J. Ibrahim, G. Melinda and T. Trong, European Patent application. High-purity granular silicon composition EP 1 900 684 A1, (43) Date of publication: 19.03.2008 Bulletin 2008/12 (21) Application number: 07120114.9 by SUN-Edison.
68. W. O. Fieltvedt, M. Javidi, M. C. Melaanen, E. Martein, H. Tatghar and P. A. Ramachandran, Development of fluidized bed reactors for silicon production, *Sol. Energy Mater. Sol. Cells*, 2010, **94**(12), 1980.
69. M. Osborn, CGL-poly touts FBR silicon matching Siemens process on purity, *Semiconductors*, March 1, 2021.
70. K. Nakajima and N. Usami, *Crystal growth of Si for solar cells*, Springer Verlag, 2009.
71. W. C. Dash, Silicon crystals free of dislocations, *J. Appl. Phys.*, 1958, **29**, 736.
72. W. C. Dash, Growth of silicon crystals free of dislocations, *J. Appl. Phys.*, 1959, **30**, 459.
73. F. Shimura, Single-Crystal Silicon: Growth and Properties, in *Springer Handbook of Electronic and Photonic Materials*, ed. S. Kasap and P. Capper, 2017, DOI: 10.1007/978-3-319-48933-9_13.
74. X. Huang, T. Taishi, I. Yonenaga and K. Hoshikawa, Dash necking in Czochralski method: influence of B concentration, *J. Cryst. Growth*, 2000, **213**, 283.
75. A. V. Kalaev, D. Lukanin, V. Zabelin, Y. N. Makarov, J. Virbulis, E. Dornberger and W. Von Ammon, Calculation of bulk defects in CZ Si growth: Impact of melt turbulent fluctuations, *J. Cryst. Growth*, 2003, **250**, 203.

76. R. Menzel, H.-J. Rost, F. M. Kiessling and L. Sylla, Float-zone growth of silicon crystals using large-area seeding, *J. Cryst. Growth*, 2019, **515**, 32.
77. J. Luo, A. Alateeqi, L. Liu and T. Sinno, Carbon solubility in liquid silicon: A computational analysis across empirical potentials, *J. Chem. Phys.*, 2019, **150**, 144503.
78. Z. Liu and T. Carlberg, Reactions between liquid silicon and vitreous silica, *J. Mater. Res.*, 1992, 7, 353.
79. T. Narushima, K. Matsuzawa, Y. Mukai and Y. Iguchi, Oxygen solubility in liquid silicon, *Materials Trans. JIM*, 1994, **35**(8), 522.
80. X. Huang, T. Nakazawa, K. Terashima and K. Hoshikawa, Silicon Crystal Growth under Equilibrium Condition of $SiO_2$-Si–SiO System: Equilibrium Oxygen Segregation Coefficient, *Jpn. J. Appl. Phys.*, 1988, **37**(12b), L1504.
81. J. G. Li and H. Hausner, Reactive Wetting in the Liquid-Silicon/Solid-Carbon System, *J. Am. Ceram. Soc.*, 1996, **79**, 873.
82. J. G. Li, Wetting of ceramic materials by liquid silicon, aluminium and metallic melts containing titanium and other reactive elements: A review, *Ceram. Int.*, 1994, **20**, 391.
83. A. Ciftja, M. Tangstad and T. A. Engh, *Wettability of silicon with refractory materials: A review*, Norwegian University of Science and Technology, Faculty of Natural Science and Technology, Department of Materials Science and Engineering Trondheim, February 2008.
84. G. Müller, A. Mühe, R. Backofen, E. Tomzig and W. Von Ammon, Study of oxygen transport in Czochralski growth of silicon, *Microelectron. Eng.*, 1999, **1**, 135.
85. Y. Endo, Y. Yatsurugi, V. Terai and V. Nozaki, Equilibrium of carbon and oxygen in silicon with carbon monoxide in ambient atmosphere, *J. Electrochem. Soc.*, 1979, **126**(8), 1422.
86. S. M. Schnurre, J. Gröber and R. Schmid-Fetzer, Thermodynamics, and phase stability in the Si-O system, *J. Non-Cryst. Solids*, 2004, **336**, 1.
87. H. Okamoto, O-Si (Oxygen-Silicon), *J. Phase Equilib. Diffus.*, 2007, **28**, 309.
88. A. Borghesi, B. Pivac, A. Sassella and A. Stella, Oxygen precipitation in silicon, *J. Appl. Phys.*, 1995, **77**(9), 4169.
89. R. Li, M.-W. Li, N. Imaishi, Y. Akiyama and T. Tsukada, Oxygen-transport phenomena in a small silicon Czochralski furnace, *J. Cryst. Growth*, 2004, **267**, 466.
90. L. Liu, S. Nakano and K. Kakimoto, Three-dimensional global modeling of a unidirectional solidification furnace with square crucibles, *J. Cryst. Growth*, 2007, **303**, 165.
91. A. D. Smirnov and V. V. Kalaev, Development of oxygen transport model in Czochralski growth of silicon crystals, *J. Cryst. Growth*, 2008, **310**, 2970.
92. B. Gao, S. Nakano and K. Kakimoto, Global simulation of coupled Carbon and oxygen transport in a unidirectional solidification furnace for solar cells, *J. Electrochem. Soc.*, 2010, **157**, H153.
93. Y.-Y. Teng, J.-C. Chen, C.-W. Lu, H.-I. Chen, C. Hsu and C.-Y. Chen, Effects of the furnace pressure on oxygen and silicon oxide distributions during the growth of multicrystalline silicon ingots by the directional solidification process, *J. Cryst. Growth*, 2010, **318**, 224.
94. L. O. Wilson, The effect of fluctuating growth rates on segregation in crystals grown from the melt: I. No backmelting, *J. Cryst. Growth*, 1980, **48**, 435.
95. L. O. Wilson, The effect of fluctuating growth rates on segregation in crystals grown from the melt: II. Backmelting, *J. Cryst. Growth*, 1980, **48**, 451.
96. K. Kakimoto and H. Ozoe, Oxygen distribution at a solid-liquid interface of silicon under transverse magnetic field, *J. Cryst. Growth*, 2000, **212**, 429.
97. M. G. Williams, J. S. Walker and W. E. Langlois, Melt motion in a Czochralski puller with a weak transverse magnetic field, *J. Cryst. Growth*, 1990, **100**, 233.

98. J. S. Walker and M. G. Williams, Centrifugal pumping during Czochralski silicon growth with a strong transverse magnetic field, *J. Cryst. Growth*, 1994, **137**, 32.

99. R. Falster, V. V. Voronkov and F. Quast, On the properties of intrinsic point defects in silicon: a perspective from crystal growth and wafer processing, *Phys. Status Solidi B*, 2000, **222**, 219.

100. M. S. Kulkarni, Lateral incorporation of vacancies in Czochralski silicon crystals, *J. Crystal. Growth.*, 2008, **310**, 3183.

101. V. V. Voronkov, The mechanism of swirl defects formation in silicon, *J. Cryst. Growth*, 1982, **59**, 625.

102. M. S. Kulkarni, J. C. Holzer and L. W. Ferry, The agglomeration of self-interstitials in growing Czochralski silicon crystals, *J. Cryst. Growth*, 2005, **284**, 353.

103. M. S. Kulkarni, Defect dynamics in the presence of oxygen in growing Czochralski silicon crystals, *J. Cryst. Growth*, 2007, **303**, 438.

104. D. Gilles, E. R. Weber and S. K. Hahn, Mechanism of internal gettering of interstitial impurities in Czochralski-grown silicon, *Phys. Rev. Lett.*, 1990, **64**, 196.

105. S. A. McHugo, E. R. Weber, M. Mizuno and F. G. Kirscht, A study of gettering efficiency and stability in Czochralski silicon, *Appl. Phys. Lett.*, 1995, **66**, 2840.

106. Y.-H. Kim, K.-S. Lee, H.-Y. Chung, D.-H. Hwang, H.-S. Kim, H.-Y. Cho and B.-Y. Lee, Internal Gettering of Fe, Ni and Cu in Silicon Wafers, *J. Korean Phys. Soc.*, 2001, **39**, S348–S351.

107. A. Haarahiltunen, Heterogeneous precipitation and internal gettering efficiency of iron in silicon, Doctoral Dissertation, TKK Dissertations 64, Espoo 2007, University of Helsinki, 2007.

108. A. Y. Liu, D. Walter, S. P. Phang and D. Macdonald, Investigating Internal Gettering of Iron at Grain Boundaries in Multicrystalline Silicon via Photoluminescence Imaging, *IEEE J. Photovoltaics*, 2012, **2**, 479.

109. V. V. Voronkov and R. Falster, Vacancy and self-interstitial concentration incorporated into growing silicon crystals, *J. Appl. Phys.*, 1999, **86**, 5975.

110. L. Mulè-Stagno, A technique for delineating defects in silicon, *Proc. Electrochem. Soc.*, 2002, **2**, 297.

111. A. Ourmazd and W. Schröter, Oxygen-related thermal donors in silicon: A new structural and kinetic model, *J. Appl. Phys.*, 1984, **56**, 1670.

112. R. Singh and P. B. Nagabalasubramanian, A review on thermal donors in CZ silicon and their impact on electronic industry, *Int. J. Appl. Eng. Res.*, 2009, **14**(2), 43.

113. T. F. Kuech, III-V compound semiconductors: Growth and structures, *Prog. Cryst. Growth Charact. Mater.*, 2016, **62**(2), 352.

114. P. Rudolph, Non-stoichiometry related defects at the melt growth of semiconductor compound crystals – a review, *Cryst. Res. Technol.*, 2003, **38**(7–8), 542.

115. T. Kimoto, Bulk and epitaxial growth of silicon carbide, *Prog. Cryst. Growth Charact. Mater.*, 2016, **62**, 329.

116. G. Dhanaraj, X. R. Huang, M. Dudley, V. Prasad, R.-H. Ma, Silicon Carbide Crystals—Part I: Growth and Characterization, in *Crystal Growth Technology*, ed. K. Byrappa, W. Michaeli, H. Waarlimont and E. Weber, William Andrew Inc., 2003.

117. V. G. Sevast'yanov, P. Ya Nosatenko, V. V. Gorski, Yu. S. Ezhoc, D. V. Sevast'yanov, E. P. Simonenko and N. T. Kuznetsov, Experimental and Theoretical Determination of the Saturation Vapor Pressure of Silicon in a Wide Range of Temperatures, *Russ. J. Inorg. Chem.*, 2010, **55**, 2073.

118. B. M. Epelbaum, D. Hofmann, M. Muller and A. Winnacker, Top-seeded Solution Growth of Bulk SiC: Search for the Fast Growth Regimes, *Mater. Sci. Forum*, 2000, **338–342**, 107.

119. T. Tomooka, Y. Shoji and T. Matsui, High temperature vapor pressure of Si, *J. Mass Spectrom. Soc. Jpn.*, 1999, **47**(1), 49.
120. V. Yakimova, M. Syväjärvi, C. Lockowandt, M. K. Linnarsson, H. H. Radamson and E. Janzén, Silicon carbide grown by liquid phase epitaxy in microgravity, *J. Mater. Res.*, 1998, **13**(7), 1812.
121. R. Yakimova and E. Janzen, Current Status and Advances in the Growth of SiC, *Diamond Relat. Mater.*, 2000, **9**, 432.
122. S. Nakamura, M. Senoh, S. Nagahama, N. Iwasa, T. Yamada, T. Matsushita, H. K. H. Kiyoku and Y. Sugimoto, Superbright Green InGaN single-quantum-well structure light emitting diodes, *Jpn. J. Appl. Phys.*, 1995, **34**(10b), L1332.
123. S. Nakamura, M. Senoh, S. Nagahama, N. Iwasa, T. Yamada, T. Matsushita, H. K. H. Kiyoku and Y. Sugimoto, InGaN-Based Multi-Quantum-Well-Structure Laser Diodes, *Jpn. J. Appl. Phys.*, 1996, **35**, L74.
124. S. Nakamura, M. Senoh, S. Nagahama, N. Iwasa, T. Yamada, T. Matsushita and T. Mukai, InGaN/GaN/AlGaN-based Leds and Laser diodes, *J. Semicond. Res.*, 1999, **4**(S1), 1.
125. C. D. Thurmond and R. A. Logan, The Equilibrium Pressure of $N_2$ over GaN, *J. Electrochem. Soc.*, 1972, **119**, 622.
126. V. Porowski and I. Grzegory, Thermodynamic properties of III-V nitrides and crystal growth of GaN at high $N_2$ pressure, *J. Cryst. Growth*, 1997, **178**, 174.
127. S. Porowski, Growth and properties of single crystalline GaN substrates and homoepitaxial layers, *Mater. Sci. Eng., B*, 1997, **44**, 407.
128. S. Porowski, Bulk and homoepitaxial GaN-growth and characterisation, *J. Cryst. Growth*, 1998, **189–190**, 153.
129. B. Lucznik, M. Bockowski, M. Wroblewski, I. Grzegory, S. Krukowski, G. Nowak, M. Leszczynski, G. Teisseyre, K. Pakula, J. M. Baranowski, T. Suski and S. Porowski, GaN Single Crystals Grown by High Pressure Solution Method, *Rev. High Pressure Sci. Technol.*, 1998, **7**, 760.
130. I. Grzegory, S. Krukowski, M. Lesczynski, P. Perlin, T. Suski and S. Porowski, in *High pressure crystallization of GaN in Nitride Semiconductors*, ed. P. Ruterana, M. Albrecht and J. Neugebauer, Wiley-VCH, 2003.
131. J. Karpinski, J. Jun and S. Porowski, Equilibrium pressure of $N_2$ over Gan and High pressure solution growth of GaN, *J. Cryst. Growth*, 1984, **66**, 1.
132. J. Karpinski and S. Porowski, High pressure thermodynamics of GaN, *J. Cryst. Growth*, 1984, **66**, 11.
133. S. Porowski, Growth and properties of single crystalline GaN substrates and homoepitaxial layers, *Mater. Sci. Eng., B*, 1997, **44**, 407.
134. F. Kawamura, M. Morishita, T. Iwahashi, M. Yoshimura, Y. Mori and T. Sasaki, Growth of Transparent, Large Size GaN Single Crystal with Low Dislocations Using Ca-Na Flux Systems, *Jpn. J. Appl. Phys.*, 2003, **42**(7A), L729.
135. F. Kawamura, M. Morishita, M. Tanpo, M. Imade, M. Yoshimura, Y. Kitaoka, Y. Mori and T. Sasaki, Effect of carbon additive on increases in the growth rate of 2 in GaN single crystals in the Na flux method, *J. Cryst. Growth*, 2008, **310**, 3946.
136. T. Kawamura, H. Imabayashi, Y. Yamada, M. Maruyama, M. Imade, M. Yoshimura, Y. Mori and Y. Morikawa, Structural Analysis of Carbon-Added Na–Ga Melts in Na Flux GaN Growth by First-Principles Calculation, *Jpn. J. Appl. Phys.*, 2013, **52**(85), 08JA04.
137. T. Kawamura, H. Imabayashi, M. Maruyama, M. Imade, M. Yoshimura, Y. Mori and Y. Morikawa, Mechanism for enhanced single-crystal GaN growth in the C-assisted Na-flux method, *Appl. Phys. Express*, 2015, **9**, 015601.
138. G. Glockler, Estimated Bond Energies in Carbon, Nitrogen, Oxygen, and Hydrogen Compounds, *J. Chem. Phys.*, 1951, **19**, 124.
139. X. Wu, H. Hao, Z. Li, S. Fan and Z. Xu, GaN crystals growth in the Na-Li-Ca flux by liquid phase epitaxy (LPE) technique, *J. Cryst. Growth*, 2019, **521**, 30.

140. J. B. Mullin, W. R. Macewan, C. H. Holliday and A. E. V. Webb, Pressure balancing: A technique for suppressing dissociation during the melt-growth of compounds, *J. Cryst. Growth*, 1972, **13–14**, 629.

141. R. Rudolph and M. Jurisch, Bulk growth of GaAs. An overview, *J. Cryst. Growth*, 1999, **188–189**, 325.

142. P. H. Yannakopoulos, G. E. Zardas, G. J. Papaioannou, Ch. I. Symeonides, M. Vesely and P. C. Euthymiou, Behavior of semiinsulating GaAs energy levels, *Rev. Adv. Mater. Sci.*, 2009, **22**, 52.

143. D. T. J. Hurle, A thermodynamic analysis of native point defect and dopant solubilities in zinc-blende III–V semiconductors, *J. Appl. Phys.*, 2010, **107**, 121301.

144. T. Morishita, Crystal growth and wafer processing of 6″ GaAs substrates for lasers, Open Sumitomo document, 2020.

145. P. S. Dutta, Bulk Growth of Crystals of III–V Compound Semiconductors, *Compr. Semicond. Sci. Technol.*, 2011, **3**, 36.

146. L. Fischer, U. Lambert, G. Nagel and H. Rüfer, Influence of pyrolytic boron nitride crucibles on GaAs crystal growth process and crystal properties, *J. Cryst. Growth*, 1995, **153**(3–4), 90.

147. M. Kaminska, EL2 Defect in GaAs, *Phys. Scr.*, 1987, **T19**, 551.

148. D. Vázquez-Cortés, E. Cruz-Hernández, V. H. Méndez-García, S. Shimomura and M. López-López, Optical and electrical properties of Si-doped GaAs films grown on (631)-oriented substrates, *J. Vac. Sci. Technol., B: Nanotechnol. Microelectron.: Mater., Process., Meas., Phenom.*, 2012, **30**, 02B125.

149. R. P. Bult, T. E. Schroeder and J. G. Needham, Method of growing gallium arsenide using boron oxide encapsulant, *US Pat.*, 4585511, 1986.

150. O. Pätzold, G. Gärtner and G. Irmer, Boron Site Distribution in Doped GaAs, *Phys. Status Solidi B*, 2002, **232**(2), 314.

151. V. Lagowski, D. G. Lin, T. Aoyama and H. C. Gatos, Identification of oxygen related midgap level in GaAs, *Appl. Phys. Lett.*, 1984, **44**, 336.

152. S. S. Hulluvarad, M. Naddaf and S. V. Bhoraskar, Detection of oxygen-related defects in GaAs by exo-electron emission spectroscopy, *Nucl. Instrum. Methods Phys. Res., Sect. B*, 2001, **183**, 432.

153. R. C. Newman, F. Thompson, M. Hyliands and R. F. Peart, Boron and carbon impurities in gallium arsenide, *Solid State Commun.*, 1972, **10**(6), 505.

154. T. M. Schmidt, P. P. M. Venezuela, M. J. Caldas and A. Fazzio, Carbon doping of GaAs: Compensation effects, *Appl. Phys. Lett.*, 1995, **66**, 2715.

155. L. Pavesi and P. Giannozzi, Atomic and molecular hydrogen in gallium arsenide: A theoretical study, *Phys. Rev. B: Condens. Matter Mater. Phys.*, 1992, **46**, 4621.

156. V. Omeljanovsky, A. V. Pakhomov and A. Y. Polyakov, Hydrogen passivation of defects and impurities in GaAs and InP, *J. Electron. Mater.*, 1989, **18**, 659.

157. J. Korb, T. Flade, M. Jurisch, A. Köhler, Th. Reinhold and B. Weinert, Carbon, oxygen, boron, hydrogen and nitrogen in the LEC growth of SI GaAs: a thermochemical approach, *J. Cryst. Growth*, 1999, **198–199**, 343.

158. V. Van Artsdalen and K. P. Anderson, The Molar Heats of Solution of Boric Oxide and Boric Acid, *J. Am. Chem. Soc.*, 1951, **73**(2), 579.

159. U. Lambert and U. Wiese, Chemical interactions in GaAs-LEC crystal growth, *Adv. Mater.*, 1991, **3**(9), 429.

160. Fr. J. Bruni, Crystal growth of sapphire for substrates for high-brightness, light emitting diodes, *Cryst. Res. Technol.*, 2015, **50**(1), 133.

161. C. P. Chattak and F. Schmid, Growth of the world's largest sapphire crystals, *J. Cryst. Growth*, 2001, **225**, 572.

162. E. R. Dobrovinskaya, L. A. Litvinov and V. Pischik, *Sapphire: Material, Manufacturing, Applications*, Springer, Berlin, 2009.

163. M. Wu, L. Liu, Y. Yang, W. Zhao and W. Ma, Effect of Crucible Location on Heat Transfer in Sapphire Crystal Growth by Heat Exchanger Method, *Heat Transfer Eng.*, 2016, **37**(3–4), 332.

164. V. A. Izhevskyi, L. A. Genova, J. C. Bressiani and A. H. A. Bressiani, Review article: Silicon Carbide. Structure, Properties and Processing, *Ceramica*, 2000, **46**, 297.

165. V. Bechstaedt, V. Furthmüller, V. Grossner and C. Raffy, Zero-and two-dimensional native defects, in *Silicon Carbide: Recent Major Advances*, ed. M. J. Choyke, H. Matsumami and H. Pensl, Springer, 2003.

166. J. A. Lely, Darstellung von Eincrystallen von Silicium carbid und Beherschung von Art und Menge der eingebaute Verunreinigungen, *Ber. Dtsch. Keram. Ges.*, 1955, **32**, 229.

167. Y. M. Tairov and V. F. Tsvetkov, Growth of Bulk Silicon Carbide Single Crystals, in *Growth of Crystals*, ed. E. I. Givargizov and S. A. Grinberg, Springer, Boston, MA, 1993, vol. 19.

168. M. Pons, E. Blanquet, J. M. Dedulle, I. Garcon, R. Madar and C. Bernard, Thermodynamic Heat Transfer and Mass Transport Modeling of the Sublimation Growth of Silicon Carbide Crystals, *J. Electrochem. Soc.*, 1996, **143**(11), 3727.

169. R. Yakimova, M. Syvjarvi, M. Tuominen, T. Yakimov, R. Råback, A. Vehanen and E. Janzen, Seeded Sublimation Growth of 6H and 4H-SIC, *Cryst. Mater. Sci. Eng. B*, 1999, **61–62**, 54.

170. R. Yakimova and E. Janzen, Current Status and Advances in the Growth of SiC, *Diamond Relat. Mater.*, 2000, **9**, 432.

171. T. Kimoto, Bulk and epitaxial growth of silicon carbide, *Prog. Cryst. Growth Charact. Mater.*, 2016, **62**, 329.

172. K. Ariyawong, C. Chatillon, E. Blanquet, J.-M. Dedulle and D. Chaussende, A first step toward bridging silicon carbide crystal properties and physical chemistry of crystal growth, *Cryst. Eng. Commun.*, 2016, **18**, 2119.

173. S. Yu. Karpov, Yu. N. Makarov, M. S. Ramm and R. A. Talalaev, Control of SiC growth and graphitization in sublimation sandwich system, *Mater. Sci. Eng., B*, 1997, **46**, 340.

174. E. N. Mokhov, M. G. Ramm, A. D. Roenkov and Yu. A. Vodakov, Growth of silicon carbide bulk crystals by the sublimation sandwich method, *Mater. Sci. Eng., B*, 1997, **46**, 317.

175. R. Vasiliauskas, Sublimation Growth and Performance of Cubic Silicon Carbide, Dissertation 1435, Semiconductor Materials Division, Department of Physics, Chemistry and Biology (IFM), Linköping University, Linköping, Sweden, 2012.

176. R. Yakimova, M. Syvajarvi, T. Yakimov, H. Jacobsson, R. Raback, A. Vehanen and E. Janzen, Polytype stability in seeded sublimation growth of 4H-SiC boules, *J. Cryst. Growth*, 2000, **217**, 255–262.

177. E. N. Mokhov, M. G. Ramm, A. D. Roenkov and Yu. A. Vodakov, SiC growth in tantalum containers by sublimation sandwich method, *J. Cryst. Growth*, 1997, **181**(3), 254–258.

178. P. G. Neudeck and J. A. Powell, Performance Limiting Micropipe Defects in Silicon Carbide Wafers, *IEEE Electron Device Lett.*, 1994, **15**(2), 63.

179. E. N. Mokhov, Doping of SiC crystals during sublimation growth and diffusion, in *Crystal Growth*, IntechOpen, 2018, DOI: 10.5772/intechopen.82346.

180. V. P. Rastagaev, D. D. Avrov, S. Areshanov and A. D. Lebedev, Features of SiC single-crystals grown in vacuum using the LETI method, *Mater. Sci. Eng., B*, 1999, **61–62**, 77.

181. G. Deffrennes, B. Gardiola, M. Alla, D. Chaussende, A. Pisch, J. Andieux and R. Schmid-Fetzer, Critical assessment and thermodynamic modelling of the Al-C silicon, *Calphad*, 2019, **66**, 101648.

182. T. L. Straubinger, M. Bickermann, R. Weingartner, P. J. Wellmann and A. Winnacker, Aluminum p-type doping of silicon carbide crystals using a modified physical vapor transport growth method, *J. Cryst. Growth*, 2002, **240**, 117.
183. A. R. G. Costa, Lattice location of impurities in Silicon Carbide, CERN-Thesis, 2018, 2018-07205/06/2018.
184. G. Pensl and W. J. Choyke, Electrical and optical characterization of SiC, *Phys. B*, 1993, **185**, 264.
185. S. R. Smith, A. O. Evwaraye, W. C. Mitchel and M. A. Capano, Shallow acceptor levels in 4H- and 6H-SiC, *J. Electron. Mater.*, 1999, **28**, 190.
186. C. Raynaud, C. Richier, P. N. Brounkov, F. Ducroquet, G. Guillot, L. M. Porter, R. F. Davis and T. Billon, Determination of donor and acceptor level energies by admittance spectroscopy in 6H SiC, *Mater. Sci. Eng., B*, 1995, **29**, 122.
187. M. Masanori, Y. Higashiguchi and Y. Hayafuji, *Ab Initio* study of substitutional impurity atoms in 4H-SiC, *J. Appl. Phys.*, 2008, **104**, 123702.
188. J. S. Hartman, B. Berno, P. Hazendonk, C. W. Kirby, E. Ye, J. Zwanziger and A. D. Bain, NMR Studies of Nitrogen Doping in the 4H Polytype of Silicon Carbide: Site Assignments and Spin–Lattice Relaxation, *J. Phys. Chem C*, 2009, **113**, 15024.
189. P. Friedrichs, T. T. Kimoto, L. Ley and G. Pensl, *Silicon Carbide, Vol. 1: Growth, Defects and Novel Applications,* Wiley-VCH, Weinheim, 2009.
190. J. Chen, S. C. Lien, Y. C. Shin, Z. C. Feng, C. H. Kuan, J. H. Zhao and W. J. Lu, Occurrence of Polytype Transformation during Nitrogen Doping of SiC Bulk Wafer, *Mater. Sci. Forum*, 2009, **600–603**, 39.
191. D. Peters, Ammonothermal synthesis of aluminium nitride, *J. Cryst. Growth*, 1990, **104**, 411.
192. D. Ehrentraut, Y. Kagamitani, C. Yokoyama and T. Fukuda, Physico-chemical features of the acid ammonothermal growth of GaN, *J. Cryst. Growth*, 2008, **310**, 891.
193. Q.-S. Chen, J.-Y. Yan, Y.-N. Jiang and W. Li, Modelling on ammonothermal growth of GaN semiconductor crystals, *Prog. Cryst. Growth Charact. Mater.*, 2012, **58**, 61.
194. T. Hashimoto, F. Wu, J. S. Speck and S. Nakamura, A GaN bulk crystal with improved structural quality grown by the ammonothermal method, *Nat. Mater.*, 2007, **6**, 568.
195. T. Hashimoto, F. Wu, J. S. Speck and S. Nakamura, Ammonothermal growth of bulk GaN, *J. Cryst. Growth*, 2008, **310**, 3907.
196. R. Dwilinski, R. Doradzinski, J. Garczynski, L. P. Sierzputowski, A. Puchalski, Y. Kanbara, K. Yagi, H. Minakuchi and H. Hayashi, Excellent crystallinity of truly bulk ammonothermal GaN, *J. Cryst. Growth*, 2008, **310**, 3911.
197. T. Hashimoto, E. Letts and S. Hoff, Current Status and Future Prospects of Ammonothermal Bulk GaN Growth, *Sens. Mater.*, 2013, **25**(3), 155.
198. K. Grabianska, R. Kucharski, A. Puchalski, T. Sochacki and M. Bockowski, Recent progress in basic ammonothermal GaN crystal growth, *J. Cryst. Growth*, 2020, **547**, 125804.
199. R. Kucharski, T. Sochacki, B. Lucznik and M. Bockowski, Growth of bulk GaN crystals, *J. Appl. Phys.*, 2020, **128**, 050902.
200. A. Yoshikawa, E. Ohshima, T. Fukuda, V. Tsuji and K. Oshima, Crystal growth of GaN by ammonothermal method, *J. Cryst. Growth*, 2004, **260**, 67.
201. J. Hu, H. Y. Wei, S. Y. Yang, C. M. Li, H. J. Li, X. L. Liu, L. S. Wang and Z. G. Wang, Hydride vapor phase epitaxy for gallium nitride substrate, *J. Semicond.*, 2019, **40**(10), 101801.
202. V. Voronenkov, N. Bochkareva, A. Zubrilov, Y. Lelikov, R. Gorbunov, P. Latyshev and Y. Shreter, Hydride Vapor-Phase Epitaxy Reactor for Bulk GaN Growth, *Phys. Status Solidi A*, 2019, **217**(3), 1900629.

203. R. J. Molnar, W. Gotz, L. T. Romano and N. M. Johnson, Growth of gallium nitride by hydride vapor-phase epitaxy, *J. Cryst. Growth*, 1997, **178**, 147.

204. T. Shibata, H. Sone, K. Yahashi, M. Yamaguchi, K. Hiramatsu, N. Sawaki and N. Itoh, Hydride vapor-phase epitaxy growth of high-quality GaN bulk single crystal by epitaxial lateral overgrowth, *J. Cryst. Growth*, 1998, **189–190**, 67.

205. M. K. Kelly, R. P. Vaudo, V. M. Phanse, L. Gorgens, O. Ambacher and M. Stutzmann, Large Free-Standing GaN Substrates by Hydride Vapor Phase Epitaxy and Laser-Induced Liftoff, *Jpn. J. Appl. Phys.*, 1999, **38**, L217.

206. Y. Kumagai, H. Murakami, H. Seki and A. Koukitu, Thick and high-quality GaN growth on GaAs substrates for preparation of freestanding GaN, *J. Cryst. Growth*, 2002, **246**, 215.

207. Y. Oshima, T. Eri, M. Shibata, H. Sunakawa, K. Kobayashi, T. Ichihashi and A. Usui, Preparation of Freestanding GaN Wafers by Hydride Vapor Phase Epitaxy with Void-Assisted Separation, *Jpn. J. Appl. Phys.*, 2003, **42**(1A), L1.

# 4 Chemistry of Semiconductor Impurity Processing

## 4.1 Introduction

We already know that deep level impurities limit or totally degrade, depending on their amount, the efficiency of a semiconductor device, due to carrier recombination effects. There is, therefore, the need not only to use ultra-pure semiconductor materials for microelectronic and photonic applications, but to reduce the almost unavoidable impact of impurity pick-up during high temperature device manufacturing processes, and to develop technological processes addressed either at minimizing the impurity concentration, or stabilising electrical deactivation, carried out during the device manufacturing process. The electrical deactivation, obviously, must remain stable during the routine operation of the device, which could imply electron, ion, or radiation impact in the case of radiation detectors.

A prior example of impurity engineering, already fully discussed in Section 2.5, is that of intrinsic gettering of TM impurities at heterogeneous sinks, consisting of oxide micro-precipitates.

Here we intend to rely on two different kinds of impurity engineering processes, presenting an essential physico-chemical character, as is the case of the hydrogenation and electrical deactivation of impurities (and of surface and interface defects) in single crystal and polycrystalline semiconductors, which will be discussed in Section 4.3, and of the extrinsic gettering (EG) processes at P-rich surface layers or at Al sinks, which will be discussed in Section 4.4 of this chapter.

Thermal annealing is another type of defect engineering, used to induce the relaxation of the mechanical stress generated in a semi-conductor sample by previous technological processes, as well as the release of impurities[†] segregated at extended defects, which will be discussed in Section 4.2.

Thermal annealing is also applied to recover the irradiation damage of ion-implanted layers or of ion implanted np junctions, to enhance the dopant diffusion from an implanted- or deposited-surface layer, and for alloying heterogeneous multi-component matrixes fabricated by sequential sublimation processes, as is the case of copper indium gallium selenide-based solar cells.[1]

The aim of this chapter is to discuss the physico-chemical background of these defect/impurity engineering processes in semi-conductors, limiting, however, most of the attention to silicon.

## 4.2 Thermal Annealing Processes

Thermal annealing processes, applied for millennia in iron metallurgy for metal hardening purposes, are currently applied to several semiconductor processes, either to induce the formation of a doped layer at the surface of a semiconductor and the dissolution of an unwanted precipitate in a semiconductor matrix, or to activate a phase transition, just to mention some important applications.

A thermal annealing process is substantially addressed at supplying the thermal work needed to drive or enhance the rate of a temperature-dependent transformation, in the form of an amount $Q = \int_0^t q \, dt$ of heat.

Heat could be supplied by a conventional or an advanced electric furnace, or by irradiation of the sample using a flash lamp, a laser, or a microwave source, this last potentially able to induce a localized heating of specific regions of the sample presenting selective microwave absorbances.[2,3]

Microwave (MW) heating of semiconductors depends on dielectric losses associated with dipole and electronic polarization,[3] which cause the onset of conditions of partial coupling of the sample with the applied electromagnetic field.

---

[†] Also dopants.

The sample heating is caused by the damped vibration of a polarized dipole, which induces a change of its dielectric constant $\varepsilon^*$

$$\varepsilon^* = \varepsilon_0(\varepsilon' - i\varepsilon'') \tag{4.1}$$

(where $\varepsilon_0$ is the dielectric constant of vacuum, and $\varepsilon'$ is the real part and $\varepsilon''$ the imaginary part of the dielectric constant) and a dielectric loss, enabling the conversion of a fraction of the electromagnetic energy to thermal energy.

The fraction $dP_{th}$ of MW power converted by a dielectric loss to thermal energy in a volume fraction $dV$ of the sample is given by the following equation

$$\frac{dP_d}{dV} = \omega \varepsilon_0 \varepsilon'' [E]^2 \tag{4.2}$$

where $\omega$ is the microwave excitation frequency and $|E|$ is the electromagnetic field energy.[4]

Only when the excitation frequencies of charged particles and dipoles in a semiconductor lies in the microwave frequency range does MW heating occur.

As an example, MW heating was shown to operate well also at temperatures lower than 300 °C on CZ silicon, not on FZ-Si as shown by Lojec,[4] who was, in fact, unable to achieve any interaction of the FZ material with the microwave field, possibly due[‡] to the negligible concentration, in FZ silicon, of dopant–oxygen dipolar species, as B–O and As–O, which could interact with the MW field.

Furnace annealing has been, instead, traditionally used in semiconductor technology for all the processes requiring thermal activation, *i.e.*

- For reaction-limited processes, as is the case of the decomposition of thermal donors (TDs) nucleated in large diameter CZ silicon ingots[§] during the post-growth, whose rate $k$ $(s^{-1})$ is given by the following equation

$$k \ (s^{-1}) = \omega \exp - \frac{E_R}{kT} \tag{4.3}$$

where $E_R$ is the activation energy of the decomposition process.

---

[‡] This is the opinion of S. Pizzini, not of B. Lojec.
[§] The large thermal capacity of these ingots favours a long stage at temperatures around 450–500 °C during the post-growth cooling. In this temperature range the formation rate of TDs is particularly high.

- For thermally activated diffusion-limited processes, as is the case of the thermal diffusion of dopants from a surface layer,[¶] where the diffusion rate depends on the diffusion coefficient

$$D = D_0 \exp - \frac{E_D}{kT} \qquad (4.4)$$

of the species involved, with $E_D$ as the activation energy of the diffusion process.

Conventional furnace annealing has also been used for more complex processes, such as the thermal oxidation of silicon for MOS fabrication technologies, and is both reaction- and diffusion-limited.[5]

With the introduction of ultra-large-scale integration (ULSI) in modern semiconductor silicon technology, and of a cluster approach including all the steps of the chip production process, a single wafer is vacuum-transferred from the chamber addressed at its dry cleaning, to those dedicated to wafer-oxidation and doping, including ion implantation and epitaxial deposition with molecular beam epitaxy. In this scheme, thermal processes (wafer oxidation, dopant diffusion and p–n junction formation) are not carried out using conventional furnace heating, but rapid thermal processes (RTP)[6] that allow ultra-clean operation conditions,[‖] faster production processes and lower thermal budgets in comparison with furnace annealing processes.

The heating source for RT processes is a flash lamp or a laser source, whose spectral range is selected as a function of the temperature dependence of the absorbance and emissivity of the silicon wafer,[**] displayed in Figures 4.1 and 4.2.[7,8]

One can see that the absorption of silicon in the IR is negligible below 600 K, excellent in the visible at every temperature, while the emission[8] has a minimum at 1100 K, followed by a modest increase at higher temperatures.

These values favour RTP conditions at temperatures around 800 °C, although the resistivity and the surface quality (roughness) influence the optical behaviour of the silicon samples and the optimal RTP conditions.

The main advantages of RTP *vs.* furnace processes are summarized in Table 4.1, which shows that RTP is favoured by short cycle times, and uniformity and repeatability of the process conditions.[6]

---

[¶] Where they are deposited at the beginning of the process.
[‖] Including the transfer of matter from one step to the other.
[**] Also the optical properties of the chamber walls contribute to the overall process yield.

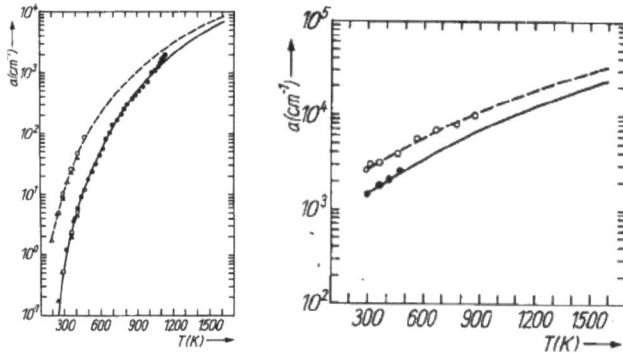

**Figure 4.1**  Temperature dependence of the optical absorption of silicon in the near IR (left figure) at $\lambda = 1152$ nm (solid curve) and 1064 nm (broken curve) and in the visible at $\lambda = 750$ nm (solid curve) and 694 nm (broken curve). Reproduced from ref. 7 with permission from John Wiley & Sons, Copyright © 1984 Wiley-VCH Verlag GmbH & Co. KGaA.

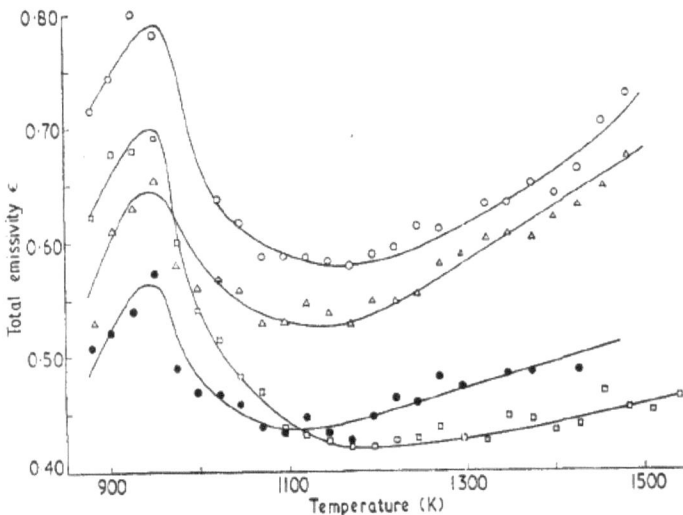

**Figure 4.2**  Temperature dependence of the total emissivity of four silicon samples: ○ 600 Ω cm, n-type; △ 0.04 Ω cm, n-type; □ 0.1 Ω cm, n-type; ● 14 Ω cm, p-type. Reproduced from ref. 8 with permission from IOP Publishing, Copyright 1971.

In the following section we will illustrate the advantages of RTP for semiconductor doping processes.

## 4.2.1  Doping Processes

Junction formation in IC and VLSI processes is accomplished by changing the local electrical properties of silicon with the introduction of dopants.

**Table 4.1** Comparison of the main properties of furnace and RTP processes. Data from ref. 6.

| Furnace | RTP |
|---|---|
| Batch of several wafers | Single wafer |
| Hot walls of the operational chamber | Cold walls |
| High cycle time | Short cycle time |
| Small $dT/dt$ | Large $dT/dt$ |

| Problems | Advantages |
|---|---|
| Low thermal budget/wafer | Excellent uniformity and repeatability |
| Presence of particles | The stress in the processed samples can be removed by a final thermal annealing |
| Environmental damage | RTP is operated in vacuum |

**Table 4.2** Diffusion properties of common dopants of silicon.

| | $D_0$ $(cm^2 s^{-1})$ | $E_A$ (eV) |
|---|---|---|
| B | 1 | 3.50 |
| P | 4.70 | 3.68 |
| As | 9.17 | 3.99 |
| Sb | 4.58 | 3.88 |

Given that dopants are substitutional impurities and slow diffusers (see Table 4.2 for the most common dopants of silicon), the doping processes in the early times of IC technologies was carried out with a thermal diffusion step in a conventional furnace, needed to drive-in the dopants deposited at the surface of a silicon wafer and create a junction of desired depth.

The full process consisted of a heating ramp to bring the sample to the desired temperature, followed by a constant temperature segment during the dopant diffusion and, finally, to a cooling stage.

When the doping process is carried out in a conventional, electrically heated furnace, the dopant concentration profile in the sample is influenced by the thermal inertia of the furnace when the sample is cooled inside the furnace, or by the thermal inertia of the sample itself when the sample is extracted from the furnace and quickly quenched to room temperature.

Instead, the dopant concentration profile is independent on the thermal inertia of the furnace if the process is carried out using the RTP, securing excellent reproducibility conditions otherwise difficult to systematically obtain.

Today, in the modern state of silicon IC technology, the junctions are so shallow that dopants are introduced at the desired depth by ion implantation followed by RTP to recover the implantation damage, this last is due to implantation-induced defects (point and extended), which would cause severe device degradation unless suitably deactivated.

As expected, the success of the RTP of ion-implanted materials strongly depends on the nature of the dopant, on the technical features of the ion implantation process and on the depth of the implanted layer.

It has been, as an example, observed that a conventional thermal annealing of B-implanted silicon is unable to remove the implantation damage, and, additionally, induces precipitation and redistribution of the dopant. Instead, pulsed laser annealing can recover the damage by irradiation just inside the implanted region, with negligible, if any, degradation of the minority carriers lifetime.[9]

The annealing problem is particularly severe in the case of the ion-implanted, ultra-shallow junctions in MOS and MOSFET devices, approaching a depth of 10 nm, since the annealing time and the process temperature must be precisely controlled to minimize changes in the junction depth due to thermal diffusion effects. It was, in fact, observed that 950 °C annealing is necessary, because at lower temperatures annealing is unable to remove the dopants distributed at depths higher than 10 nm,[10] see Figure 4.3.

A further problem occurs when silicon is n-type doped to concentrations up to $3 \times 10^{20}$ cm$^{-3}$, which is the typical case of the ultra-shallow junctions in FET and MOSFET devices. At these high P or As concentrations, free carrier saturation is observed both with junctions prepared with MBE or with ion implanted junctions,[11] due to the

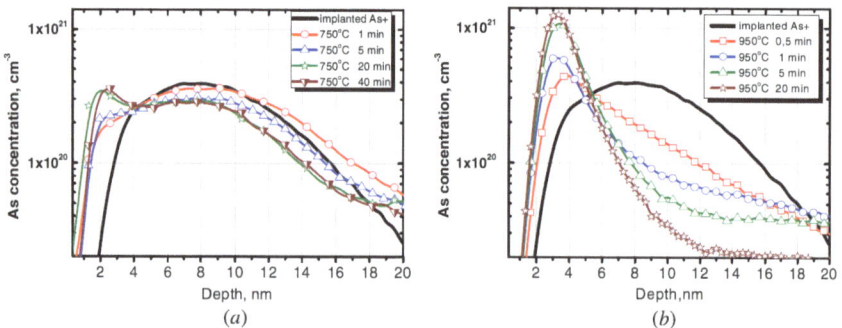

**Figure 4.3**  As-depth profiles after annealing at 750 °C (a) and 950 °C (b). Reproduced from ref. 10 with permission from IOP Publishing, Copyright 2014.

thermally induced formation of electrically neutral vacancy–donor complexes.

As an example, in the case of As-implanted junctions fabricated on CZ silicon, due to the presence of oxygen and of vacancy defects generated by irradiation, oxygen interacts with vacancies and dopants, leading to the formation of neutral As–O and O–As–V complexes, with partial dopant deactivation.

The physico-chemical backgrounds of these interaction processes will be considered in Section 4.5.4 when discussing the conditions of phosphorous diffusion gettering of metallic impurities.

Point defect-stimulated gettering of oxygen during the thermal activation of implanted As in silicon has also been observed. As an example, the dependence of the As and oxygen depth profiles on the thermal annealing temperature and duration has been investigated,[10] showing that oxygen interaction with vacancies does occur with the segregation of precipitates inside the junction, on which As segregation does occur, with electrical As deactivation.

In these cases, the annealing temperature should be kept as low as possible, and MW annealing[††] could be a potential solution.[3,4,12–15]

## 4.3 Hydrogen-assisted Impurity and Defect Passivation

Atomic hydrogen from various dry or wet sources is used in the semiconductor technology for the electrical deactivation of surface states and of shallow and deep levels arising from dopants, impurities, and point and extended defects in silicon and other semiconductors.[16–26]

Interface states are responsible for surface carrier recombination effects, which could significantly reduce, as a non-unique example, the efficiency of solar cells. These states are localized in an interlayer extended over a few Angstroms from the surface, and result from silicon dangling bonds with different back-bond configurations.

The hydrogenation processes are known in the literature as passivation processes, arising from the chemical bonding of interstitial hydrogen[‡‡] with surface states and impurities, which leads to a shift of their original gap levels out of the gap,[27,28] and with their electrical deactivation.

---

[††] At 5.8 GHz.
[‡‡] Hydrogen dissolves in silicon as an interstitial species, with a solubility given by the following equation $S(H) = 9.1 \times 10^{21} \exp(-1.80 \text{ eV}/kT) \text{ cm}^{-3}$.

Hydrogen passivation processes were largely used to reduce the carrier recombination processes occurring at grain boundaries of multi-crystalline solar cells or at silicon interfaces[30] by reaction of hydrogen with surface defects and formation of Si–H bonds.

The passivation (compensation) of donor and acceptor dopants by hydrogen in silicon could be, instead, discussed considering that atomic hydrogen is an amphoteric impurity, and behaves as a donor species in p-type silicon

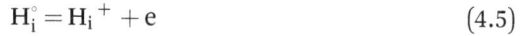

$$H_i^\circ = H_i^+ + e \qquad (4.5)$$

and as an acceptor in n-type silicon

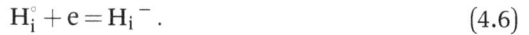

$$H_i^\circ + e = H_i^- . \qquad (4.6)$$

This property is well illustrated in Figure 4.4, where one sees that the stable condition for interstitial hydrogen in p-type silicon is the $H_i^+$ state, while it is the $H_i^-$ state in n-type silicon.[17] Apparently, the neutral state $H_i^\circ$ is never stable in silicon, but the error bar is too large in these calculations to exclude unambiguously the stability of $H_i^\circ$.[17]

Since the passivation of dangling bonds (DBs) at silicon surfaces would require the presence of neutral interstitial hydrogen $H_i^\circ$ as reactant, and the hydrogenation of DBs operates well with un-doped silicon, we suggest that $H_i^\circ$ is actually stable in undoped silicon.

Compensation, eventually, occurs with ionized shallow acceptors[25] in p-type silicon

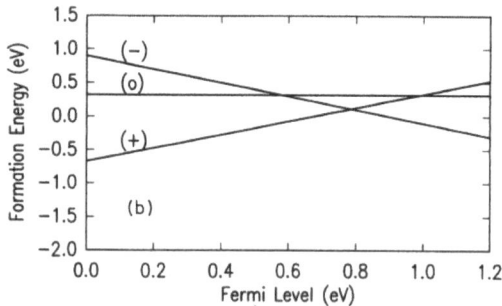

$$H_i^+ + B_{Si}^- \rightleftharpoons H - B \qquad (4.7)$$

**Figure 4.4**  Calculated formation energies of hydrogen in the different charge states in silicon as a function of doping. Reproduced from ref. 17 with permission from American Physical Society, Copyright 1989.

and n-type silicon

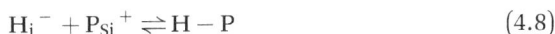

$$H_i^- + P_{Si}^+ \rightleftharpoons H - P \qquad (4.8)$$

with the formation of neutral pairs, partially covalently bonded.

The efficiency of the hydrogen passivation process depends on the energetics of hydrogen bonding with impurities and dangling bonds, determined, as an example, by Van der Walle,[31,32] with first-principles pseudopotential-density functional calculations, see Figure 4.5, showing that the average bond energy of hydrogen with B and P holds $\sim$ 2.0 eV, not far from the values of 1.2 and 1.8 for B–H and P–H given by Johson *et al.*[18] and from 2.17 to 3.6 eV for Si–H bonds, depending on their configuration.

Such high values of energy of formation of Si–H bonds (200–400 kJ mol$^{-1}$) favour the thermal stability of passivated interfaces, a property particularly relevant in solar cell applications.

Wet hydrogenation is conventionally carried out by dipping a silicon sample in an aqueous solution of HF[21] or by rinsing in water and then dipping in a NH$_4$F solution a previously oxidated silicon

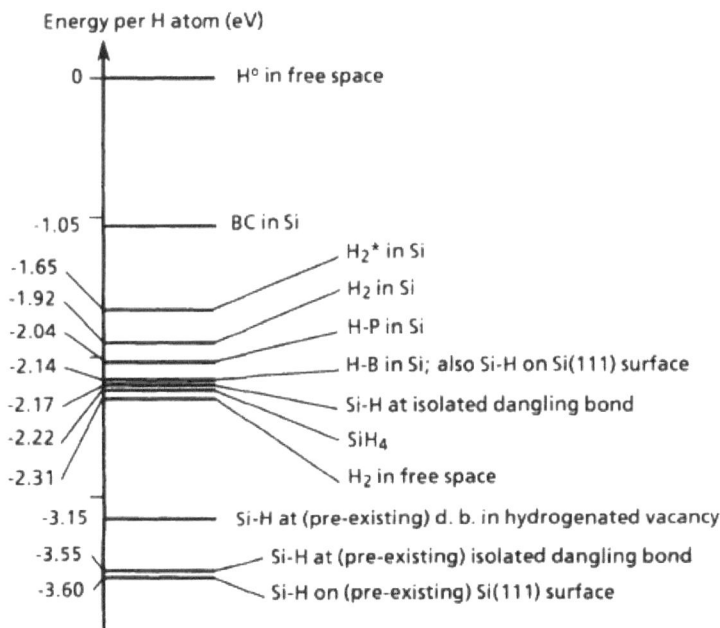

**Figure 4.5** First-principles binding energy calculations for isolated hydrogen–silicon species. Reproduced from ref. 31 with permission from American Physical Society, Copyright 1994.

surface[30] which leads to fully hydrogenated and remarkably hydrophobic surfaces.

In both cases the hydrogenation reaction involves HF as the reacting species

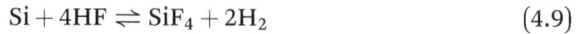

$$Si + 4HF \rightleftharpoons SiF_4 + 2H_2 \qquad (4.9)$$

and is strongly shifted to the right, given the value of the Gibbs energy $\Delta G$ (298 K) $= -479.85$ kJ mol$^{-1}$ of reaction (4.9), which is suggested to occur as a coupled electrochemical process,[§§] with the prior oxidation of silicon

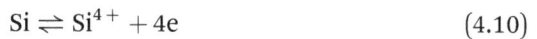

$$Si \rightleftharpoons Si^{4+} + 4e \qquad (4.10)$$

and the further reduction of HF

$$4HF + 4e \rightleftharpoons 2H_2 + 4F^- \qquad (4.11)$$

with silicon itself acting as an electrical bridge, as occurs with the electrochemical corrosion of Zn in HCl, see Figure 4.6.[29]

HF is an effective, but temporary, surface passivating agent, since storage in air of HF-passivated surfaces rapidly induce a deterioration

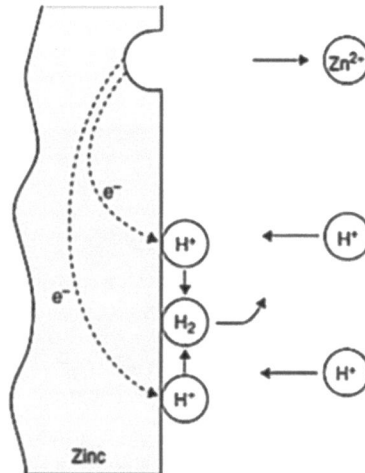

**Figure 4.6**  Electrochemical reactions occurring during corrosion of Zn in HCl.

---

§§ A process common in the corrosion of metals in aqueous solutions.

of the electrical properties and a decrease in the density of the passivated surface states,[30] as can be seen in Figure 4.7, which shows a decrease in the H-terminated Si(111) species, amounting to three orders of magnitude, when the hydrogenated sample is stored in air for 150 min.

The effective passivation depth[¶¶] and yield depend, however, on a delicate balance of inward and outward hydrogen bulk diffusion, whose diffusion coefficient holds[25] $D_H = 9.4 \times 10^{-3} \exp(-0.48 \text{ eV}/kT)$ cm$^2$ s$^{-1}$, and of trapping and de-trapping processes, both having a strong temperature dependence.

The outward hydrogen diffusion is, however, limited by a surface barrier associated with surface reconstruction as high as 2.4 eV,[33] which limits the absolute value of hydrogen losses.

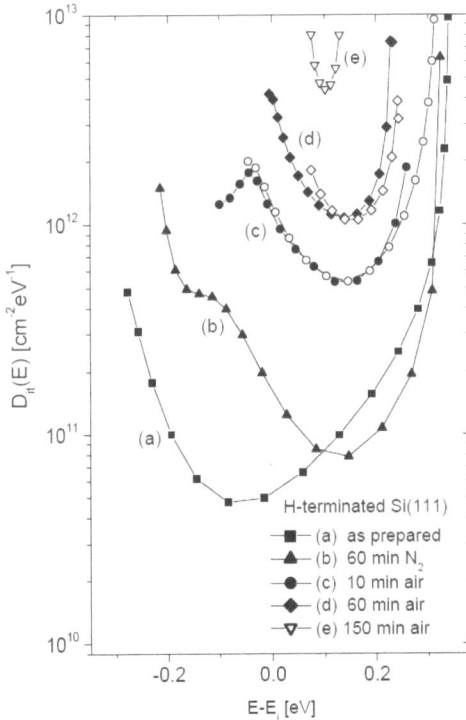

**Figure 4.7** Distribution of interface states on Si(111) surfaces passivated using a NH$_4$F solution. Reproduced from ref. 30 with permission from Elsevier, Copyright 2004.

---

[¶¶] In several cases the passivation should be extended to the entire thickness of the device, which in mc-Si solar cells is around 200 μm.

Instead, the hydrogenation stability is strongly dependent on de-trapping processes associated with the hydrogen oxidation, which explains the loss of surface passivation observed in all the samples of Figure 4.7 after storage in air.

Permanent surface and volume-defect passivation is however obtained using so-called wet-chemical oxidation,[30] which is carried out with a preliminary cleaning step with the RCA clean,[‖] followed by hydrogenation with $NH_4F$ and oxidation with ultrapure water at 80 °C. This procedure allows obtaining surfaces much more stable upon exposure to clean-room air than those prepared by conventionally prepared H-terminated surfaces and presenting the lowest density of surface states.

Excellent results are also obtained with plasma hydrogenation, which is capable of also passivating bulk vacancies with strong (0.96 eV) V–H bonds.[32]

## 4.4 Dry Etching Processes and Microfabrication

The dry etching processes of silicon are standard tools in VLSI technologies, as the substitutes of the almost isotropic wet etching processes, based on the use of aqueous mixtures of hydrofluoric acid (HF) and nitric acid ($HNO_3$), employed in the early days of IC fabrication. They grant, in fact, almost anisotropic and near-vertical etched sidewalls, and are fully compatible with modern VLSI process technologies.[34–36]

Dry etching processes also have a major impact on the solution of 3D micro-machining problems, since they have not only excellent anisotropy and vertical etch walls, but they also have the ability to etch through entire substrate thicknesses, see Figure 4.8, with high aspect ratio details, and present good selectivity with conventional masking material layers, reproducible performance, and low environmental impact.[37]

Dry etching processes operate with three main configurations, plasma etching, reactive ion etching and ion milling, and use

---

[‖] RCA clean is an etching procedure for removing organic substances from the silicon surface using a solution of $H_2O_2$ and $NH_4OH–H_2O$ (RCA1-clean). This procedure might be followed by a second clean, RCA-2, consisting of a solution of $H_2O_2–HCl–H_2O$ to further clean the surface. The silicon surface after the RCA clean is covered by a thin $SiO_2$ layer.

**Figure 4.8** SEM image of the cross section of a deep etched silicon wafer processed with a dry etch technology. Reproduced from ref. 37, https://doi.org/10.3390/mi12080991, under the terms of the CC BY 4.0 license, https://creativecommons.org/licenses/by/4.0/.

halogenated compounds as reactants, which enable the formation of volatile silicon halides.

Plasma etching of silicon could be carried-out using tetra-chloromethane ($CCl_4$), but etching of silicon dioxide and silicon nitride requires the use of trifluoromethane ($CHF_3$).

In reactive ion etching (RIE) and ion milling processes, $SF_6$ and $CF_4$ gases are employed for etching silicon and silicon carbide, while $C_3F_8$ and $C_2F_6$ are used for silicon dioxide. One of the problems encountered with fluorine-based etching processes is the fast rates of their radicalic reactions, which lead to isotropic etching unless adopting continuous- or cyclic-sidewall passivation techniques,[37] using as an example, polytetrafluoroethylene (Teflon) as the passivating agent. Room temperature operation is adopted for dry etching techniques, but modern deep reactive ion etching techniques (DRIE) operate at cryogenic (173 K) temperatures.

As can be seen in Figure 4.9, the operational pressure range varies from $10^2$–$10^{-1}$ torr for plasma etching to $10^{-3}$–$10^{-5}$ torr for ion milling processes. Under high pressure operation the constituents of the plasma experience multiple collisions, thereby not impinging normally on the substrate and making the etch configuration more

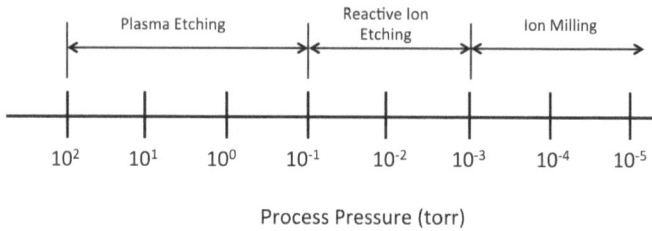

Process Pressure (torr)

**Figure 4.9**  Pressure range of dry etching processes. Reproduced from ref. 37, https://doi.org/10.3390/mi12080991, under the terms of the CC BY 4.0 license, https://creativecommons.org/licenses/by/4.0/.

isotropic than desired. For this reason, low pressure techniques are preferred for technological applications.

## 4.5 External Gettering (EXG)

### 4.5.1 Introduction

As discussed in Section 2.5, internal gettering (IG) provides the removal of TM impurities[38] from the active region of the device, with a process consisting of the localization, trapping and inactivation of impurities at oxide precipitates, deliberately created by a thermal reaction process occurring with silicon and oxygen impurities present in the material. The consequence of this process is the formation of an impurity-depleted layer[39] in correspondence with the active region of the device, where carrier recombination losses are, therefore, made to be negligible.

Different from IG processes, external gettering (EXG) processes in silicon occur[40–51] with the segregation and the electrical deactivation of impurities at the back surface of a wafer (thus the name "external"), in correspondence with a region of deliberately introduced, mechanically or chemically activated impurity sinks, leaving the entire volume of the sample impurity depleted.

In the first case, the active entities, where impurities diffuse and interact, are surface defects (dislocations) generated by mechanical abrasion or by ion implantation at the back surface of a silicon wafer, or dangling bonds at the surface of polycrystalline silicon layers, suitably deposited on the back surface of a silicon wafer. Both processes leave impurity depleted the entire volume of the wafer.

In the second case the active entity is a layer of elemental gettering species (phosphorus and aluminium, as non-exclusive examples) where impurities diffuse, dissolve or segregate, and become electrically inactivated by chemical bonding processes, possibly with the assistance of point defects.

EXG processes have been studied, developed and applied to provide a final cleaning of EG silicon wafers and of multi-crystalline silicon wafers used for PV applications[51] from residual TM impurities.

Although in today's industrial practices gettering at mechanical damaged surfaces or at polysilicon layers is the only kind of EXG applied to an electronic grade 200 mm wafer, while no EXG is applied to 300 mm wafers, EXG at chemical sinks for PV applications represent an excellent piece of chemistry in the semiconductor technology, meriting full consideration in a book devoted to the chemistry of semiconductors.

## 4.5.2 Thermodynamics of External Gettering Processes

An external gettering process could be treated as the segregation, or the dissolution and the electrical inactivation, of a single- or of multiple-chemical species dissolved in a semiconductor wafer*** to a sink located at the back surface of the wafer.

In either case, the equilibrium condition for the distribution of an impurity i between the semiconductor bulk phase ($\alpha$) and the surface sink ($\beta$) is given by the following equation

$$\mu_i^{\circ,\alpha} + RT \ln \gamma_i^{\alpha}(x) x_i^{\alpha} = \mu_i^{\circ,\beta} + RT \ln \gamma_i^{\beta}(x) x_i^{\beta} \qquad (4.12)$$

where $\mu_i^{\circ,\alpha}$ and $\mu_i^{\circ,\beta}$ are the standard chemical potentials of the impurity i in the bulk phase and in the surface sink, $\gamma_i^{\alpha}(x)$ and $\gamma_i^{\beta}(x)$ are the concentration-dependent activity coefficients of i, and $x_i^{\alpha}$ and $x_i^{\beta}$ are the concentrations[†††] of i in the two phases.

Since impurities in semiconductors are normally present in diluted solutions, the activity coefficient of i in the bulk phases could be approximated to unity, although their interaction with other impurities could bring the activity coefficient $\gamma_i^{\alpha}(x)$ far from unity, as is the

---

***On top of which should set the active region.
[†††] In molar fractions.

well-known case of Fe in B-doped silicon, which forms thermo-dynamically stable Fe–B complexes.

Eventually, the ratio of the equilibrium concentration of the impurity i segregated at the surface sinks $x_i^\beta(x)$ and that in the bulk phase $x_i^\alpha$ is given by the distribution coefficient $k$

$$k = \frac{x_i^\beta}{x_i^\alpha} = \Gamma\left(x_i^\beta, \ x_i^\alpha\right) \exp - \frac{\mu_i^{\circ,\beta} - \mu_i^{\circ,\alpha}}{kT} = A\Gamma\left(x_i^\beta, \ x_i^\alpha\right) \exp - \frac{\Delta H_B}{kT} \qquad (4.13)$$

where the difference in the chemical potentials $\mu_i^{\circ,\beta} - \mu_i^{\circ,\alpha} = \Delta G_B$ is the Gibbs energy of the interaction reaction of i at a surface sink site, $\Delta H_B$ is the reaction enthalpy of the impurity at a surface sink site, the pre-exponential $A$ has an entropic character and $\Gamma(x_i^\beta, x_i^\alpha)$ is the ratio of the activity coefficients of i, which accounts for all the interactions occurring between the impurity atoms and the atoms of the matrix in both the bulk and sink "phase".

Equilibrium conditions are achieved if the diffusion of impurities toward the gettering region across the entire width of the wafer is fast and assumed to be ruled by a Fickian process[‡‡‡]

$$\Phi_i = -D_i \Delta x_i \qquad (4.14)$$

where $\Phi_i$ is the molar flux (mol m$^2$ s$^{-1}$) of the impurity I toward the sink, $D_i$ is the diffusion coefficient of the impurity in cm$^2$ s$^{-1}$ and $x_i$ is the concentration of i.

It is immediately obvious that the absolute values of the diffusion coefficients of the impurities to be gettered have a substantial role in the success of the gettering process, fast diffusers being the best candidate for high process yields.

The temperature dependence of the diffusion coefficient of transition metals in silicon[38] is displayed in Figure 4.10. One can see that Cu and Ni diffuse much faster than Fe,[§§§] Cr and Ti, with differences of orders of magnitude, which makes titanium, chromium and iron more difficult to be gettered in comparison with Cu and Ni.

Some topical details of the EXG processes, the role of defects in the process rate, the specific configuration of the gettering sites and the nature of the species to be gettered are the issues considered in the next two sections.

---

[‡‡‡] In the absence of strain or electrical fields that could affect the diffusion process.

[§§§] $D_{Fe} = 1.3 \times 10^{-3} \exp - \dfrac{0.68 \ (eV)}{kT}$.

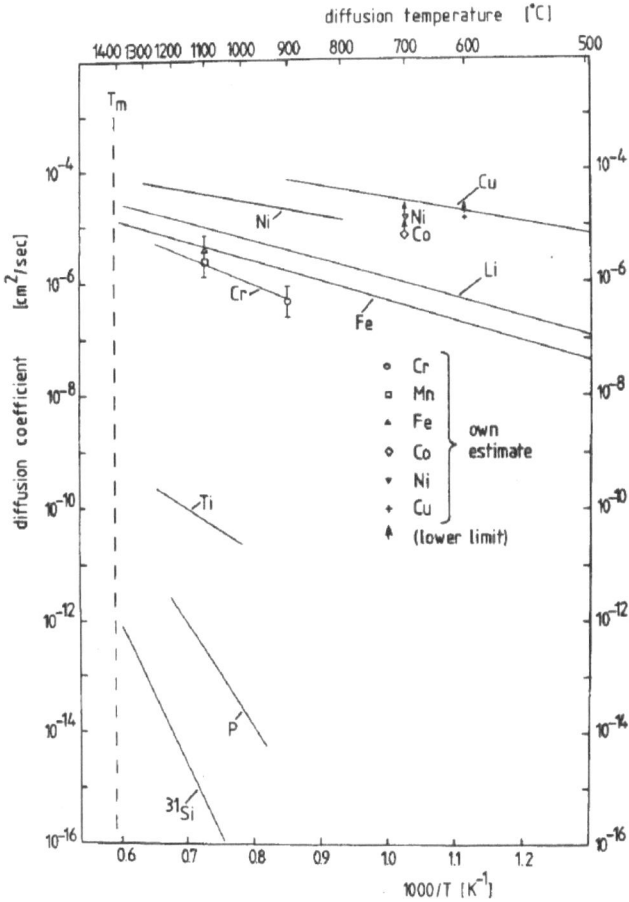

**Figure 4.10** Temperature dependence of diffusion coefficients of 3d metals in silicon. For comparison the diffusion coefficient of Li and P and the self-diffusion coefficient of $^{31}$Si are also displayed. Reproduced from ref. 38 with permission from Springer Nature, Copyright 1983.

## 4.5.3 Gettering at Mechanically Damaged Surfaces and at Polycrystalline Silicon Deposits

Gettering of TM at mechanically damaged surfaces or at polycrystalline silicon layers at the back surface of a silicon wafer is suggested to occur with two different mechanisms, *i.e.* with a conventional segregation process of the impurities at the mechanically damaged region or at the surface of polycrystalline silicon surface phase, following the conditions of an equilibrium segregation (eqn (4.13) and (4.14)) or through a relaxation gettering (RG) process, associated with the relaxation of a surface strain due to the impurity segregation.[52]

Relaxation gettering can be, at least qualitatively, understood considering that the surface strain induced by mechanical abrasion of the back surface of a silicon wafer is directly associated with the presence of dislocation arrays, which behave as powerful impurity sinks, and that the surface strain of a polycrystalline silicon layer deposited at the back surface of a silicon wafer is associated with the presence of dangling bonds.

In both cases the segregation of impurities at surface defects causes a stress relaxation, and in both cases relaxation gettering is also a defect-controlled gettering process.[52]

With respect to mechanical damaged back-surfaces, polycrystalline silicon back surface layers have the advantage[45] of withstanding high process temperatures, up to 1100 °C, *i.e.*, the temperatures at which top surface epitaxial processes are carried out, without losing their gettering properties. It is, furthermore, also known[45] that gettering of TM impurities at polycrystalline silicon back surface layers occurs with an equilibrium segregation process at the EXT gettering process temperature (eqn (4.13) and (4.14)) and with a relaxation process during the final cooling when the impurity solution becomes supersaturated (see Figure 4.11 for the solubilities of TM in silicon[38]).

The RG process yield at a polycrystalline silicon layer might be, however, very modest, as is the case of the segregation of iron on a polysilicon layer from a saturated iron solution in Si ([Fe] $> 10^{16}$ at cm$^{-3}$) at 1150 °C,[38] which leads, in fact, to a segregation coefficient of 2.6:

$$k = \frac{S_{Fe}^{poly}}{S_{Fe}^{Si}} \approx 2.6 \qquad (4.15)$$

where $S_{Fe}^{poly}$ is the solubility of iron in polysilicon and $S_{Fe}^{Si}$ is the solubility of iron in the matrix.

Considering that DBs are responsible for the gettering effects in the polycrystalline silicon back-surface phase, the use of an equilibrium relationship as that given in eqn (4.13) is, at least, improper, because polysilicon is very inhomogeneous and contains both grain boundaries where the impurities can segregate and relatively unstressed silicon within the grains which has no segregation effect. Therefore, the macroscopic segregation coefficient should depend on the microscopic structure of the polysilicon layer.[¶¶¶]

The work carried out by Kirscht *et al.*[52] adds a consistent piece of additional information concerning the effect of temperature on the Fe

---

[¶¶¶] Original text in Istratov's paper.

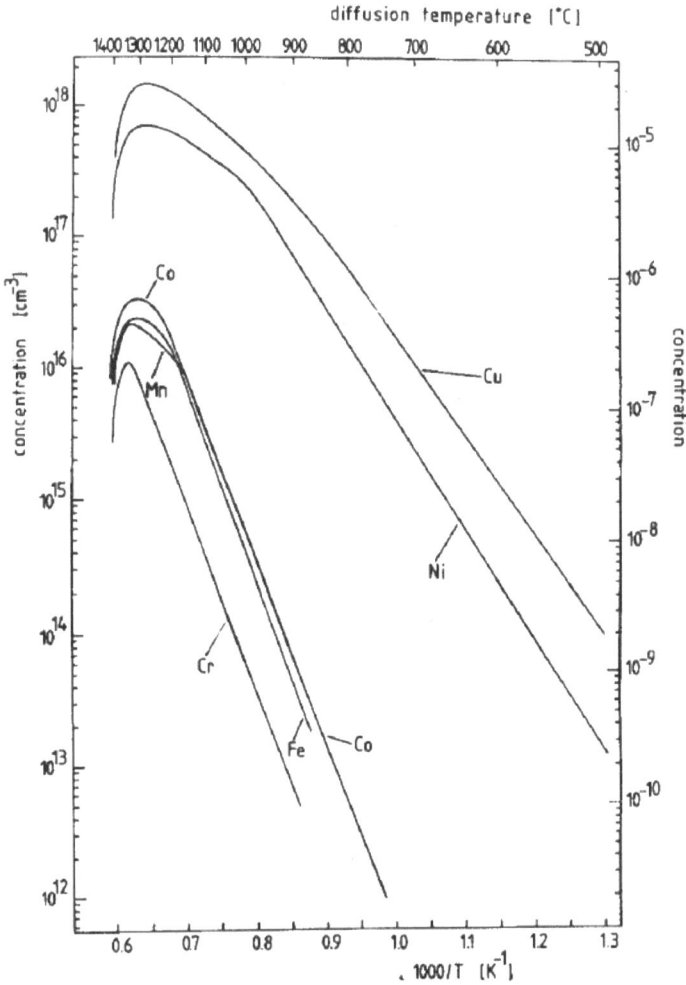

**Figure 4.11**   Temperature dependence of the solubility of 3d metals in silicon. Reproduced from ref. 38 with permission from Springer Nature, Copyright 1983.

gettering yield at a back polycrystalline silicon layer, with the simultaneous occurrence of epitaxial growth of a silicon layer at the top silicon surface of silicon wafers lightly doped with B or P, with an Fe concentration of $2 \times 10^{13}$ cm$^{-3}$.

They show, in fact, that while the backside strain due to a poly-silicon layer takes the largest values at the lowest deposition temperature, within the selected process temperature range (640–680 °C), a strong strain decrease is observed as a consequence of the growth of a top surface epitaxial layer carried out at 1100 °C.

A side-effect of this situation is a partial segregation of Fe impurities at the front interface, with a significant influence of p-type or n-type doping on the total impurity gettering balance, with a final Fe concentration of around $8 \times 10^{12}$ cm$^{-3}$ at the back surface and $5 \times 10^{11}$ cm$^{-3}$ at the front interface.

The main conclusion from these and other studies[53] is that EXG of iron at mechanically damaged surfaces is a very complex process, with most of the details still needing to be understood.

## 4.5.4 Phosphorus Diffusion Gettering (PDG)

Physico-chemical processes of EXG using phosphorus as a chemical sink of unwanted impurities allow mild, effective and reliable conditions of operation, as experimentally shown in the early studies of Ourmazd and Schröter[40] and Bourret and Schröter,[41] who showed that Ni, preliminarily deposited at the top surface of a silicon wafer and further thermally diffused in the volume of the wafer as interstitial Ni$_i$, is efficiently gettered at the back surface, in correspondence with a heavily P-doped surface layer. The gettering process was carried out with the initial deposition of P from $P_2O_5$ vapours in a nitrogen flux at 260 °C at the surface of the sample wafer, followed by a thermal annealing step at 900 °C[||||] for 130 min, and by a final quenching step, addressed at inducing Ni supersaturation conditions.

Microstructural investigation carried out with HRTEM and RX diffraction on the gettered sample at the end of the process show that an amorphous phosphosilicate glass (PSG) surface layer covers the P-doped surface layer,[****] and that silicon mono-phosphide (SiP) particles segregate at the Si/PSG interface (see Figure 4.12) from the supersaturated P-solution.

These measurements also show the presence of minute microprecipitates different form silicon phosphide, identified as NiSi$_2$ particles from X-ray microanalysis.

The author's conclusion was that the NiSi$_2$ formation is associated with a process of self-interstitial injection from the SiP particles, due to the difference in the molar volume of silicon and of SiP, and occurs according to the following equation

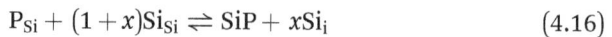

$$P_{Si} + (1+x)Si_{Si} \rightleftharpoons SiP + xSi_i \tag{4.16}$$

---

[||||] Today, PDG is carried out either using POCl$_3$ as the source gas or P-ion implantation.
[****] Because of the use of $P_2O_5$ vapours as the phosphorus source.

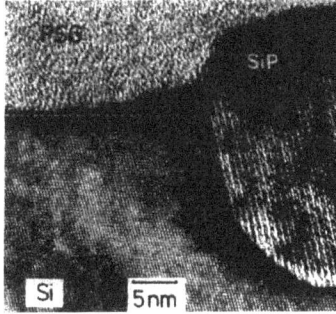

**Figure 4.12** HRTM lattice image of a P-doped Si/PSG interface. On the right side of the image, one can see a SiP particle protruding from the P-doped silicon layer. Reproduced from ref. 40 with permission from AIP Publishing, Copyright 1984.

which leads to the relaxation of the P-doped silicon substrate by the epitaxial growth of $NiSi_2$ at the Si/PSG interface[††††]

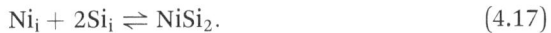

$$Ni_i + 2Si_i \rightleftharpoons NiSi_2. \tag{4.17}$$

In other words, the PDG of nickel is associated with its precipitation as nickel silicide from a supersaturated solution of Ni in the P-rich silicon phase, driven by the enthalpy of formation of nickel disilicide, which is $-78.62$ kJ mol$^{-1}$.

The metal silicide precipitation with PDG has been confirmed, for Fe, Co, Cu[51] and for Pt,[54] but further works suggest that the silicide formation is not the unique route for the PDG of TMs, since the segregation of the metal impurities in a P-rich silicon phase, at P-concentrations lower than the saturation, follows an alternative, and efficient, gettering route.[55]

This is the case of the PDG gettering of Co, Fe and Cu, which apparently works with the segregation of the metal impurities from the metal impurity-contaminated silicon phase to the heavily P-doped surface silicon phase, as shown by Shabani *et al.*[56] They demonstrated that a PDG gettering carried out at 900 °C for 90 min can remove efficiently Ni and Cu in the experimental metal impurity concentration range of $10^{12}$–$10^{15}$ cm$^{-3}$, with a reduction in their concentration of four to five orders of magnitude, also in the presence of dislocations and other bulk microdefects, which could work as secondary gettering sources. In the case of Fe, a reduction in its concentration, in the same experimental conditions, is only of two orders

---

[††††] Original sentence of the authors.

of magnitude, with a further reduction of the gettering efficiency for B-doped samples, due to the presence of stable $(Fe–B)_x$ complexes.

SIMS depth profiles after a PDG carried out at 900 °C confirm that Fe, Cu, and Ni are gettered in the highly P-concentrated near surface region, up to a depth of 0.5 μm, with a gettering efficiency that decreases from Cu to Fe.

Further PDG work was carried out by Talvitie *et al.*[49,57,58] on B-doped CZ silicon samples with an average concentration of $1.7 \times 10^{13}$ cm$^{-3}$. They could demonstrate, see Figure 4.13, that the residual iron concentration in the silicon phase depends on the PDG process temperature[‡‡‡‡] and time, and decreases with a decrease in the process temperature, leading to a bulk Fe concentration lower by two orders of magnitude than the initial one, for a process temperature of 600 °C and 60 min of process time.

An exponential relationship is followed for the temperature dependence of the residual Fe concentration $x_{Fe}$ after PDG

$$x_{Fe} \approx x_{Fe}^0 \exp - \frac{\Delta H_{segr}}{RT} \tag{4.18}$$

with an experimental value of the enthalpy of segregation of 2.5 eV or $-241$ kJ mol$^{-1}$, which should correspond to the enthalpy of formation of a Fe–X complex in the P-rich silicon surface phase.

They showed also that the Fe segregates down to a depth of 0.1 μm in the high-P concentration region,[57] as shown in Figure 4.14. for a

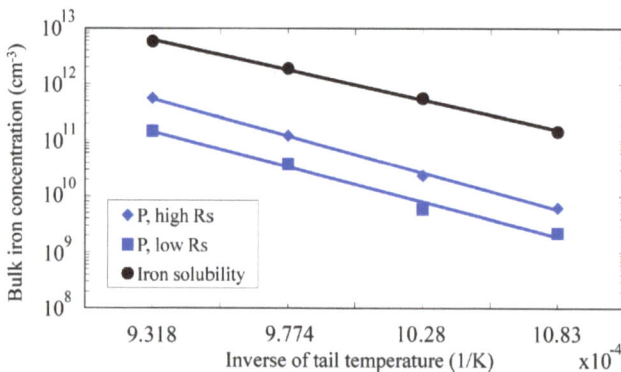

**Figure 4.13** Residual Fe bulk concentration after PDG as a function of the anneal (tail) temperature and gettering time ($R_s$ in the figure) high $R_s = 30$ min, low $R_s = 60$ min. Reproduced from ref. 57 with permission from AIP Publishing, Copyright 2011.

[‡‡‡‡] "Tail temperature" in the text.

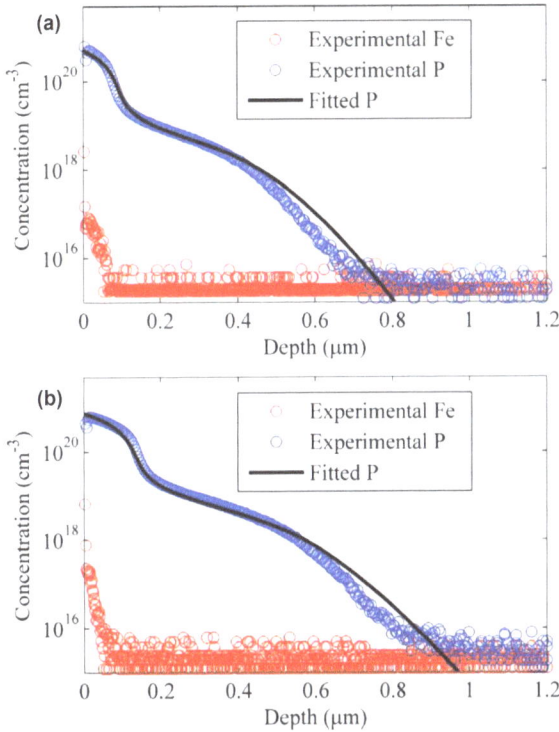

**Figure 4.14** SIMS depth profiles of P and Fe for (a) 30 min annealing at 870 °C and (b) 60 min annealing at 870 °C. Reproduced from ref. 57 with permission from AIP Publishing, Copyright 2011.

PDG process carried out at 870 °C. The depth of the P-diffused layer shows a significant dependence on the annealing time, increasing from 0.8 μm to 1 μm with the increase in the annealing time from 30 min to 60 min, while the depth of the region where Fe segregates remains substantially constant.

The physical and chemical backgrounds of PDG of metal impurities in silicon will be discussed in the following two sections, where account is given for the role of temperature and P-concentration on the segregation of impurities as phosphorus–metal complexes (segregation gettering) or as silicide precipitates (relaxation gettering).

### 4.5.4.1 Modelling the PDG Segregation Gettering: The Role of the Phosphorus–Vacancy Complexes

Different from the precipitation gettering, which occurs with the precipitation of metal silicides in the P-rich silicon back surface

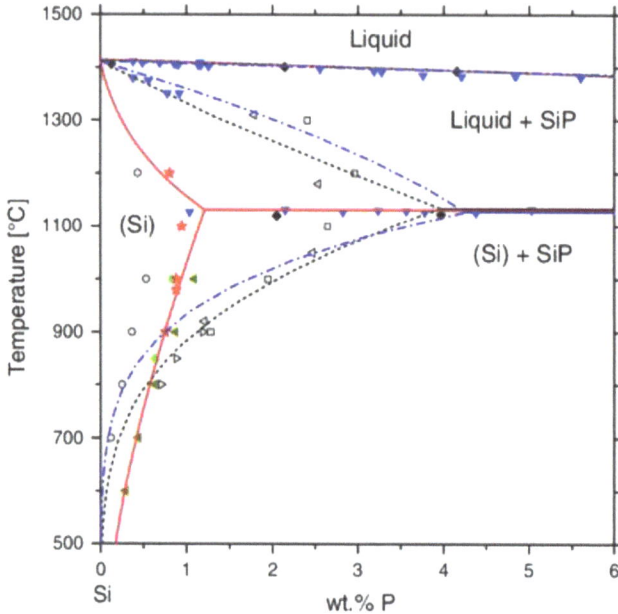

**Figure 4.15**    Section of the Si–P phase diagram, with the homogeneous P-rich silicon β phase. Reproduced from ref. 59 with permission from Springer Nature, Copyright 2014.

β phase (see the Si–P phase diagram in Figure 4.15, which shows a maximum of 1% mol of P at 1100 °C), segregation gettering involves the equilibrium between metal species presenting different chemical configuration in the bulk silicon phase (α) and in the P-rich surface phase (β)[59]

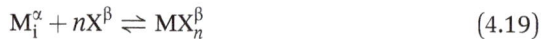

$$M_i^\alpha + nX^\beta \rightleftharpoons MX_n^\beta \tag{4.19}$$

where $MX_n$ is a thermodynamically stable complex or a bonded configuration of the species M in the environment of the β phase. Assuming the formation of a truly molecular species $MX_n$ in the P-rich (β) phase, its concentration would depend on its Gibbs energy of formation $\Delta G_f^{MX_n}$

$$c_{MX_n}^\beta = Ac_M^\alpha \exp - \frac{\Delta G_f^{MX_n}}{RT} = ABc_M^\alpha \exp - \frac{\Delta H_f^{MX_n}}{RT} \tag{4.20}$$

where $A$ is a term that accounts for the ratio of the activity coefficients, $B$ is an entropic term and $\Delta H_f^{MX_n}$ is the enthalpy of formation of the $MX_n$ species in the P-rich (β) phase.

The molecular species $MX_n$ that could be supposed to form in a $P_xSi$ phase is a metal atom bound to a phosphorus–vacancy complex $P_nV$, whose presence in phosphorus–silicon solutions is supported by copious experimental and theoretical results.

As an example, according to Watkins and Corbett,[60] the main product of the irradiation with 6 MeV electrons of oxygen-lean, lightly P-doped ($10^{15}$–$10^{16}$ cm$^{-3}$) FZ silicon, is the E-centre, a defect that ESR spectroscopy measurements allow to identify as a P atom neighbour to a silicon vacancy, or a phosphorus–vacancy complex PV (see Figure 4.16), which introduces an acceptor level at $\sim E_c - 0.43$ eV, making it electrically active and detectable by PL and DLTS measurements.

Markevich *et al.*[61] additionally showed that not only the E-centre is introduced by 6 MeV electron irradiation of lightly P-doped FZ silicon, but also another defect, consisting of interstitial-substitutional carbon pairs $[C_i-C_{Si}]$, arising from the systematic presence in FZ silicon of carbon impurities at a concentration of $10^{14}$–$10^{15}$ cm$^{-3}$ that presented a PL emission at 969.5 meV.

They also show that with the anneal of the E-centre in the temperature range 125–175 °C a decrease in the DLTS signal associated with the E-centre and in the PL intensity of the emission at 969.5 meV associated with the $[C_i-C_{Si}]$ defects is measured, together with an increase in the DLTS signal associated with a level at $E_c - 0.21$ meV, characteristic of the substitutional dicarbon complex $[C_{Si}-C_{Si}]$.[62]

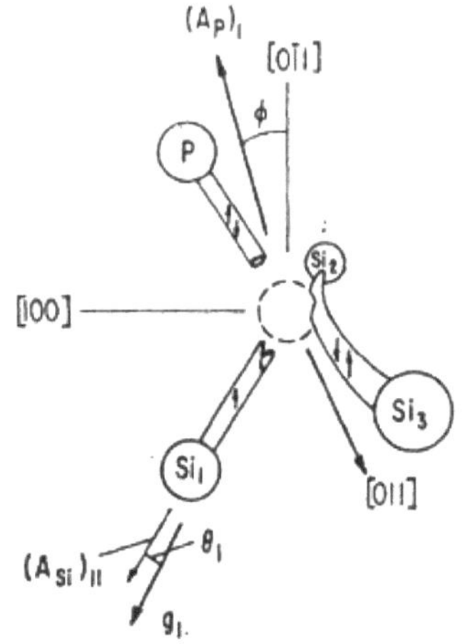

**Figure 4.16** Structural model of the E-centre. Reproduced from ref. 60 with permission from American Physical Society, Copyright 1964.

We could, therefore, conclude that in lightly P-doped FZ Si the E-centre thermally dissociates

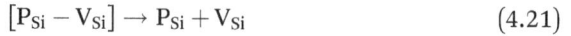

$$[P_{Si} - V_{Si}] \rightarrow P_{Si} + V_{Si} \tag{4.21}$$

accompanied by the simultaneous formation of a $C_{Si}$–$C_{Si}$ complex *via* the following defect reaction

$$V_{Si} + [C_i - C_{Si}] \rightleftharpoons [C_{Si} - C_{Si}]. \tag{4.22}$$

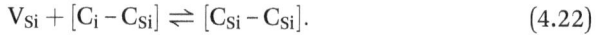

Since the monophosphorus–vacancy complex PV dissociates at temperatures much lower than those operated for PDG (800–900 °C), it could not be supposed to be involved in the gettering process of TM impurities.

An important contribution to the understanding of the problem arises from a work of Kveder *et al.*,[51,55] who showed that the P-diffusion in highly P-doped Si, which is the condition occurring in PDG processes, cannot be associated with the diffusion of phosphorus interstitial $P_i$,[§§§§] generated by a kick-out mechanism involving phosphorous in substitutional positions and silicon self-interstitials

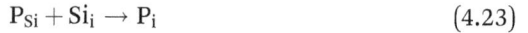

$$P_{Si} + Si_i \rightarrow P_i \tag{4.23}$$

as it occurs in lightly P-doped silicon, since at high P concentrations ($\geq 3 \times 10^{19}$ cm$^{-3}$) the diffusivity takes a $\propto P^2$ dependence, which could be explained by supposing that it occurs *via* mobile vacancy-diphosphorus (VP$_2$) complexes.

The results of the modellization carried out with these criteria[51] show, in fact (see Figure 4.17a), that the deviations of the experimental concentration profile of substitutional phosphorus ($P_{Si}$) from that expected by the sole $P_i$ diffusion (dotted line) in the first 0.4 μm, could be reasonably well accounted for if one considers an additional contribution of the vacancy–diphosphorus VP$_2$ complexes.

From the same figure it is also quite evident that the concentration of isolated (naked) vacancies (V) fits with the literature values displayed in Figure 4.18 for intrinsic silicon[66–68] and is four orders of magnitude lower than that of VP$_2$ complexes. Eventually, the excellent fit of the experimental concentration profiles of substitutional $P_{Si}$ ($\bigcirc$) with the simulation profiles (solid lines) for the four different surface

---

§§§§ $P_{Si}$ in substitutional sites is supposed not be mobile.

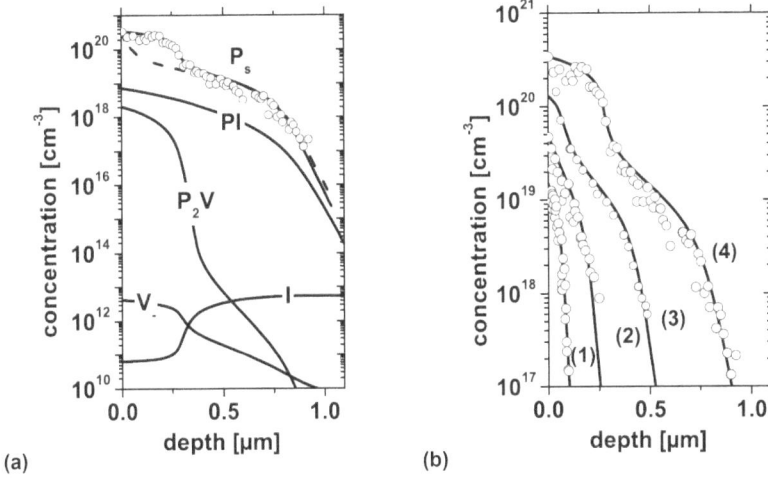

(a)

(b)

Figure 4.17 (a) Calculated and experimental (○) depth concentration profiles for substitutional $P_s$; P interstitials (PI), $P_2V$ complexes, vacancies (V) and silicon self-interstitials (I) for a P in-diffusion process at 900 °C for 60 minutes. (b) Comparison of experimental P concentration profiles (○) with the simulation profiles at various surface P concentrations (1) $1 \times 10^{19}$ cm$^{-3}$, (2) $5 \times 10^{19}$ cm$^{-3}$, (3) $1.5 \times 10^{-20}$ cm$^{-3}$, (4) $3.5 \times 10^{-20}$ cm$^{-3}$. Reproduced from ref. 51 with permission from John Wiley & Sons, Copyright © 2012 John Wiley & Sons, Ltd.

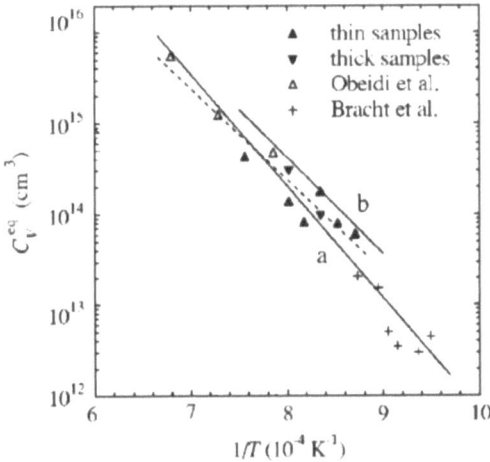

Figure 4.18 Equilibrium concentration of vacancies in intrinsic silicon. Reproduced from ref. 66 with permission from AIP Publishing, Copyright 2004.

P concentrations in Figure 4.17b is supplementary proof of the validity of the model and of the contribution of $VP_2$ complexes to the P diffusion.

The question that remains open is whether only the vacancy-diphosphorus complex [VP$_2$] is responsible of the metal gettering activity of PDG, due to a bonding reaction with metal impurities

$$VP_2 + Me_i \rightleftharpoons [M - PV_2] \tag{4.24}$$

since, in principle, all the VP$_n$ complexes with $2 \leq n \geq 4$ could contribute, having already agreed about the thermodynamic instability of the PV complex at the PDG temperatures.

The problem was considered by Chen *et al.*[64] and Lee *et al.*[69] who carried out the *ab initio* calculation of the formation energies $E_f$ of all the VP$_n$ complexes, assuming that phosphorus atoms occupy substitutional silicon P$_{Si}$ positions and that its reaction path takes the form

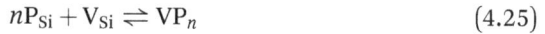

$$nP_{Si} + V_{Si} \rightleftharpoons VP_n \tag{4.25}$$

for $n > 1$, although we argue that the actual reaction paths could occur *via* the following reaction

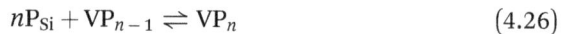

$$nP_{Si} + VP_{n-1} \rightleftharpoons VP_n \tag{4.26}$$

which consists of the addition of a P atom to a pre-existing VP$_{n-1}$ complex.

The results are displayed in Table 4.3, which shows that the formation energies $E_f$ of VP$_n$ complexes[¶¶¶¶] decrease in absolute values with the increase of the $n$ value, leading to negative values for the complexes with $3 \leq n \geq 4$, which then behave as thermodynamically stable species, different from the complexes with $n < 3$.

**Table 4.3** *Ab initio* calculated formation energies $E_f$ (eV) of VP$_n$ complexes (LDA: local density functional approximation, CGA: generalized gradient approximation) at 0 K.

| Method | V | VP | VP$_2$ | VP$_3$ | VP$_4$ | Ref. |
|---|---|---|---|---|---|---|
| CGA | 3.50 | 2.31 | 0.97 | −0.18 | −1.68 | 64 |
| LDA | 3.49 | 2.38 | 0.99 | −0.18 | −1.29 | 69 |
| CGA | 3.62 | 2.34 | 0.78 | −0.26 | −1.89 | 69 |

---

¶¶¶¶ Calculated as the difference between the energy of formation of a silicon vacancy co-ordinated with $n$P atoms and that of a system consisting of $n$ isolated P atoms.[64]

Apparently, the calculated $E_f$ value for the naked vacancy (3.5–3.62 eV) in Table 4.3 fits well with the *ab initio* value (3.5 eV) calculated with the Car–Parrinello method,[70] giving significant support to the reliability of these calculations.

Since the Gibbs formation energies $\Delta G_f$ of a $VP_n$ complex can be calculated at any temperature by adding to the calculated enthalpy $E_f$ value at 0 K, an entropic $-T(S_{th} + S_{conf})$ term, where the configurational[71] $-TS_{conf}$ term should play the major role, one could foresee that the thermodynamic stability of complexes with $n \leq 3$, and then also of the $VP_2$ complex, is enhanced at high temperatures by the entropic term, leading them to behave as thermodynamically stable species.

A further insight into the question comes from the works of Ranki *et al.*[72,73] who carried-out positron annihilation spectroscopy (PAS)[||||||] measurements[73] of the temperature dependence of the vacancy concentration in highly P-doped silicon, whose results are displayed in Figure 4.19, showing an exponential dependence on the temperature.

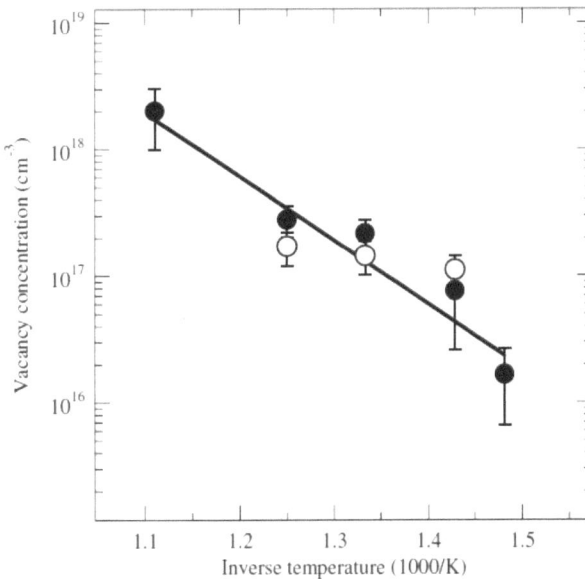

**Figure 4.19** Temperature dependence of the equilibrium vacancy concentration in highly P-doped silicon ($10^{20}$ cm$^{-3}$). Reproduced from ref. 72 with permission from American Physical Society, Copyright 2004.

[||||||] That is sensitive to voids in the lattice, and therefore to vacancies and vacancy complexes.

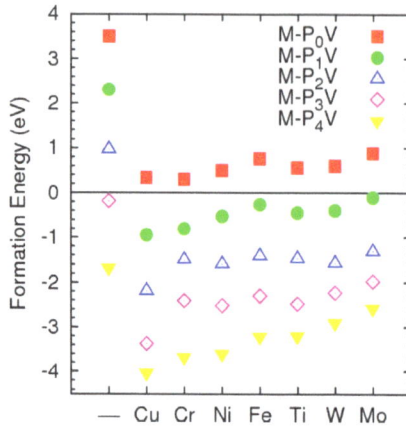

**Figure 4.20** Calculated formation energies of MP$_n$V complexes. The first column displays the formation energies of the pure VP$_n$ complexes. Reproduced from ref. 64 with permission from AIP Publishing, Copyright 2014.

Apparently, the Ranki's experimental vacancy concentration at 920 K ($2 \times 10^{18}$ cm$^{-3}$) is orders of magnitude higher than that in undoped silicon[66] ($10^{12}$ cm$^{-3}$), see Figure 4.18, and corresponds well with the concentration of VP$_2$ complexes calculated by Seibt and Kveder,[51] and displayed in Figure 4.17a.

Furthermore, the experimental values of the vacancy formation enthalpy ($\Delta H_f = 1.1$ eV) determined from the slope of the Arrhenius plot of Figure 4.19 are lower than the literature values of the silicon vacancy formation in intrinsic silicon (2.44 eV)[66,68] or of the *ab initio* value (3.5 eV) calculated with the Car–Parrinello method,[70] but fits well with the formation enthalpy of the VP$_2$ complex, $E_f = 0.97$ eV calculated by Chen *et al.*[64] or $E_f = 0.99$ eV calculated by Lee *et al.*[69] and reported in Table 4.3.

Thus, the vacancy-defect determined by PAS measurements at temperatures compatible with PDG processes is a vacancy decorated by two P atoms, *i.e.* it is the VP$_2$ complex.

The question now open is whether the VP$_2$ complex is, in fact, involved in the gettering activity of phosphorus in a PDG process. This view has been, again, considered by Chen *et al.*,[64] who suggest that metals interact with VP$_n$ complexes leading to M–PV$_n$ complex species, with the calculated formation energies displayed in Figure 4.20. On that basis, all the M–PV$_n$ complexes are thermodynamically stable, including the M–VP$_2$ complex, which is, therefore, a good candidate for the chemical species involved in the gettering activity of phosphorus in a PDG process.

### 4.5.4.2 Modelling the PDG Relaxation Gettering

Relaxation gettering of metallic impurities with the PDG process occurs with the formation of metal silicide precipitates when

- The metals diffuse in the P-rich silicon back-surface layer and concentrate as $MVP_n$ complexes.
- The metal impurity concentration in the P-rich layer is brought to supersaturated conditions $c_M(T) > c_M^\circ(T)$ by a dedicated cooling step process, which brings the system to an equilibrium between the solution of the metal in silicon, the metal silicide phase and the SiP phase, as is the case of Ni in the isothermal section (800 °C) of the ternary phase Ni–Si–P diagram of Figure 4.21,[75] where the P-rich silicon phase contaminated with Ni is evidenced in red.
- The phosphorus concentration overcomes its solubility in silicon $[c_P(T) > c_P^\circ(T)]$ with the formation of SiP precipitates or/and of dense dislocation networks.

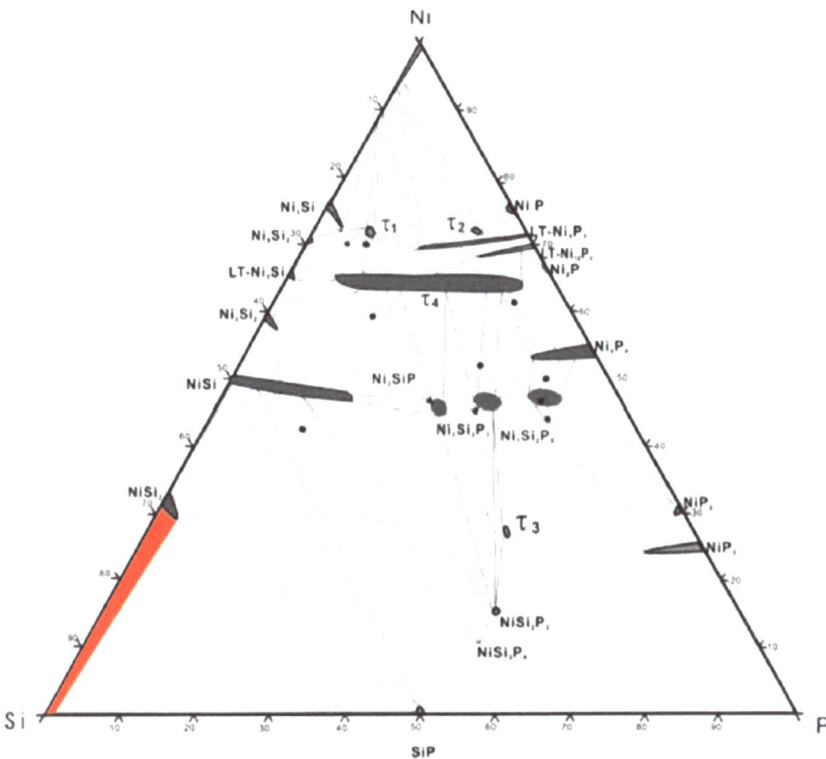

**Figure 4.21**  Isothermal section at 800 °C of the ternary Ni–P–Si. Reproduced from ref. 75 with permission from Elsevier, Copyright 2013.

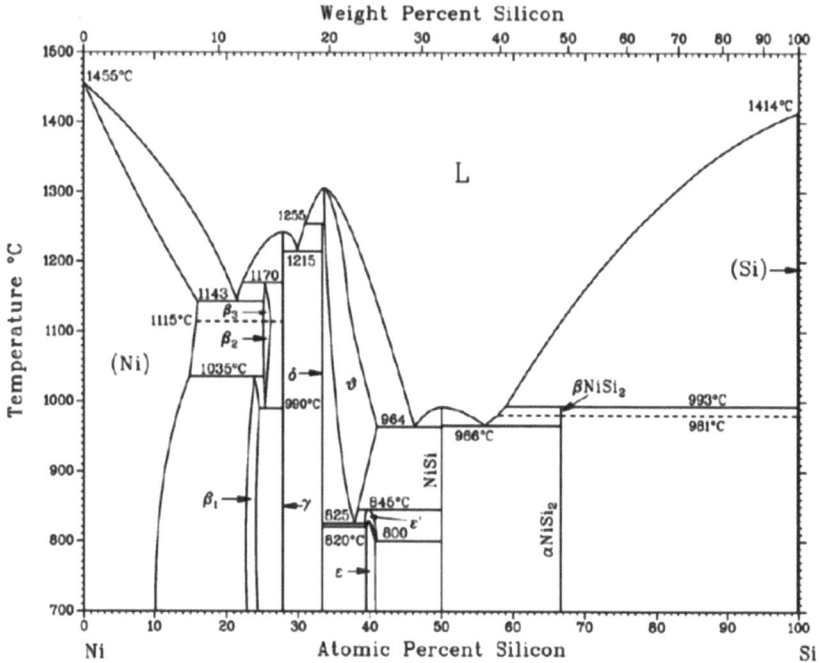

**Figure 4.22** Phase diagram of the binary Ni–Si system. Reproduced from ref. 76 with permission from Elsevier, Copyright 2010.

The practical situation is better seen for the case of Ni impurities in the binary Ni–Si phase diagram of Figure 4.22,[76] which shows that, below the eutectic temperature at 993 °C, the nickel-saturated silicon (α) phase is in equilibrium with the $NiSi_2$ (β) phase

$$\mu_{Ni}^{\alpha} = \mu_{Ni}^{\beta}. \tag{4.27}$$

The schematic diagram of Figure 4.23 illustrates the Ni concentration changes during an appropriate process cycle, consisting of a heating step AB, an isothermal step BC, and a cooling step. The AB step brings the sample to a process temperature $T > 933$ °C, with the concentration of Ni present in the solid solution remaining almost constant, the isothermal BC step enriches the sample in Ni with a segregation process in the P-rich Si layer, and the step CD brings the sample to a temperature where the relaxation occurs, with the precipitation of $NiSi_2$.

It is apparent that to design and profitably operate a PDG relaxation process, it is necessary to know the temperature dependence of the solubility of the metal in the silicon–phosphorus alloy, *i.e.* the

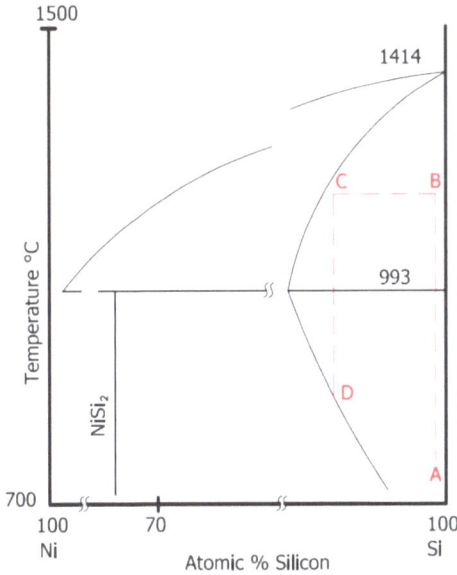

**Figure 4.23** Schematic diagram of a PDG relaxation cycle (in red). The range of the solid solution phase of Ni in silicon is strongly enlarged in size.

width of the Me–Si–P ($\beta$) phase in the ternary Me–Si–P system, in the full process temperatures range: an option not easily available. Additional design requirements are the knowledge of the kinetics of the segregation process at the chosen temperature, to evaluate the metal enrichment in the isothermal step, and the temperature of the relaxation step, where the metal segregates as a metal silicide phase.

Under these strict requirements, not only the design of a PDG process of multiple impurities is very difficult, but even that of single impurities, as shown, as an example, by Istratov *et al.*,[45] who showed that in the case of iron both the segregation and relaxation processes may simultaneously operate, with potentially severe consequences on the behaviour of the electronic components.[77]

Dissolution of P in Si, with its atomic radius very different from that of silicon, induces, in fact, the conditions for the set-up of local mechanical stress $\sigma$

$$\sigma = \frac{E}{1-\nu}\gamma C_s \tag{4.28}$$

where $E$ is the Young modulus, $\nu$ is the Poisson coefficient, $\gamma = \dfrac{r_{imp} - r_{Si}}{r_{Si}}$ (with $r_{imp}$ as the impurity radius and $r_{Si}$ as the silicon radius), and $C_s$ is the dopant concentration at the surface of the sample at the beginning of the P diffusion process.

When $\sigma$ exceeds the critical stress necessary for the generation of dislocation,[78] the system relaxes with the set-up of a network of misfit dislocations, which act as a very efficient impurity gettering system.

An additional condition for dislocation generation occurs when the supersaturated dopant P segregates as an intermediate compound (SiP), leading to the local onset of elastic stress due to molar volumes misfit between the silicon host and SiP, which relaxes with self-interstitial emission and extrinsic dislocation generation. In both cases dislocations will work as beneficial nucleation centres for silicide precipitation and as impurity gettering centres in the final relaxation step, but will also run upward to the surface device region.

Since impurity-decorated dislocations behave as recombination centres for minority carriers (see Section 2.4.5), their presence should be avoided in correspondence with the device junction, requiring delicate attention to the management of the PDG process.

On the other side, self-interstitial emission is responsible of the enhanced diffusion toward the P-rich surface of transition metal impurities sitting in interstitial positions, which mainly diffuse with a kick-off mechanism[79]

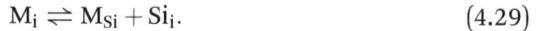

$$M_i \rightleftharpoons M_{Si} + Si_i. \tag{4.29}$$

Thus, PDG involves self-interstitial-enhanced diffusion of impurities toward the P-rich surface layer, where are trapped at dislocations or segregate as solutes in the $Si_{1-x}P_x$ solution with a segregation co-efficient $k_i > 1$

$$k_i = \frac{x_i^{Si_{1-x}P_x}}{x_i^b} > 1 \tag{4.30}$$

since the solubility of the impurity in the P-diffused surface layer is larger than in the bulk.

Several hypotheses have been proposed to explain the enhanced solubility of metal impurities in P-diffused regions, which increases with the increase in the P-doping.[80]

The formation of ternary complexes $M_x P_{1-x} Si$[81] or of electrostatically stabilized pairs $M_{Si}^- P_{Si}^+$ for metallic impurities sitting in substitutional positions,[80] was considered in the past the most physically sensible explanation of the enhanced solubility. We suggest that the formation of ternary $MP_n V$ complexes, which we discussed in the previous section,[64] could be the alternative explanation.

## 4.5.5 Aluminium Gettering

Aluminium gettering is the simplest process among those discussed in this section, as it consists of the equilibrium segregation of impurities from a bulk silicon phase to a diluted (liquid) $Al_x-Si_{1-x}$ phase, which is obtained by depositing a thin Al layer on the Si back surface. The process is carried out with excellent gettering yield at 700–900 °C, above the eutectic temperature of the binary Al–Si system (577 °C),[82] see Figure 4.24.

Al has a relatively low solubility in solid silicon at any temperature (the maximum solubility is $4.3 \times 10^{-4}$ in molar fraction or $2.9 \times 10^{19}$ at cm$^{-3}$ at 1450 K). In addition, the solubility of TM impurities in Al is higher than in silicon. As an example, at 1093 K, the segregation coefficient of Co in Al, $k_{Co}^{Al} = x_{Co(Si)}/x_{Co(Al)} = 10^4$, and therefore the solubility of Co $x_{Co(Al)}$ in Al is $10^4$ times higher than the solubility $x_{Co(Si)}$ of Co in silicon.

The segregation coefficients of Fe and Ti are even larger ($k_{Fe}^{Al}$ (1073 K) = $1.7 \times 10^{11}$, $k_{Fe}^{Al}$ (1273 K) = $5.9 \times 10^9$: $k_{Ti}^{Al}$ (1073 K) = $3.8 \times 10^9$; $k_{Ti}^{Al}$ (1273 K) = $1.6 \times 10^7$), but mention is made by Abdelbarey *et al.*[48] that large deviations from these values could be expected when accounting for the M–Al–Si ternary system involved.

In fact, in ternary systems involving TM elements as the third component, see Figure 4.25 for the case of Fe,[83] the composition range of the Al-rich liquid phase strongly depends on the nature of the

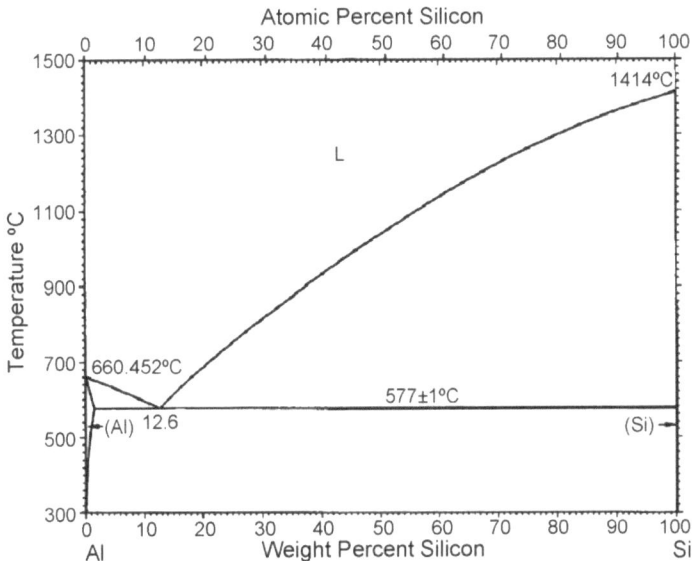

**Figure 4.24** Phase diagram of the Al–Si system. Reproduced from ref. 82 with permission from Elsevier, Copyright 2007.

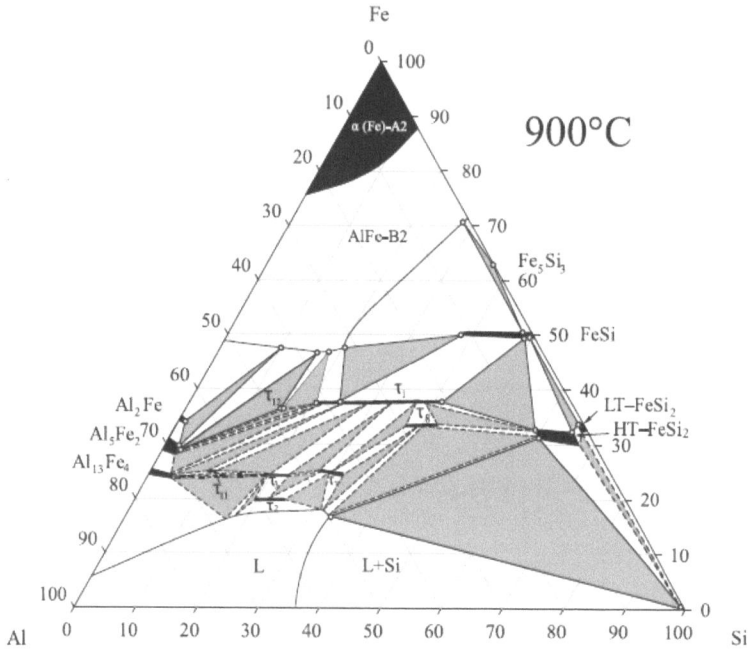

**Figure 4.25** Phase diagram of the system Al–Fe–Si at 900 °C. Reproduced from ref. 83 with permission from Elsevier, Copyright 2011.

TM impurity and the Al liquid phase could be in equilibrium with a metal silicide phase, *i.e.* the FeSi$_2$ phase in the case of iron.

Therefore, the design of a Al-gettering process could be more complicated than that of the PDG relaxation gettering, even for a single impurity, as shown by Abdelbarey *et al.*[48] for the case of iron, and as is immediately apparent from simple thermodynamic considerations.

In the approximation that the solution of the TM impurity M in aluminium behaves as a regular solution,[84] the thermodynamic stability and, therefore, the free energy $G^s$ of the ternary solution phase could be described by the following equation

$$G^s = G^{0,s}_{Al-M}y_{Al-M} + G^{0,s}_{Si-M}x_{Si-m} + RT(y_{Al-M}\ln y_{Al-M} + x_{Si-M}\ln x_{Si-M})$$
$$+ \Omega_{[Al-M]-[Si-M]}y_{Al-M}x_{Si-M}$$

$$(4.31)$$

where $\Omega_{[Al-M]-[Si-M]}$ is the interaction coefficient.[84]

Here, the deviations from the ideality are dominated by the term $\Omega_{[Al-M]-[Si-M]}y_{Al-M}x_{Si-M}$, which accounts for the binary Al–M and Si–M interactions.

The $\Omega_{[Si-M]}$ values could be deduced by the formation enthalpies of the silicides, which are $-20.4$ kJ mol$^{-1}$) for FeSi$_2$, $-78.62$ kJ mol$^{-1}$ for NiSi$_2$ and $-148$ kJ mol$^{-1}$ for CoSi$_2$, while the $\Omega_{[Al-M]}$ values can be evaluated from the respective values of the aluminides, which are $-25$ kJ mol$^{-1}$ for FeAl, $-55.2$ kJ mol$^{-1}$ for CoAl and $-58.99$ for NiAl. On this basis, the $\Omega_{[Al-M]-[Si-M]}$ term is $-4.6$ kJ mol$^{-1}$ for Fe, $-19.63$ kJ mol$^{-1}$ for Ni and $-92.8$ kJ mol$^{-1}$ for Co, showing a progressive increase in impurity interactions and, consequently, of a progressive deviation of the impurity solubility in the ternary M–Al–Si systems with respect to that in the binary M–Al systems.

The success of Al gettering depends, therefore, on the structural and thermodynamic details of the ternary systems involved.

# References

1. J. Ramaunjam and U. P. Sing, Copper indium gallium selenide based solar cells – a review, *Energy Environ. Sci.*, 2017, **10**, 1306.
2. K. V. T. Maremyanin, V. V. Parshin, E. A. Serov, V. V. Rumyantsev, K. E. Kudryavtsev, A. A. Dubinov, S. Fokin, A. P. Morosov, V. Y. Aleshkin, M. Y. Glyavin, G. G. Denisov and S. V. Morozov, Investigation into Microwave Absorption in Semiconductors for Frequency-Multiplication Devices and Radiation-Output Control of Continuous and Pulsed Gyrotrons, *Semiconductors*, 2020, **54**, 1069.
3. T. Sameshima, T. Hayasaka and T. Haba, Analysis of microwave absorption caused by free carriers in silicon, *Jpn. J. Appl. Phys.*, 2009, **48**, 021204.
4. B. Lojec, Low Temperature microwave annealing of source-drain (S/D) extensions, Proceedings 16th IEEE International Conference on Advanced Thermal Processing of Semiconductors, 2008, RTP2008 978-1.
5. K. Kageshima, M. Uematsu, T. Akiyama and T. Ito, Microscopic Mechanism of silicon Thermal oxidation Process, *ECS Trans.*, 2007, **6**(3), 449.
6. R. B. Fair, *Rapid Thermal Processing: Science and Technology*, Academic Press, 2000.
7. E. H. Sin, C. K. Ong and H. S. Tan, Temperature dependence of interband optical absorption of silicon at 1152, 1064, 750 and 694 nm, *Phys. Status Solidi A*, 1984, **85**, 199.
8. S. C. Jain, S. K. Agarwal, W. N. Borle and S. Tata, Total emissivity of silicon at high temperatures, *J. Phys. D: Appl. Phys.*, 1971, **4**, 1207.
9. J. Narayan, R. T. Young and C. W. White, A comparative study of laser and thermal annealing of boron implanted silicon, *J. Appl. Phys.*, 1978, **49**, 3912.
10. O. Oberemok, V. Kladko, V. Litovchenko, B. Romanyuk, V. Popov, V. Melnik, A. Sarikov, O. Gudymenko and J. Vanhellemont, Stimulated Oxygen Impurity Gettering Under Ultra-Shallow Junction Formation in Silicon, *Semicond. Sci. Technol.*, 2014, **29**, 055008.
11. V. Ranki, *Vacancies in highly doped silicon studied by positron annihilation spectroscopy*, PhD thesis, Helsinki University of Technology, Espoo (Finland), 2005.
12. M. Kowalski, J. E. Kowalski and B. Lojek, Microwave annealing for low temperature activation of As in Si, Proceedings 15th IEEE International Conference on Advanced Thermal Processing of Semiconductors, RTP 2007, Catania, Italy, 2007.
13. T. L. Alford, D. C. Thompson, J. W. Mayer and N. D. Theodore, Dopant activation in ion implanted silicon by microwave annealing, *J. Appl. Phys.*, 2009, **106**, 114902.

14. P. Xu, C. Fu, C. Hu, D. W. Zhang, D. Wu, J. Luo, C. Zhao, Z.-B. Zhang and S.-L. Zhang, Ultra-shallow junctions formed using microwave annealing junctions formed using microwave annealing, *Appl. Phys. Lett.*, 2013, **02**, 122114.

15. C. Fu, *et al.*, Understanding the microwave annealing of silicon, *AIP Adv.*, 2017, 7, 035214.

16. J. I. Pankove, D. E. Carlson, J. E. Berkeyheiser and H. O. Wance, Neutralization of Shallow Acceptor Levels in Silicon by Atomic Hydrogen, *Phys. Rev. Lett.*, 1983, **51**, 2224.

17. C. G. Van de Walle, P. J. H. Denteneer, Y. Bar-Yam and S. T. Pantelides, Theory of hydrogen diffusion and reactions in crystalline silicon, *Phys. Rev. B: Condens. Matter Mater. Phys.*, 1989, **39**, 10791.

18. N. M. Johnson, C. Doland, F. Ponce, J. Walker and G. Anderson, Hydrogen in crystalline semiconductors, *Physica B*, 1991, **170**, 3.

19. S. M. Myers, M. I. Baskes, H. K. Birnbaum, J. W. Corbett, G. G. DeLeo, S. K. Estreicher, E. E. Haller, P. Jena, N. M. Johnson, R. Kirchheim, S. J. Pearton and M. J. Stavola, Hydrogen interactions with defects in crystalline solids, *Rev. Mod. Phys.*, 1992, **64**, 559.

20. S. K. Estreicher, Hydrogen-related defects in crystalline semiconductors: a theorist's perspective, *Mater. Sci. Eng., R*, 1995, **14**, 319.

21. S. Pizzini, M. Acciarri, S. Binetti, D. Narducci and C. Savigni, Recent achievements in semiconductor defect passivation, *Mater. Sci. Eng., B*, 1997, **45**, 126.

22. S. Binetti, S. Basu, C. Savigni, M. Acciarri and S. Pizzini, Passivation of extended defects in silicon by catalytically dissociated molecular hydrogen, *J. Phys. III*, 1997, 7, 1487.

23. G. Van de Walle and J. Neugebauer, Universal alignment of hydrogen levels in semiconductors, insulators and solutions, *Nature*, 2003, **423**, 626.

24. C. G. Van de Walle, Universal alignment of hydrogen levels in semiconductors and insulators, *Physica B*, 2006, **376–377**, 1.

25. S. K. Estreicher, M. Stavola and J. Weber, *Hydrogen in Si and Ge in Silicon, Germanium and Silicon-germanium alloys: growth, defects, impurities and nanocrystals*, ed. G. Kissinger and S. Pizzini, CRC Press, 2015.

26. S. Pizzini, *Physical Chemistry of Semiconductor Materials and Processes*, Wiley & Sons, 2015.

27. S. J. Pearton, J. W. Corbett and M. Stavola, *Hydrogen in Crystalline Semiconductors*, Springer, Berlin, 1992.

28. C. H. Seager, R. A. Anderson and J. K. G. Panitz, The diffusion of hydrogen in silicon and mechanism for unintentional hydrogenation during ion beam processes, *J. Mater. Res.*, 1987, **2**, 96.

29. M. Fontana, *Corrosion Engineering*, McGraw-Hill Books, 1987, ISBN 0-07-100360-6.

30. H. Angermann, W. Henrion, M. Rebien and A. Röseler, Wet-chemical passivation and characterization of silicon interfaces for solar cell applications, *Sol. Energy Mater. Sol. Cells*, 2004, **83**(4), 331.

31. C. G. Van der Walle, Energies of various configurations of hydrogen in silicon, *Phys. Rev. B: Condens. Matter Mater. Phys.*, 1994, **49**(7), 4579.

32. C. G. Van de Walle and R. A. Street, Structure, energetics, and dissociation of Si-H bonds at dangling bonds in silicon, *Phys. Rev. B: Condens. Matter Mater. Phys.*, 1994, **49**, 14766.

33. M. Dürr and U. Höfer, Hydrogen diffusion on silicon surfaces, *Prog. Surf. Sci.*, 2013, **88**(1), 61.

34. G. C. Schwartz and P. M. Schaible, Reactive ion etching of silicon, *J. Vac. Sci. Technol.*, 1979, **16**, 410.

35. S. Tachi, K. Tsujimoto and S. Okudaira, Low-temperature reactive ion etching and microwave plasma etching of silicon, *Appl. Phys. Lett.*, 1998, **52**, 616.

36. A. J. van Roosmalen, J. A. G. Baggerman and S. J. H. Brader, *Dry Etching for VLSI*, Springer Science & Business Media, 2013.
37. M. Huff, Recent Advances in Reactive Ion Etching and Applications of High-Aspect-Ratio Microfabrication, *Micromachines*, 2021, **12**, 991.
38. E. R. Weber, Transition metals in silicon, *Appl. Phys.*, 1983, **A30**, 1.
39. R. J. Falster and V. V. Voronkov, Rapid Thermal Processing and the Control of Oxygen Precipitation Behaviour in Silicon Wafers, *Mater. Sci. Forum*, 2008, **573–574**, 45–60.
40. A. Ourmazd and W. Schröter, Phosphorus gettering and intrinsic gettering of nickel in silicon, *Appl. Phys. Lett.*, 1984, **45**, 781.
41. A. Bourret and W. Schröter, HREM of SiP precipitates at the (111) silicon surface during phosphorus predeposition, *Ultramicroscopy*, 1984, **14**(1–2), 97.
42. M. Apel, I. Hanke, R. Schindler and W. Schröter, Aluminum gettering of cobalt in silicon, *J. Appl. Phys.*, 1994, **76**(7), 4432.
43. S. Kusanagi, T. Sekiguchi, B. Shen and K. Sumino, Electrical activity of extended defects and gettering of metallic impurities in silicon, *Mater. Sci. Technol.*, 1995, **11**, 685.
44. C. del Canizo and A. Luque, A comprehensive model for the gettering of lifetime-killing impurities in silicon, *J. Electrochem. Soc.*, 2000, **147**(7), 2685.
45. A. A. Istratov, W. Huber and E. R. Weber, Experimental evidence for the presence of segregation and relaxation gettering of iron in polycrystalline silicon layers on silicon, *Appl. Phys. Lett.*, 2004, **85**(19), 4472.
46. T. Buonassisi, A. A. Istratov, M. D. Pickett, M. A. Marcus, T. F. Ciszek and E. R. Weber, Metal precipitation at grain boundaries in silicon: Dependence on grain boundary character and dislocation decoration, *Appl. Phys. Lett.*, 2006, **89**, 042102.
47. A. Haarahiltunen, H. Vainola, O. Anttila, M. Yli-Koski and J. Sinkkonen, Experimental and theoretical study of heterogeneous iron precipitation in silicon, *J. Appl. Phys.*, 2007, **101**(4), 043507.
48. D. Abdelbarey, V. Kveder, W. Schroter and M. Seibt, Aluminum gettering of iron in silicon as a problem of the ternary phase diagram, *Appl. Phys. Lett.*, 2009, **94**, 061912.
49. A. Haarahiltunen, H. Savin, M. Yli-Koski, H. Talvitie and J. Sinkkonen, Modeling phosphorus diffusion gettering of iron in single crystal silicon, *J. Appl. Phys.*, 2009, **105**(2), 023510.
50. S. P. Phang and D. Macdonald, Direct comparison of boron, phosphorus, and aluminum gettering of iron in crystalline silicon, *J. Appl. Phys.*, 2011, **109**(7), 073521.
51. M. Seibt and V. Kveder, Gettering processes and the role of extended defects, in *Advanced silicon materials for photovoltaic applications*, ed. S. Pizzini, John Wiley & Sons, 2012.
52. F. G. Kirscht, M. B. Shabani, T. Yoshimi, S.-B. Kim, B. Snegirev, C. Wang, L. Williamson, T. Takashima, P. Taylor and D. Lange, Strain and gettering in epitaxial silicon wafers, *Solid State Phenom.*, 1997, **57–58**, 355.
53. C. S. Chen and D. K. Schröder, Kinetics of gettering in silicon, *J. Appl. Phys.*, 1992, **71**, 5858.
54. M. Seibt, A. Döller, V. Kveder, A. Sattler and A. Zozime, Platinum silicide precipitate formation during Phosphorous Diffusion gettering of Platinum, *Solid State Phenom.*, 2002, **82–84**, 411.
55. V. Kveder, W. Schröter, A. Sattler and M. Seibt, Simulation of Al and phosphorus diffusion gettering in Si, *Mater. Sci. Eng., B*, 2000, **71**, 175.
56. M. B. Shabani, T. Yamashita and E. Morita, Study of Gettering Mechanisms in Silicon: Competitive Gettering Between Phosphorus Diffusion Gettering and Other Gettering Sites, *Solid State Phenom.*, 2008, **131–133**, 399.

57. H. Talvitie, V. Vähänissi, A. Haarahiltunen, M. Yli-Koski and H. Savin, Phosphorous diffusion gettering of iron in monocrystalline silicon, *J. Appl. Phys.*, 2011, **109**, 093505.

58. J. Schön, V. Vähänissi, A. Haarahiltunen, M. C. Schubert, W. Warta and H. Savin, Main defect reactions behind the phosphorous diffusion gettering of iron, *J. Appl. Phys.*, 2014, **116**, 244503.

59. S.-M. Liang and R. Schmid-Fetzer, Modeling of Thermodynamic Properties and Phase Equilibria of the Si–P System, *J. Phase Equilib. Diffus.*, 2014, **35**, 24.

60. G. D. Watkins and J. W. Corbett, Defects in Irradiated Silicon: Electron Paramagnetic Resonance and Electron–Nuclear Double Resonance of the Si–E Center, *Phys. Rev.*, 1964, **134**, A1359.

61. V. P. Markevich, O. Andersen, I. F. Medvedeva, J. H. Evans-Freeman, I. D. Hawkins, L. I. Murin, L. Dobaczewski and A. R. Peaker, Defect reactions associated with the dissociation of the phosphorus–vacancy pair in silicon, *Phys. B*, 2001, **308–310**, 513.

62. A. Nylandsted Larsen, A. Mesli, K. Bonde Nielsen, H. Kortegaard Nielsen, L. Dobaczewski, J. Adey, R. Jones, D. W. Palmer, P. Briddon and R. S. Öberg, E Center in Silicon Has a Donor Level in the Band Gap, *Phys. Rev. Lett.*, 2006, **97**, 106402.

63. S. Dannefaer, G. Suppes and V. Avalos, Evidence for a vacancy-phosphorus-oxygen complex in silicon, *J. Phys. Condens. Matter.*, 2009, **21**(1), 015802.

64. R. Chen, B. Trzynadlowski, T. Scott and S. T. Dunham, Phosphorus vacancy cluster model for phosphorus diffusion gettering of metals in Si, *J. Appl. Phys.*, 2014, **115**, 054906.

65. X. Zhu, Y. X. Xuegong, P. Chen, Y. Liu, J. Vanhellemont and D. Yang, Effect of Dopant Compensation on the Behaviour of Dissolved Iron and Iron-Boron Related Complexes in Silicon, *Int. J. Photoenergy*, 2015, 154574.

66. L. Lerner and N. A. Stolwijk, Vacancy concentration in silicon determined by the indiffusion of iridium, *Appl. Phys. Lett.*, 2004, **86**, 011901.

67. S. Obeidi and N. A. Stolwijk, Diffusion of iridium in silicon: Changeover from a foreign-atom-limited to a native-defect-controlled transport mode, *Phys. Rev. B: Condens. Matter Mater. Phys.*, 2001, **64**, 113201.

68. H. Bracht, J. Fage Pedersen, N. Zangenberg, A. Nylandsted Larsen, E. E. Haller, G. Lulli and M. Posselt, Radiation enhanced silicon self-diffusion and the silicon vacancy at high temperatures, *Phys. Rev. Lett.*, 2003, **91**, 245502.

69. M. Lee, H.-Y. Ryu, E. Ko and D.-H. Ko, Effect of Electrical, and Chemical Properties of Phosphorus-Doped Silicon Films, *ACS Appl. Electron. Mater.*, 2019, **1**(3), 288.

70. R. Car and E. Smargiassi, First-principles free energy calculations on condensed-matter systems: Lattice vacancy in silicon, *Phys. Rev. B: Condens. Matter Mater. Phys.*, 1996, **53**(15), 9760.

71. C. Sutton and S. V. Levchenko, First-principles Atomistic Thermodynamics and Configurational Entropy, *Front. Chem. B.*, 2020, **8**, 757.

72. V. Ranki and K. Saarinen, Formation of thermal vacancies in highly As and P doped silicon, *Phys. Rev. Lett.*, 2004, **93**, 255502.

73. V. Ranki, *Vacancies in highly doped silicon studied by positron annihilation spectroscopy*, PhD thesis, Helsinki University of Technology, Espoo (Finland), 2005.

74. R. V. Siegel, Positron annihilation spectroscopy, *Annu. Rev. Mater. Sci.*, 1980, **10**, 393.

75. Y. Liu, D. Cao, T. H. Hao, X. Su, J. Wang, C. Wu and H. Peng, Phase equilibria of the Ni–Si–P system at 800 °C, *J. Alloys Compd.*, 2013, **577**, 643.

76. E. Çadırlı, D. M. Herlach and E. Davydov, Microstructural, mechanical, electrical, and thermal characterization of arc-melted Ni–Si and Co–Si alloys, *J. Non-Cryst. Solids*, 2010, **356**, 1735.

77. M. Seibt, Relaxation-Induced Gettering of Metal Impurities in Silicon: Microscopic Properties of Effective Gettering Sites, *MRS Online Proceeding Library (OPL)*, 1992, **262**, 957.
78. S. Prussin, Generation and Distribution of Dislocations by Solute Diffusion, *J. Appl. Phys.*, 1961, **32**, 1876.
79. U. Gösele, W. Frank and A. Seager, Mechanism and kinetics of the diffusion of gold in silicon, *Appl. Phys.*, 1980, **23**, 361.
80. L. Baldi, G. F. Cerofolini, G. Ferla and G. Frigerio, Gold solubility in silicon and gettering by phosphorus, *Phys. Status Solidi A*, 1978, **48**, 523.
81. R. W. Wilcox, T. J. La Chapelle and D. H. Forbes, Gold in Silicon: Effect on Resistivity and Diffusion in Heavily-Doped Layers, *J. Electrochem. Soc.*, 1964, **111**, 1377.
82. W.-Y. Li, C. Zhang, X. P. Guo, G. Zhang, H. L. Liao and C. Coddet, Deposition characteristics of Al–12Si alloy coating fabricated by cold spraying with relatively large powder particles, *Appl. Surf. Sci.*, 2007, **253**(17), 7124.
83. M. C. J. Marker, B. Skolyszewska-Kuhberger, H. S. Effermberg, C. Schmetterer and K. W. Richter, Phase equilibria and structural investigations in the system Al-Fe-Si, *Intermetallics*, 2011, **19**, 1919.
84. S. Pizzini, *Physical chemistry of semiconductor Materials and processes*, Wiley, 2015, ch. 1.

# 5 Semiconductor Nanomaterials

## 5.1 Introduction

Zero-dimensional, one-dimensional, two-dimensional and three-dimensional nanostructured materials are a family of "small" systems, which have already found widespread technological applications in chemical catalysis, and in drug delivery for medical applications,[†] but are also credited with novel application in advanced electrochemistry (lithium and sodium batteries, fuel cells) and solar cells,[1] of strategic impact for the planet's survival of the climate crisis.

The class of nanostructured materials which we intend to discuss in this chapter is that of nanosized elemental and compound semiconductors, for their potential impact on advanced photonic applications,[2–4] and on chemical sensor technologies, granted using wide-gap metal oxide semiconductors.[5]

The advantage of semiconducting nanomaterials, presently manufactured as nanometric grains (dots), wires, rods and films, is that their chemical and physical properties strongly depend on their size, morphology and crystalline structure, with important differences from the properties of the corresponding macroscopic bulk phases.

Figure 5.1 displays, as an example, the morphology of micro- and nano-rods of ZnO,[‡] prepared by wet techniques, and the effect of the morphology on their photoluminescence (PL) emission.[6]

---

[†] For these applications the use of fatty acids nanospheres to protect the mNRA in Covid vaccines is the last, important innovation.
[‡] A semiconducting oxide already used as an active sensor material.

---

Chemistry of Semiconductors
By Sergio Pizzini
© Sergio Pizzini 2024
Published by the Royal Society of Chemistry, www.rsc.org

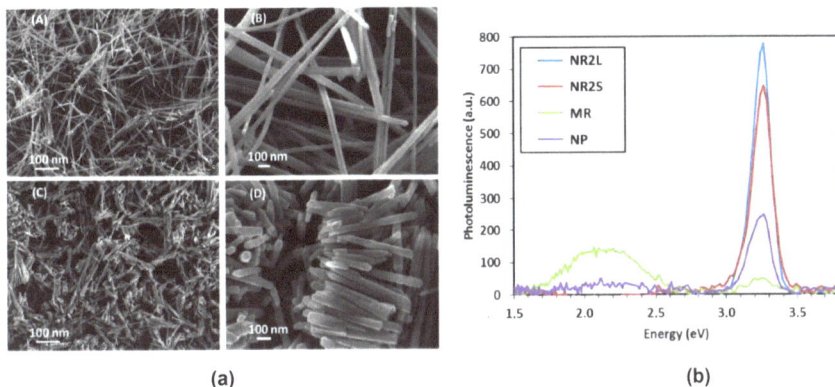

Figure 5.1 (a) SEM images of ZnO nano (A and C) and microrods (B and D), (b) the corresponding photoluminescence (PL) spectra. Reproduced from ref. 6 with permission from the Royal Society of Chemistry.

The effect of the morphology on PL is very clear, since only nanorods (NRs) present a strong excitonic emission at ~3.26 eV, in close correspondence with the value of the energy gap of bulk ZnO at 3.3 eV. This emission is, instead, very feeble in microrods (MR), which exhibit an additional emission at 2.13 eV, possibly due to defects that behave as secondary emission centres. These features seem to be common to other ZnO nanostructures, whose band gap depends on the morphology and ranges between 3.33 and 3.19 eV for undoped mixtures of ZnO nanorods and spherical particles,[7] with significant band gap narrowing or widening.

Even more critical is the effect of the nano-structuration on silicon nanowires, since they may crystallize both with the diamond-cubic and with the hexagonal structure of wurtzite,[§] this last presenting a cathodoluminescence emission at 1.5 eV, blue-shifted at 0.4 eV with respect to the emission of diamond cubic silicon,[8] see Figure 5.2, where a comparison is given between the emissions of diamond cubic- and hexagonal-silicon NWs.

One of the most important consequences of nano-structuration is, in fact, a change in the electronic structure of the material with size, which leads to the development of a distribution of bonding and antibonding orbitals at nanometric sizes, with the set-up of discrete energy levels, as is schematically shown in Figure 5.3. This gives to a nanostructured material physical properties that are intermediate between molecules and bulk semiconductors,[9] strongly depending on their morphology.

---

[§] See Appendix 5.1 for details about the structure of silicon allotropes.

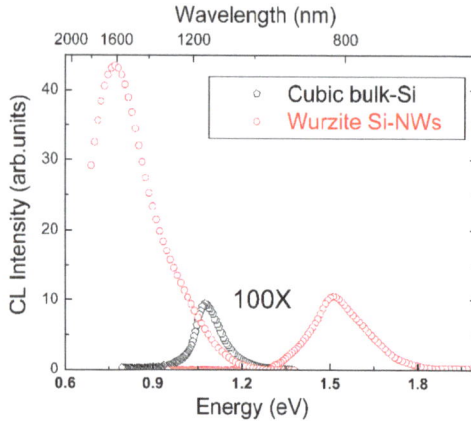

**Figure 5.2**  Cathodoluminescence spectra of cubic silicon and of wurtzite nanowires. Reproduced from ref. 8 with permission from Springer Nature, Copyright 2014.

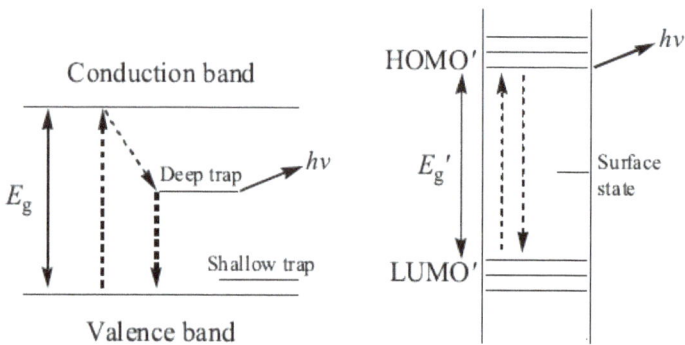

**Figure 5.3**  A comparison of the configuration of the electronic diagram of a bulk semiconductor with that of a nanosized one. Reproduced from ref. 9 with permission from MDPI, Copyright 2003.

The physical and chemical properties of nano-structured solids[¶] are not only influenced by the modification of their electronic structure, but may also be controlled by quantum-size effects, by surface and interface effects, and by size-induced modifications to their lattice symmetry,[3,10–13] which lead to the stabilization of structural properties that are not present in the bulk reference phases.

The main consequence of quantum size conditions is the localization (confinement) of carriers due to a reduction in the degrees of freedom of the mobile carrier particles, or to a reduction in their phase space.[14]

---

[¶] Metallic and semiconducting solids.

Quantum confinement (QC) appears when one or more spatial dimensions of a nanoparticle compare with the Bohr radius $a_B$ of the excitons in the semiconductor[15] (see Table 5.1)

$$a_B = \frac{h^2 \varepsilon}{e^2} \left( \frac{1}{m_e^*} + \frac{1}{m_h^*} \right) = \frac{h^2 \varepsilon}{e^2 \mu} = a_0 \frac{\varepsilon}{\varepsilon^\circ} \frac{m_e}{\mu} \qquad (5.1)$$

where $\varepsilon$ is the dielectric constant of the material, $a_0 = \dfrac{\varepsilon^\circ h^2}{m_e e^2} = 0.0529$ nm is the Bohr radius of hydrogen, $\varepsilon^\circ$ is the permittivity of the free space, $m_e$ is the rest mass of the electron, $\mu = \dfrac{m_e^* m_h^*}{m_e^* + m_h^*}$ is the reduced mass of the exciton, and $m_e^*$ and $m_h^*$ are the effective masses of the electron and hole, which depend on the semiconductor nature.

Accordingly, a nanometric grain (a dot) might be confined in three dimensions, a nanometric wire in two dimensions and a nanometric film (or a well) in one dimension.

One of the main effects of QC on indirect gap semiconductors (Ge and Si, as an example) is to allow the onset of optical transitions without the aid of a phonon. In fact, the breaking of the momentum conservation rule[15] occurs because of the reduction of the system dimensions, which implies a spread of the electron/hole wavefunctions in the momentum space. This would allow the onset of pseudo-direct transitions in nanometric Ge and Si crystals, which will, however, differ from the direct transition for their lifetimes that are of the order of microseconds, as compared with lifetimes of picoseconds for direct transitions.[14]

**Table 5.1** Exciton Bohr radius of selected semiconductors.

| Material | Exciton Bohr radius (nm) |
| --- | --- |
| ZnO | 0.9 |
| Si | 4.2 |
| CdSe | 6 |
| CdTe | 8 |
| GaAs | 15 |
| PbS | 20 |
| Ge | 24 |
| InAs | 34–74 |
| PbSe | 46 |
| InSb | 86–138 |
| PbTe | 1700 |

Besides these effects, a dot radius-depending blue shift $E(R)$ of the energy gap $E_G$, and thus of the optical emission, is observed in nanometric semiconductors

$$E(R) = E_G + \frac{\hbar^2 \pi^2}{2\mu R^2} - \frac{1.786 e^2}{\varepsilon R} - 0.248 E_R \qquad (5.2)$$

where $R$ is the radius of the semiconducting nanoparticle, $\mu = \dfrac{m_e^* m_h^*}{m_e^* + m_h^*}$ is again the reduced mass of the exciton, $\varepsilon$ is the dielectric constant and $E_R$ is the effective Rydberg energy, making absorption and PL spectroscopies the favoured investigation tools in nanostructural research.

Since also size-depending strain induces modifications in the band structure,[16] its effect cannot be neglected because it becomes predominant in arrays of close dots, where it is shown to modify also their structural self-organization.[17,18]

Pure quantum confinement effects, like the blue shifts of the PL, might be, however, experimentally difficult to obtain for some of the semiconductors reported in Table 5.1 because of the value of the size scale, in one or more directions, needed to fulfil the condition $L \leq a_B$. And even if this condition is fulfilled, it cannot be excluded that the experimentally observed changes in the physical properties of the nanostructured material *vs.* that of the bulk is not be due to other causes.

As an example, a dot of 1 nm of silicon, with a relative Bohr size of 0.24 Bohr excitons, has almost 90% of atoms as surface atoms, and it is disputable whether the luminescence[||] of this silicon dot depends only on QC effects, or on surface defects, which might behave as secondary light emission centres.

Extended defects are also systematically present in nanophases. Twins and stacking faults, as an example, are typical defects of silicon carbide (SiC) nanowires, but also of nanocrystalline silicon films,[19] see Figure 5.4. Their presence not only influences the mechanical properties of the Si and SiC nanocrystals, but also their optoelectronic properties, as was shown, among many others, by Wang *et al.*[20]

We can, therefore, foresee that surface, interface and extended defects, together with size-induced modifications of the lattice symmetry, would influence, or dominate the physical properties of nanostructures.

---

[||] With respect to a macroscopic sample.

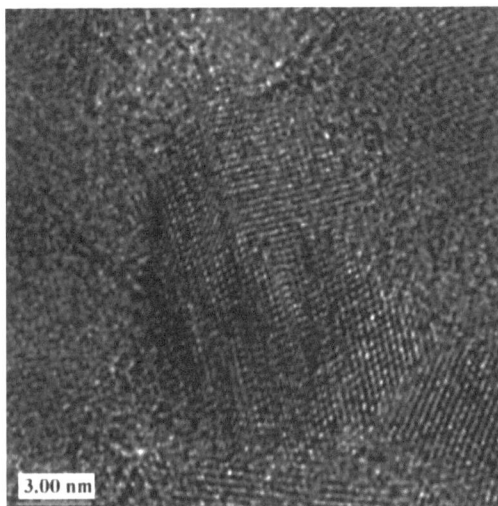

3.00 nm

**Figure 5.4**  A stacking fault decorating a grain of a nanocrystalline silicon film. Reproduced from ref. 19 with permission from Elsevier, Copyright 2006.

Besides optoelectronic properties, the thermodynamic properties are also expected to change at the nanometric length scale, as is the case of the melting temperature $T_m$ (and of the eutectic temperature) of (metallic and) semiconductor nanocrystals,[21,22] which decreases with the decrease in size, as can be seen in Figure 5.5[23] for the typical case of Ge nanostructures.

It is, in fact, easy to demonstrate that thermodynamics should fail when applied to nanoparticles, with their atomic mass density much less than that of the corresponding bulk phase and the large surface to volume ($S/V$) ratio, because the intensive variables ($T$, $P$, $\mu$), which are independent of each other in a bulk solid, are interrelated in a nanometric solid,[24,25] and also because the surface energy term cannot be neglected.

Therefore, the basic Gibbs equilibrium equation of a macroscopic system ($N \to \infty$)

$$dE = TdS - pdV + \sum_i \mu_i dN_i \qquad (5.3)$$

cannot be directly applied, as such, to a nanoscopic system because an additional surface energy term $E^*$ proportional to $N^{2/3}$ should be added to eqn (5.3).

This additional term should be calculated with respect to a macroscopic ensemble, consisting of $N$ equivalent non-interacting

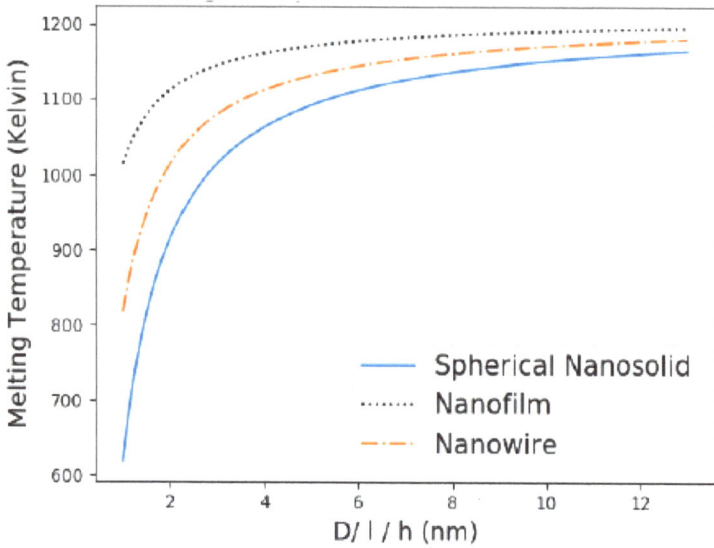

**Figure 5.5**   Evolution of the melting temperature of nano-germanium dots, wires and films as function of their size. Reproduced from ref. 23, http://www.ijnnonline.net/article_245843.html, under the terms of the CC BY 4.0 license, https://creativecommons.org/licenses/by/4.0/legalcode.

systems, rather than a single system, to be formally compatible with the condition of eqn (5.3) derived for a macroscopic system.[24]

On that basis, the equilibrium condition for a nano-system is given by the following equation, with the surface energy $E^*$ as an intensive property of the nano-system

$$dE_i = T dS - p dV + \sum_i \mu_i N_i + E^* dN \tag{5.4}$$

which, after integration and division by $N$, gives the total energy $E$ of a nano-system

$$E = TS - pV + \sum_i \mu_i N_i + E^*. \tag{5.5}$$

A significant example of the failure of thermodynamics applied to nano-systems is given when using the Thomson equation[26] to account for the decrease in the melting temperature $T_m$ of a silicon nano-crystal with decrease in the size

$$\frac{T_0 - T_m}{T_0} = \frac{2\sigma_{sl}}{R\lambda_0} v_s \tag{5.6}$$

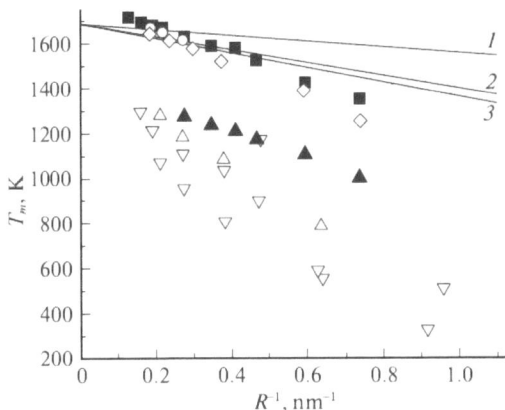

**Figure 5.6** Size-dependence of the melting point of silicon. $\nabla$ and $\triangle$ are experimental points. Other symbols are calculated values using molecular dynamics. Straight lines 1, 2 and 3 display the results obtained using the Thomson equation for three different values of $\sigma_{sl}$. Reproduced from ref. 22 with permission from Springer Nature, Copyright 2019.

where $T_0$ is the melting temperature of the bulk phase, $\sigma_{sl}$ is the surface tension at the interface solid/melt, $v_s$ is the specific volume of the solid bulk crystalline phase, $\lambda_0$ is the macroscopic melting heat and $R$ is the radius of the nanoparticle.

As can be seen in Figure 5.6,[22] the experimental dependence of the melting temperature of silicon nanoparticles on the particles radius $R$ follows the $1/R$ behaviour predicted by the Thomson equation, but the experimental values of melting temperatures strongly deviate from the values calculated using the Thomson equation when applied with values of surface tension $\sigma_{sl}$ characteristic of the bulk case, and from the prediction of molecular dynamics (MD) simulations.

Considering that the Thomson equation holds for a thermodynamic equilibrium involving a surface melt and a core consisting of a crystalline solid with the characteristics of its bulk phase, the deviations from the Thomson equations are understandable, if we consider that the core of a nanocrystal consists of a nanophase with a large $S/V$ ratio, which will lead to the dominance of the surface properties on the properties of the whole.[25]

The deviations of the experimental values of melting temperatures *vs.* those of the MD simulations are also understandable if we follow the arguments of Ercolessi *et al.,*[27] who mention *that the surfaces of nanophases are far being ideal, that surface reconstruction takes place, that the assumption of a size-independent surface energy is questionable, and that also the very concept of surface energy is ill defined.*[**]

The consequences of these limits can be observed in Figure 5.7, which shows that the calculated temperature $T_m$ of gold nanoparticles decreases with the decrease in the cell size from the melting temperature of bulk gold (1377 K), but that the typical signatures of melting, a jump of the internal energy $\varepsilon(T)$, which corresponds to the melting enthalpy $\Delta H_m$, can be observed only with the largest cell sizes ($N = 477$ and 879), while melting occurs without a jump for a cell size of $N = 219$.

This means that the melting enthalpy decreases with the size, that a critical diameter $d_c$ exists at which the melting enthalpy vanishes, but that $T_m$ is still finite.

The self-assembly of metallic and semiconductor nanocrystals into super-crystals (SC) has also been demonstrated by Coropceanu *et al.*[28] They showed that colloidal solutions of metallic and semiconductor NCs capped with conductive inorganic metal chalcogenide complex ligands (such as $Sn_2S_6^{4-}$, $Sn_2Se_6^{4-}$, $In_2Se_4^{2-}$, $AsS_4^{3-}$, and $Cu_6S_4^{2-}$), using alkali metal complexes as charge balance, undergo a phase transition with the formation of supercrystals, which combine long-range order, see Figure 5.8, with strong electronic coupling.

Eventually, non-stoichiometry, which is an intrinsic feature of compound semiconductors, leads to the presence of wide homogeneity regions in their $x - T$ domains, shifts the melting point from that of the corresponding bulk phases and leads to the presence of point defects that compensate the non-stoichiometry and may behave as

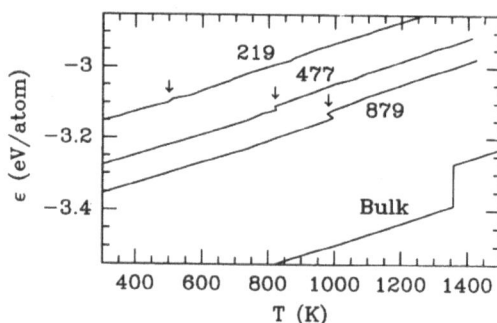

**Figure 5.7** Temperature dependence of the internal energy of gold clusters of $N = 219$, 477, 879 and for the infinite bulk. Reproduced from ref. 27 with permission from American Physical Society, Copyright 1961.

---

[**] In italics Ercolessi's sentence.

**Figure 5.8** TEM micrographs of ordered domains of Pd, Ni, PbS and PbSe. Reproduced from ref. 28 with permission from the American Association for the Advancement of Science, Copyright 2022.

carrier recombination centres.[29] As mentioned by Richard Feynman more than half a century ago *"There is plenty of room at the bottom"*.

## 5.2 Preparation, and Structural and Physical Properties of Nanocrystalline Silicon

Silicon nanocrystals, nanowires and nanocrystalline films represent a class of materials of relevant fundamental and technological interest because nano-crystallinity promotes novel photonic properties on silicon, that, due to its poor optical properties in the bulk phase, is unable to compete with compound semiconductors for light emitting devices and laser applications, despite years of almost unsuccessful research work on Er impurities,[30] dislocations and oxide precipitates[31–34] as potential sources of improved optical properties.

In the nanocrystalline configurations, instead, as individual nanocrystallites, as nanocrystalline films, consisting of a distribution of silicon nanocrystals in a dielectric matrix of amorphous silicon (a-Si) or of silicon oxide ($SiO_2$), as silicon nanocrystals/$SiO_2$ superlattices, and as Si nanowires, nanocrystalline silicon (Si NC) presents not only very promising photonic properties[2] but also a chance for its full integration with silicon microelectronics.

It will be shown, however, that the chemical contaminations and surface oxidation arising from impurities in the preparation process, the presence of surface amorphous phases and of extended defects (stacking faults and twins), and the set-up of phase transitions induced by surface strain, play a critical role on the properties of these nanomaterials, preventing so far their full technological application.

### 5.2.1 Nanocrystalline Silicon Dots

Nanocrystalline silicon dots can be prepared by several dry and wet methods, as a function of the final application, since their properties strongly depend on the process used. As an example, mechanical milling, centrifugation and spry drying were used to fabricate silicon nanoparticles with an average size of 62 nm, employed for high performance Si-electrodes in Li-ion batteries,[35] a process that certainly could not be used for photonic applications, where dedicated wet and dry processes are instead applied.

It is, however, easy to demonstrate that the fabrication of Si NC for microelectronic and photonic application with both dry and wet processes is very demanding in terms of product reproducibility, as can be seen when looking to the photoluminescence (PL) emission wavelength $\lambda_{em}$ of Si NC with a size compatible with QC emission, see Table 5.2, which is shown to depend not only on the synthetic process used, but also on the surface treatment adopted to protect the Si NC from atmospheric degradation, with variations of the optical emission not compatible with their own size.[36]

In the remainder of this section we will deal only with dry processes, for their almost immediate compatibility with conventional microelectronic processes, which will enable future integration.

This is the case of the fabrication, at mbar pressures, of nanocrystalline silicon dots using a radiofrequency (RF)-activated plasma of silane and argon, confined in the inner space of the two electrodes of the continous-flow, non-thermal plasma reactor shown in Figure 5.9.[37]

Under these process conditions, a powder consisting of individual silicon nanocrystals, which present a core-ordered structure and a defective surface, is created inside the plasma phase,[38,39] whose size and rate of formation depends on the residence time and on the gas mixture composition.

**Table 5.2** Photoluminescence emission $\lambda_{em}$ of silicon nanocrystals as a function of the process route employed for their preparation and of the surface coverage. Data from ref. 36.

| Synthetic process | Average size (nm) | Surface coverage | $\lambda_{em}$ (nm)/eV |
|---|---|---|---|
| Anodic etching of porous Si | 1 | H-capped | 320–380/3.87–3.3 |
| Ar/silane in a continuous flow reactor | 1.6 | Octyl-capped | 420/2.95 |
| Wet $SiCl_4$ reduction in the presence of micelles | $1.6 \pm 0.2$ | Alkyl functionalized | 280–290/4.4–4.3 |
| Wet $SiCl_4$ reduction in the presence of micelles | $1.8 \pm 0.2$ | 1-Heptyl capped | 335/3.7 |
| Reaction of sodium silicide with $NH_4Br$ | $3.9 \pm 1.3$ | Octyl capped | 438/2.8 |

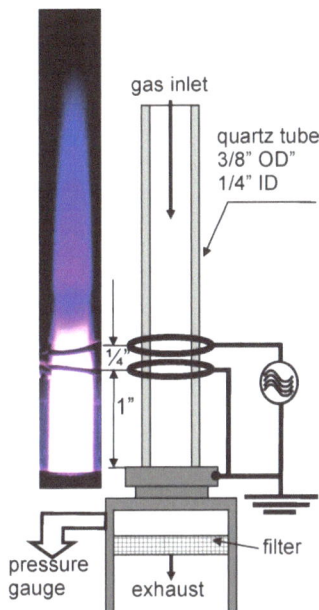

**Figure 5.9** Schematic diagram of a continuous-flow non-thermal plasma reactor. Reproduced from ref. 38 with permission from American Chemical Society, Copyright 2005.

Individual silicon nanocrystallites might behave, depending on their size, as single quantum dots and, thus, as single photo-luminescence sources, as shown by Foitik *et al.*,[40,41] who prepared nanometric Si crystallites, see Figure 5.10, in a silane ($SiH_4$)–argon plasma, which were subsequently suspended in cyclohexane to protect them from irreversible atmospheric oxidation.

It was shown that the luminescence emission of the as-grown particles, after ageing the suspension in air, lies in the yellow-red range (1.5–1.9 eV) for dot sizes ≤3 nm, while those etched in an HF solution (dot size less than 1 nm) emit in the blue (2.7 eV). While it is apparent that a strong blue shift is present in both the samples,[††] the problem is why the dots submitted to a mild oxidation in solution are optically active in the red-yellow, whereas the dots etched in HF are optically active in the blue.

Considering the size and the morphological properties of the two samples, see Figure 5.10, it could be suggested that that the red luminescence is due to a recombination process at an oxide shell

---

[††] The energy gap of bulk silicon is 1.12 eV at 300 K.

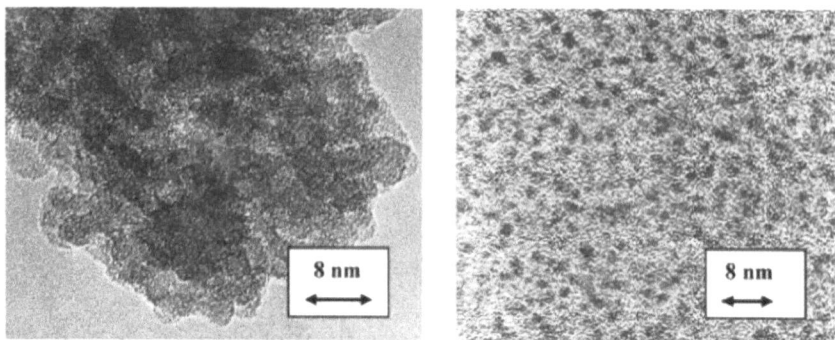

**Figure 5.10** TEM micrographs of red luminescent (left) and blue luminescent (right) Si nanoparticles. Reproduced from ref. 41 with permission from American Chemical Society, Copyright 2006.

covering the dot surface, grown as the consequence of the mild oxidation, while the blue luminescence is compatible with exitonic recombination at hydrogen-passivated Si dots, substantially free of surface defects.[42]

We recall here that hydrogenation with dry or wet methods is a common defect passivation treatment for bulk silicon surfaces, which makes the surface defects, mostly dangling bonds, electronically inactive.[3]

In the dry case, surfaces are passivated by the direct absorption of hydrogen in a plasma phase on silicon dangling bonds[43] with the formation of stable Si–H bonds

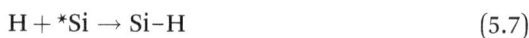

$$H + {^*}Si \rightarrow Si\text{–}H \tag{5.7}$$

where $^*$Si is a dangling bond at the silicon surface.

In the wet case, using an aqueous solution of HF, the process proceeds[‡‡] with the initial etching of the oxide layer[44]

$$4HF + SiO_2 \rightarrow SiF_4 + 2H_2O \tag{5.8}$$

which leaves a hydrophilic surface on which the following reaction occurs

$$Si(OH)^-_{surf} + HF \rightleftharpoons SiF^-_{surf} + H_2O \tag{5.9}$$

which leaves a fluorinated hydrophobic surface.

---

[‡‡] Simplified reactions.

Hydrogen passivation is eventually obtained after a long HF etch, with the formation of stable Si–H bonds at the silicon surface

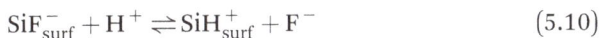

$$SiF_{surf}^- + H^+ \rightleftharpoons SiH_{surf}^+ + F^- \tag{5.10}$$

which saturate the surface defects.

The conclusions concerning the Foitik works,[40,41] which show the effect of size and surface passivation on the PL emission of Si NC, are validated by a work of Beard *et al.*,[45] who demonstrate that Si NC with a size of 9.5 nm, dispersed in hexane or tetrachloroethylene, with well passivated surfaces, maintain the properties of an indirect gap semiconductor, because no direct gap transition is detectable in this size range, while a QC emission, with a blue-shift of the photo-luminescence, is observed for NC with a diameter of 3.8 nm, see Figure 5.11a, with an emission peak at 1.8 eV.

These results are also supported by Hannah *et al.*,[46] who were able to show that the PL emission of silicon nanocrystals with sizes within 2.6 and 4.6 nm covers the range 1.3–1.9 eV, see Figure 5.11b, with QC emissions. It should be therefore concluded that QC effects in Si NCs are only observed for crystal sizes ≤3.8 nm, but that oxidation and passivation processes also strongly influence their PL emission energy.

## 5.2.2 Nanocrystalline Germanium Dots

Ge is a particularly attractive material for optoelectronic applications at the nanosize, since it combines a large exciton Bohr radius (24 nm)

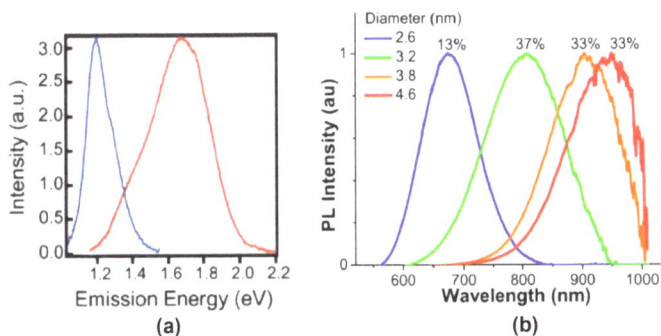

Figure 5.11 Size dependence of the photoluminescence of Si NCs. (a) Reproduced from ref. 45 with permission from American Chemical Society, Copyright 2007. (b) Reproduced from ref. 46 with permission from American Chemical Society, Copyright 2012.

with a narrow band gap (0.661 eV) and high carrier mobilities ($\mu_e = 3900$ cm$^2$V$^{-1}$s$^{-1}$, $\mu_h = 1900$ cm$^2$V$^{-1}$s$^{-1}$). Its application would be compatible with existing CMOS processing methods, but for the majority of applications considered, which include chemical sensors, solar cells and photodetectors, there is the need for high production yield, narrow size distribution and reproducible surface properties; conditions that are not yet achievable.

Both dry and wet methods have been used for the fabrication of Ge dots.

A radio frequency magnetron sputtering from a Ge target in Ar–H$_2$ atmosphere at a pressure of 1.5 Torr was used, as an example, by Shiratani et al.[47] At this pressure, the mean free pass of Ge atoms in the gas phase is short, allowing the formation of Ge nanoparticles of 10 nm in diameter, that are then deposited on a glass substrate at 180 °C. These Ge dots have been used for quantum-dot sensitized solar cells, with very poor preliminary results, that only show the potentialities of the technology but require important supplementary work to reduce the carrier recombination losses due to still unknown defects.

Solution-phase synthesis of Ge NCs was, instead, carried out by Carolan et al.[48] by reduction of Ge halide salts (GeCl$_4$ or GeBr$_4$) by hydride reducing agents, as lithium aluminium hydride, LiAlH$_4$, which allows Ge nanoparticles to be obtained with a diameter of $3.0 \pm 0.4$ nm, much below the Bohr exciton radius and thus strongly quantum confined.

High resolution TEM patterns of Ge NCs confirm the diamond cubic structure of the nanocrystals, while the photoluminescence spectra displayed in Figure 5.12 show that Ge nanocrystals prepared by reduction of GeBr$_4$ present a maximum of intensity of emission at 320 nm (3.9 eV), largely blue-shifted above the energy gap of Ge, due to quantum confinement.

A seedless heteroepitaxial process has also been exploited, using a square or hexagonal pattern of pits, approximately 40 nm deep, on an oxidized Si$\langle 001 \rangle$ substrate, prepared by electron beam lithography and plasma etching.[17]

Ge NC nucleation occurs directly inside the pits, see Figure 5.13a and c, but a self-ordered nucleation of Ge nano-islands arrays also occurs at the strained periphery of the pits with square or hexagonal arrangement, see Figure 5.13b and d. When the distance between the pits is sufficiently large to make negligible their interaction, the number of NCs nucleated around each pit is six. It is, therefore,

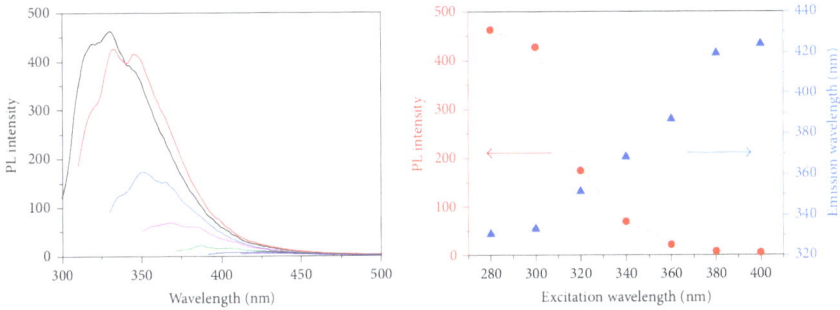

**Figure 5.12** Left: Photoluminescence spectra of Ge nanocrystals prepared by reduction of GeBr$_4$ as a function of the excitation wavelength. Right: Influence of the excitation wavelength on the PL intensity. Reproduced from ref. 48, https://doi.org/10.1155/2015/506056, under the terms of the CC BY 3.0 license, https://creativecommons.org/licenses/by/3.0/.

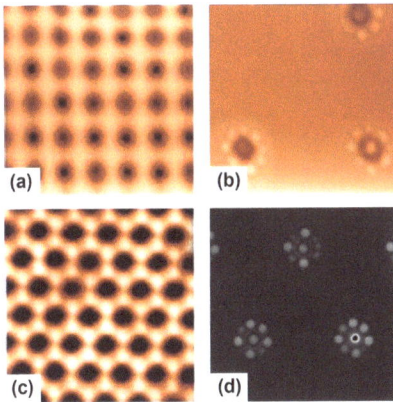

**Figure 5.13** AFM images (3.0 × 3.0 μm$^2$) of arrays of laterally ordered Ge(Si) QDs on pit-patterned SOI substrates: (a), (b) pits arranged to a square lattice with a period of 0.5 and 2 μm, respectively; (c), (d) pits arranged to a hexagonal lattice with a period of 0.5 and 2 μm, respectively. Reproduced from ref. 17 with permission from Springer Nature, Copyright 2020.

supposed that each pit creates a distribution of preferential nucleation sites at its periphery, as a Monte Carlo simulation based on strain effects qualitatively confirms.

Considering that the whole process is carried out in the frame of CMOS technologies, it is of potential interest for future photonic applications of Ge NCs.

## 5.2.3 Silicon Nanowires

Among silicon nanostructures, silicon nanowires (NWs) are considered the resource most compatible with industrial applications[49,50] for their facile transfer to common micro- and opto-electronic processes.

Silicon nanowires have been grown with two main type of processes of which the first is a metal-seeded vapor–liquid–solid (VLS) process, carried out either with CVD or plasma enhanced CVD technology. The second is a metal-assisted electrochemical process (MACE), although unseeded processes have also been experimented by growing arrays of Si nanowires in a vapour phase of Si sulfides,[51] where the metal catalyst is replaced by a molten $SiS_2$ phase.

As is well known, the main difference between CVD and PECVD technologies consists of the nature of the reacting species and of the reaction phase. In a conventional CVD process the reactants are directly the precursor molecules [silane ($SiH_4$) or tetrachlorosilane ($SiCl_4$)], while radicalic species are the reactants in PECVD processes. In turn, the reaction phase is a gaseous phase with conventional CVD processes, while the reaction phase is a plasma phase in PECVD processes, with intense high energy ions bombardment of the reaction interface.

### 5.2.3.1 Growth of Silicon Nanowires with CVD Processes

With the metal-seeded VLS process, silicon nanowires (Si NWs) of various diameters and lengths can be grown using silane ($SiH_4$) or tetrachlorosilane ($SiCl_4$) as the precursors, although, due to lower process temperatures, silane is mostly employed. As catalysts and seeds for the NW growth, nano-size metal particles deposited on the surface of a single crystal silicon substrate[52,53] are used, with Au as the preferred catalyst, though Al, Zn, In, Sn and Pt have also been used.[52,54]

The metallic seeds could be deposited on the surface of a crystalline silicon substrate either as a thin (0.6 nm) metallic layer of Au,[55] which segregates, by heating, a disordered array of metal nanoparticles (NPs), or as an ordered distribution of preformed metallic dots. In both cases, at temperatures above the Si–Au eutectic, randomly distributed or ordered arrays of nano-droplets of a liquid Au–Si alloy spontaneously form on the silicon surface.

Once silane is introduced in the process atmosphere, the liquid Au–Si alloy droplets catalyse its decomposition to elemental silicon and hydrogen on their surface and behave as silicon sinks

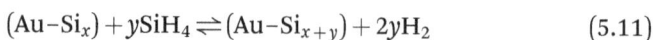

$$(\text{Au--Si}_x) + y\text{SiH}_4 \rightleftharpoons (\text{Au--Si}_{x+y}) + 2y\text{H}_2 \qquad (5.11)$$

with the formation of a saturated and super-saturated metal–silicon liquid alloy.

Thermodynamic equilibrium is eventually achieved with the segregation of the excess Si from the super-saturated Au–Si alloy in the form of a wire that holds the gold drop on its tip, see Figure 5.14, with a process that implies a wire diameter expansion until the process driven by reaction (5.11) is allowed to proceed, see Figure 5.15, with the growth of a wire.

The Si NW growth occurs with a metal-seeded vapor–liquid–solid (VLS) process at process temperatures above the eutectic temperature of the binary Me–Si systems, while a metal-seeded vapor–solid–solid (VSS) process occurs at temperatures below the eutectic. In both cases the eutectic temperature $T_e$ depends on the size of the seed.

It has been demonstrated that not only the melting temperature $T_m$, but also the eutectic temperature $T_e$ of binary Me–Si alloys

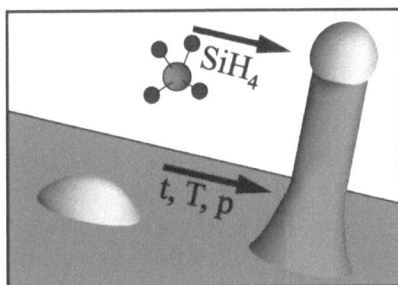

**Figure 5.14** Schematic illustration of the metal-assisted VLS growth of a silicon nanowire. Reproduced from ref. 52 with permission from John Wiley & Sons, Copyright © 2009 Wiley-VCH Verlag GmbH & Co. KGaA, Weinheim.

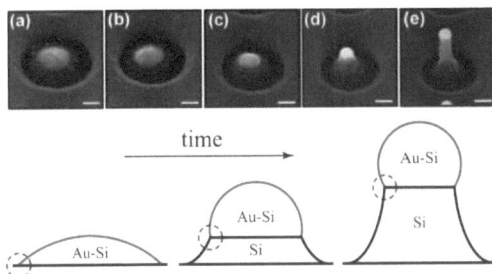

**Figure 5.15** SEM images showing the evolution of the metal droplet shape during the initial growth phase. Bottom: Schematic illustration of the changes of the droplet and wire during the initial growth phase. Reproduced from ref. 53 with permission from American Chemical Society, Copyright 2010.

depends on the seed size, and decreases with the decrease in the diameter of the catalyst droplet, as is shown in Figure 5.16 (where $\Delta T = T_e - T_e^\circ$, and $T_e^\circ$ is the eutectic temperature of the bulk phase) for the case of the Al–Si, Au–Si and Ag–Si systems.[53]

As suggested by Hourlier *et al.*,[56,57] a more complex behaviour could be expected to hold, since the eutectic temperature depends also on the morphology of the nanostructure grown from it. They calculated, for the Au–Si system, a eutectic temperature $T_e$ of 250 °C in the case of the segregation of spherical nanocrystals of 5 nm radius, but an eutectic temperature of 450 °C in the case of the segregation of a nanowire of 5 nm radius, both substantially different from the bulk eutectic temperature of 363 °C of the Au–Si system.

We will see that the Hourlier[56,57] calculation is rarely respected in practice, and that many additional parameters influence the growth thermodynamics, the morphology, structure and defectivity of the wires, like the nature, size, and distribution of the metal catalyst on the substrate, the concentration and flow rate of the gaseous precursors, and the details of the growth and post-growth process, as experimentally shown by Hyvl *et al.*[58] Using plasma-enhanced CVD (PECVD) process, with $SiH_4$ and $H_2$ as the gaseous precursors, they could show that the growth of metal (Au, Sn, In) catalysed NWs on silicon substrates could be experimentally observed much below the (bulk) eutectic temperature of the corresponding Si alloys (363, 232 and 156 °C, respectively) and found to occur even at 113 °C below the bulk eutectic temperature of the Au–Si system. Furthermore, the

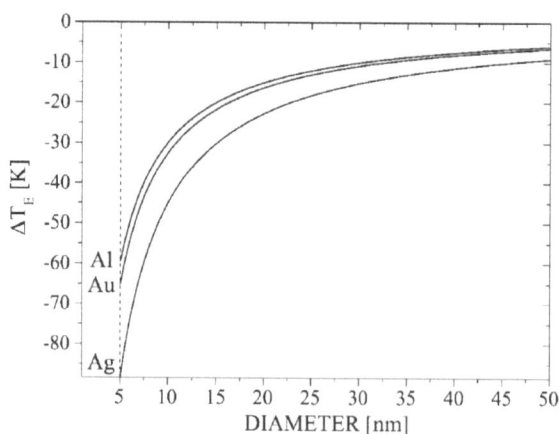

**Figure 5.16**   Dependence of the eutectic temperature for the Al–Si, Au–Si and Ag–Si systems as a function of the diameter of a nanocrystalline seed. Reproduced from ref. 53 with permission from American Chemical Society, Copyright 2010.

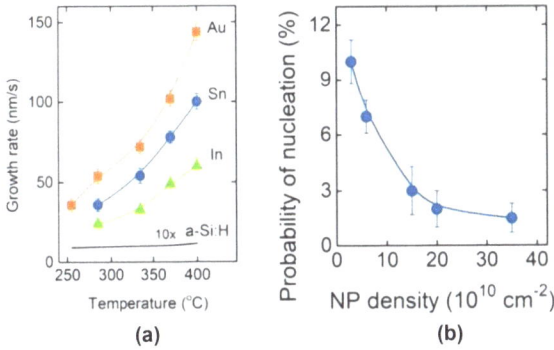

**Figure 5.17**   (a) Effect of temperature on the growth rate of Si NWs for In, Sn and Au catalysts. (b) Effect of metal seed (NP) density on the nucleation probability of SiNWs catalysed by Sn. Reproduced from ref. 58 with permission from IOP Publishing, Copyright 2020.

critical temperature at which the NWs start to grow at an appreciable rate ($>20$–$25$ nm s$^{-1}$) for Au, Sn and In catalysts has been found to be around 250–270 °C, see Figure 5.17a. In all these cases the NWs consists of a crystalline core and of an amorphous silicon shell, with the crystallographic structure of the crystalline core phase not explicitly mentioned.

Apparently, Au favours the highest growth rates, and thus nucleation rates, but the nucleation probability is shown to strongly decrease with the increase in the density of the metallic seeds at the surface of the silicon substrate, see Figure 5.17b, showing that only isolated metal nanoparticles of 10–20 nm in size behave as effective seeds, while larger NPs are inactive, inducing non-equilibrium process conditions.

Higher critical growth temperatures are credited by Puglisi *et al.*,[59] who successfully grew crystalline[§§] Si NWs at a process temperature of 395 °C with a PECVD process using silane as the precursor and Au dots of 1.6 nm in diameter as metallic seeds. This growth temperature is higher than the calculated[53] eutectic temperature for a dot of 1.6 nm (see again Figure 5.16), but lower than that calculated by Hurlier for a liquid seed/solid NW configuration. They, additionally, demonstrate that the NW size distribution, for a seed density of $2 \times 10^{12}$ cm$^{-2}$, is a factor of ten higher than that of Hyvl,[58] and exhibits an almost exponential decrease with an increase in the wire diameter, see Figure 5.18, above a maximum that depends on the plasma excitation power.

---

[§§] No explicit information is given on the crystalline structure of these wires.

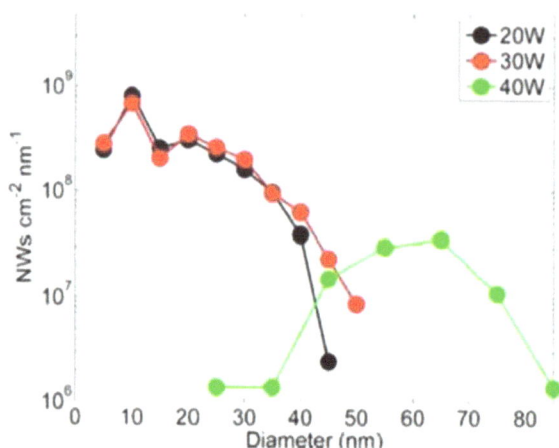

**Figure 5.18** Experimental size distribution of Si NWs as a function of the wire diameter and of the excitation power. Reproduced from ref. 59 with permission from American Chemical Society, Copyright 2019.

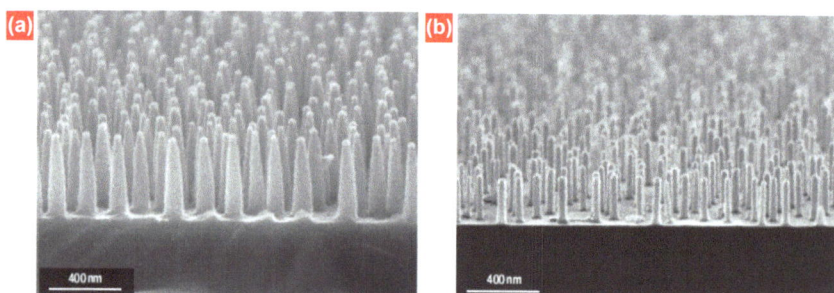

**Figure 5.19** SEM cross sectional images of arrays of silicon nanowires grown at different temperatures on a silicon substrate using Al as catalyst: (a) 490 °C, (b) 430 °C. Reproduced from ref. 54 with permission from Springer Nature, Copyright 2006.

Apparently, low plasma power favours the nucleation rate and the growth of small diameter (less than 10 nm) nanowires.

Depending on the process temperature and catalyst nature, the growth cannot occur uniquely by segregation of the excess silicon from the silicon-saturated seed, but also by decomposition of the precursor in the plasma phase on the wire sidewalls, causing a radial growth and a tapering effect, as shown by Wang *et al.*[54] who grew Si NWs using Al as the catalyst, and observed that the growth is only catalyst activated at 430 °C, while tapering occurs at a growth temperature of 490 °C, as is seen in Figure 5.19.

Given the number of process variables involved, the optimization of growth conditions by empirical methods is, therefore, very demanding, and requires the use of chemical kinetics studies based on a thermodynamic approach for a rational solution.

This approach allows,[52] as an example, the demonstration that the steady-state growth rate $\nu$ (in nm s$^{-1}$) of a cylindrical wire of radius $r$ should be equal to the crystallization rate, and is given by the following equation

$$\nu = \nu_\infty + A \frac{2\Omega\sigma(T)}{r} \qquad (5.12)$$

where $\nu_\infty$ is the steady growth rate at infinite radius, $A$ is a constant, $\Omega$ is the atomic volume of Si, and $\sigma(T)$ is the surface free energy of silicon, which foresees a $1/r$ dependence of the growth rate and the existence of a critical radius of the metallic seed at which the growth velocity falls to zero[52]

$$r_{\text{crit}} = \frac{2\Omega\sigma(T)}{kT \ln p/p_0} \qquad (5.13)$$

where $\Omega$ is again the atomic volume, $\sigma(T)$ is the surface free energy of silicon and the term $kT \ln p/p_0$ is a measure of the supersaturation ratio, with $p_0$ as the partial pressure of the silane at which the supersaturation ratio is one.

The final radius of the nanowire depends on the radius of the metallic nanoparticles used to seed the wire growth. With a well-controlled CVD growth process, using silane as the precursor and hydrogen as the carrier gas, seeding the process with Au nanoclusters having a size of few nanometers, nanowires with a constant diameter down to 3 nm can be grown.[52,60]

Experimental kinetics studies of Si NW growth in CVD conditions were carried out, as a first example, by Zhao *et al.*,[61] who used SiH$_4$ as the precursor, in a mixture of 10% SiH$_4$ and 90% He, and Au colloidal seeds (15 nm in diameter), statistically dispersed at the surface of a silicon $\langle 100 \rangle$ substrate, covered by an approximately 5 nm thick layer of native oxide.

Their results show that in the temperature range 475–600 °C the growth of single crystalline NWs is systematically observed after a period of induction and occurs with a process that can be divided into three main steps as a function of the SiH$_4$ pressure, as can be seen in Figure 5.20.

(1) No growth occurs for a SiH$_4$ pressure of ~10 Torr in the induction stage at the beginning of region A.

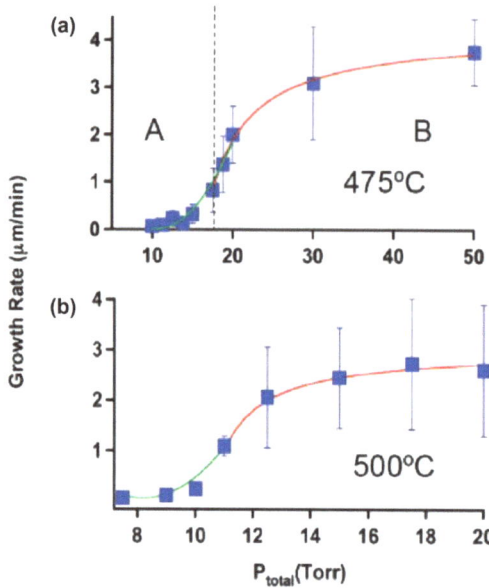

**Figure 5.20** Growth rate as a function of the pressure of a 10% $SiH_4$ and 90% He mixture at a process temperature of 475 and 500 °C. A indicates the low partial pressure range (<20 torr). B indicates the range of pressures higher than 20 torr. Reproduced from ref. 61 with permission from American Chemical Society, Copyright 2008.

(2) Start of growth, with a rate that depends on the $SiH_4$ pressure, according to a $P_{SiH_4}{}^2$ law, region A.

(3) The growth rate becomes constant and independent of the $SiH_4$ pressure, region B.

Considering that the VLS process should involve four main steps:

(1) The mass-transport of $SiH_4$ in the gas phase and its delivery at the gas phase/liquid Au interface

(2) The chemisorption of $SiH_4$ at the vapour–liquid Au interface

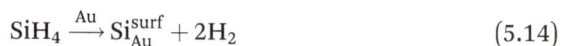

$$SiH_4 \xrightarrow{Au} Si_{Au}^{surf} + 2H_2 \tag{5.14}$$

(3) The diffusion of Si in the liquid phase, which eventually reaches supersaturation conditions, and the segregation of silicon at the liquid Au interface, when $\mu_{Si}^{Au} \geq \mu_{Si}^{Si}$

(4) The growth of a Si NW with the incorporation of Si atoms in the crystal lattice

one should conclude that, during the induction stage, mass transport and chemisorption occur in the liquid seed, with the formation of a supersaturated solution of Si in liquid Au, which provides the necessary conditions for the start of NW growth.

Once these conditions are satisfied, growth occurs with a $P^2_{SiH_4}$ law, which indicates that the delivery of nutrients at the reaction interface is rate determining, while constant-rate growth conditions are satisfied when the rate determining step is the chemisorption process.

Apparently, step 4 is never rate-determining, like step 3, because the diffusion of Si in the molten eutectic phase is fast.[62]

The kinetics of the growth process regime at higher temperatures with a conventional CVD process was also investigated, by Jeong *et al.*,[63] who studied the case of Au- and Pt[¶¶]-catalysed growth on a crystalline silicon substrate between 950 and 1100 °C, using SiCl$_4$ as the precursor, diluted in H$_2$ and Ar.

As in the previous case, the growth rates tend to saturate at high precursor flow rates, see Figure 5.21, while the growth rate dependence on temperature at a constant flow rate of 20 sccm, when chemisorption is rate determining, is displayed in the Arrhenius plot of Figure 5.22.

It is apparent that the chemisorption process is thermally activated, and that the activation energy is lower for the Pt-catalysed case, with activation energies of 80 and 130 kJ mol$^{-1}$ for Pt and Au, respectively.

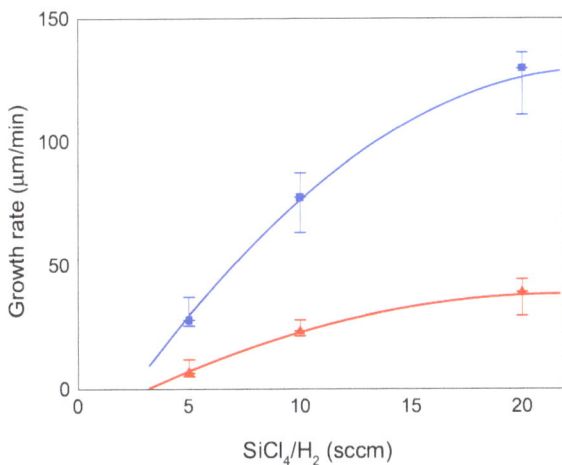

**Figure 5.21** Dependence of the growth rate of Si NWs for the Pt-catalysed case (blue line) and Au-catalysed case (red line). Reproduced from ref. 63 with permission from Elsevier, Copyright 2009.

¶¶ The bulk eutectic temperature of the Pt–Si system is 830 °C.

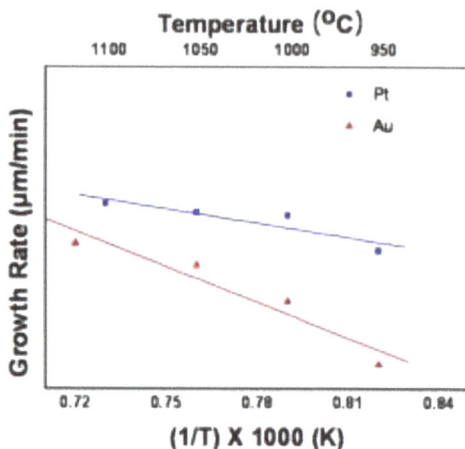

**Figure 5.22** Growth rate of silicon nanowires catalysed by Au and Pt. Reproduced from ref. 63 with permission from Elsevier, Copyright 2009.

Both activation energies are, instead, lower than the activation energy (213.38 kJ mol$^{-1}$) experimentally determined for the un-catalysed, thermally activated CVD deposition of Si from SiH$_4$,[64] when the chemisorption process occurs at the Si/gas phase interface, making the CVD growth at sidewalls kinetically inhibited.

In whole range of temperatures, TEM measurements show that the NWs are single crystalline and defect-free.

Another process regime was investigated by Wen *et al.*,[65] who carried-out kinetic studies of the Cu-catalysed, CVD growth directly in a transmission electron microscope, in process pressure conditions compatible with direct TEM use (a maximum pressure of $4 \times 10^{-5}$ Torr) using disilane (Si$_2$H$_6$) diluted in argon at a pressure of $10^{-7}$ Torr, at a growth temperature of 470–550 °C.

In this temperature range, the Cu seeds react with disilane and convert into crystalline Cu silicide, leading to a VSS growth, with a wire orientation that depends both on the silicon substrate and the crystalline Cu silicide seed. Under these process conditions the growth rate (maximum rate around 100 nm min$^{-1}$ at 530 °C) is limited by the precursor supply in the gas phase, and the wires are systematically found to be diamond cubic.

It is possible, therefore, to conclude that in a wide temperature range (475–1100 °C):

- The Me-catalysed CVD growth occurs with the mass transport of precursors at the seed surface

- The chemisorption of the precursor is the rate determining step
- Thermodynamic equilibrium conditions are established at very low growth rates.

Although Au presents several advantages with respect to other metal catalysts, among which a relatively low eutectic temperature,[|||] and thus, the possibility of systematic growth from a liquid alloy, this does not avoid the Au contamination of the Si NW, with Au concentrations up to $10^{18}$ at cm$^{-3}$,[66] making the Au catalysis incompatible with microelectronic or optoelectronic applications, and the Al catalysis the most viable.

The gold contamination of CVD-grown, B- and P-doped Si NWs grown at 600 °C using silane as the precursor was also demonstrated by Sato *et al.*[67] They showed that DLTS measurements carried out in both types of NWs indicate the presence of a donor level (0/+) at $E_v + 0.36$ eV, corresponding to a Au–H complex.

The concentration of this complex in B-doped NWs is $5 \times 10^{15}$ cm$^{-3}$, two orders of magnitude higher than the gold solubility in bulk silicon at 600 °C ($3 \times 10^{13}$ cm$^{-3}$). No conclusive explanation of this excess solubility of gold in B-doped Si NWs is reported in the literature, but an excess Au concentration due to the nanometric size can be excluded, since the solubility of impurities in silicon decreases with the decrease in size, as shown by Bulyariskiv *et al.*[68] for the case of P and Sn in the 1200–600 °C temperature range.

Also the crystal structure that determines the optical properties of the NWs and the vertical wire alignment, which is a highly desirable NW morphology, are shown to be strongly process-condition dependent.

This was observed, among others, by Ho *et al.*[69] who carried out Si NW growth at low growth rates using SiCl$_4$ as the precursor, hydrogen as the carrier gas and gold as the catalyst, and demonstrated that the NW quality and orientation depend on the precursor concentration and on the process temperatures, showing that a process temperature of 850 °C enables the growth of well aligned and epitaxial NWs, 150–200 nm in diameter, oriented along the ⟨111⟩ direction.

The same conclusion was proposed by Lieber *et al.*[70–73] who carried out Si NW growth at 450–460 °C, using SiH$_4$ as the precursor, He as the carrier gas and Au nanoclusters of different radius (5, 10, 20, 30 nm) as catalysts, and obtained NWs with a diamond cubic crystalline core

---

[|||] The bulk eutectic temperatures of the Zn–Si, Al–Si and Pt–Si systems are 419, 577 and 830 °C, respectively.

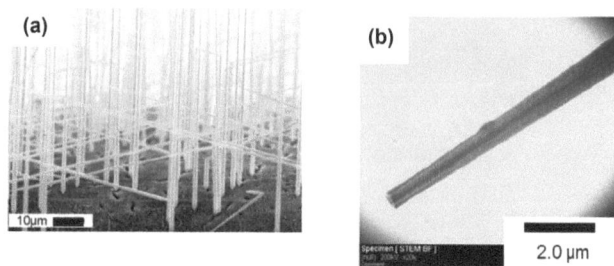

**Figure 5.23** (a) SEM image of a vertical array of silicon nanowires grown at 1230 °C with a metal catalyst-free process. (b) TEM image of a nanowire. Reproduced from ref. 74, https://doi.org/10.1038/srep30608, under the terms of the CC BY 4.0 license, https://creativecommons.org/licenses/by/4.0/.

and a shell of amorphous $SiO_x$. Depending on the seed radius, the NW diameter ranged from 6 to 31 nm. They, additionally, demonstrated that the NW orientation depends on the wire diameter, with preferential ⟨100⟩ orientation for the small diameter NWs and exclusively along the ⟨111⟩ direction for the large diameter NWs.

Higher growth temperatures were used by Ishiyama *et al.*,[74] who could grow on a silicon ⟨111⟩ substrate arrays of epitaxial, vertically aligned, ⟨111⟩ oriented, diamond-cubic Si NWs, see Figure 5.23a, with a diameter of 26 nm using a catalyst-free process, carried out at 1230 °C, from a molten $SiS_2$ phase. As can be seen in Figure 5.23b, the wires are tapered, an indication that silicon deposition also occurs on the sidewalls in the absence of a catalyst on the top of the wires.

Different from the previous cases, Foncuberta I. Morral *et al.*[75] demonstrated, using standard CVD on an oxidized silicon substrate with $SiH_4$ as the precursors, $H_2$ as the carrier gas and Au as the catalyst, that at temperatures between 500 and 650 °C and at a growth rate of 1 $\mu m\,min^{-1}$, a large percentage of the Si NWs crystallize with hexagonal structures, called wurtzite A and B, whose properties are shown in Table 5.3, both belonging to the same space group but with important differences in their lattice parameters.

From Table 5.3 it can be observed that the lattice constants of the wurtzite type A samples fit well with those of the metastable 4H Si hexagonal silicon allotrope*** (lonsdaleite) ($a = 0.383$ nm, $c = 0.634$ nm),[76–78] which could be prepared by decompression from a static pressure application of 11.7 GPa of diamond-cubic Si crystals[79] and is granted an indirect gap of 0.95 eV and an optically forbidden

---

*** Details on the extended allotropism of silicon are reported in Appendix 5.1.

**Table 5.3** Comparison of the lattice parameters of the CVD-grown Si NWs with those of the diamond-cubic silicon (Si I) and of the metastable unstrained and strained 4H Si (HD silicon) allotrope.

| Symbol/space group | Lattice constants | | Ref. |
| --- | --- | --- | --- |
| | $a$ (nm) | $c$ (nm) | |
| Wurtzite A | 0.380 | 0.627 | 75 |
| $P6_3/mcc$ | | | |
| Wurtzite B | 0.404 | 0.660 | 75 |
| $P6_3/mcc$ | | | |
| 4H Si | 0.3824 | 0.6323 | 76 |
| $P6_3/mcc$ | | | |
| 4H Si | 0.3837 | 1.2586 | 77 |
| $P6_3/mcc$ | | | |
| 4H Si | 0.3837 | 0.6317 | 78 |
| $P6_3/mcc$ | | | |
| 4H Si (5% biaxially strained) | 0.402 | 0.6234 | 80 |
| Si I | 0.54310205 | | 81 |
| $Fd3m$ | | | |

direct gap at 1.65 eV, which becomes allowed under biaxial strain.[80] Furthermore, a comparison of the Foncuberta's wurtzite B lattice constants with the DFT-calculated[80] values of biaxially strained (+5%) hexagonal silicon given in Table 5.3,[†††] suggests that the wurtzite B samples are, possibly, strained variants of hexagonal silicon.

In fact, the free-energy landscape of silicon presents so many local energy minima that, at least in principle, it allows the formation of a variety of unusual structures, some of which could be stabilized at ambient pressure.[79]

A problem, however, exists with the structural and physical characterization of the 4H allotrope in bulk conditions, since so far it has been very difficult to prepare hexagonal Si crystals of size sufficiently large to allow reliable X-ray diffraction and optical measurements.[77,78]

Therefore, a comparison of the lattice constants of the hexagonal silicon nanowires with those of the 4H allotrope should be handled with severe caution.

As can be seen from the statistical distribution of cubic (I) and hexagonal nanowires as a function of the wire diameter displayed in Figure 5.24, large diameter (>20 nm) NWs mostly exhibit the hexagonal symmetry $IV_B$, and never the cubic symmetry, while cubic wires and hexagonal wires with the Si $IV_B$ structure are equally distributed among smaller diameters (<20 nm) NWs.

---

[†††] This also displays the standard value of the lattice constant of diamond cubic silicon.[81]

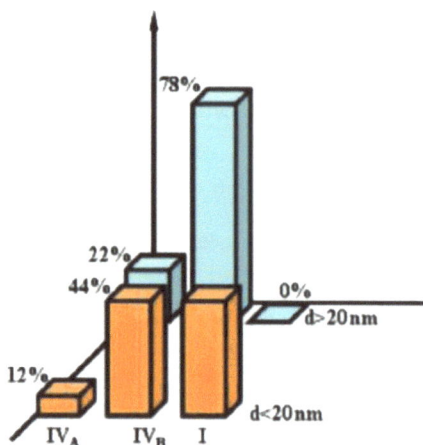

**Figure 5.24**   Statistical distribution of the structural configurations of silicon nanowires as a function of the wire diameter. Reproduced from ref. 75 with permission from John Wiley & Sons, Copyright © 2007 Wiley-VCH Verlag GmbH & Co. KGaA, Weinheim.

Apparently, surface strain due to nano-structuration plays the same role as static compression in the stabilization of the hexagonal symmetry, but additional process conditions influence the simultaneous formation of cubic, strained and un-strained hexagonal NWs, as is illustrated by the following example.

A different, morphological distribution was, in fact, observed by Fabbri *et al.*[8,82] among doped and undoped, Au-catalysed CVD-grown Si NWs. The growth was performed in an atmosphere of silane $SiH_4$ for undoped nanowires, and of silane and diborane $(B_2H_6)$ (1%) or phosphine $(PH_3)$ (1%) in $H_2$ for B-doped and P-doped nanowires, respectively, using Au nano-colloids 3 nm in diameter as metallic seeds. The growth temperature was held at 600 °C and the deposition substrate was an oxidized crystalline silicon sample.

Undoped NWs are a 90–10% mixture of the metastable hexagonal Si IV (HD Si) allotrope of Table 5.3 ($a = 0.380$ and $c = 0.627$ nm) and of diamond-cubic Si.

The B-doped NWs keep the structure of the Si IV (HD Si) allotrope, but some are so strongly twinned to lead to a macroscopic cubic structure. Eventually, the P-doped ones mostly keep the diamond cubic structure. The statistical distribution of the silicon phases in these NWs is displayed in Figure 5.25.

The cathodo-lumincence (CL) spectra of bulk diamond-cubic Si and of hexagonal NWs displayed in Figure 5.2[8] show that the hexagonal NWs present two emissions at 0.9 eV and 1.5 eV, of which that at 0.9 eV

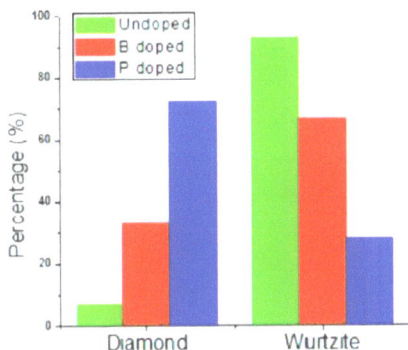

**Figure 5.25** Statistical distribution of the silicon phases in undoped and doped silicon nanowires. Reproduced from ref. 82 with permission from American Chemical Society, Copyright 2013.

is the emission from defect states and that at 1.5 eV corresponds to the direct gap of 4H-Si, optically allowed by high biaxial strain.[80,83,84]

Eventually, hexagonal Si nanowires, with diameters in the 2–4 nm range, were also grown by Tang *et al.*[85,86] with a PECVD process at 400 °C, using Sn as the catalyst. The deposition was carried out on a Cu grid, silver-attached to a Corning glass substrate used to carry-out TEM measurements on the as-grown samples. Given the negligible solubility of Sn in Si, the MP of tin (231.9 °C) corresponds to the eutectic temperature,[87] see Figure 5.26.

They observed that most of the nanowires, which are the thinnest among those considered so far, due to the negligible solubility of Si in Sn, present the crystal structure of the hexagonal silicon allotrope, with $a = 0.3828$ nm and $c = 0.6325$ nm. Dedicated thermal annealing treatments of these hexagonal wires show that they are stable up to 700 °C, in agreement with a known property of the metastable Si IV allotrope.[88]

While it seems apparent from the former results that silicon nanowires can spontaneously keep both cubic and hexagonal structure, there is no clear evidence of a specific physico-chemical reason for this attitude, like the catalyst nature and configuration, the precursor, and the process temperature, as is seen in Table 5.4.

According to Tang *et al.*,[85] at the size scale of most of the experiments the surface tension from the side-walls of silicon NWs becomes significant and will exert a strong anisotropic stress on the nanowire structure that could mimic the anisotropic stress at the origin of the formation of the metastable hexagonal phase.[‡‡‡]

---

[‡‡‡] In italics a sentence of Tang *et al.*[85]

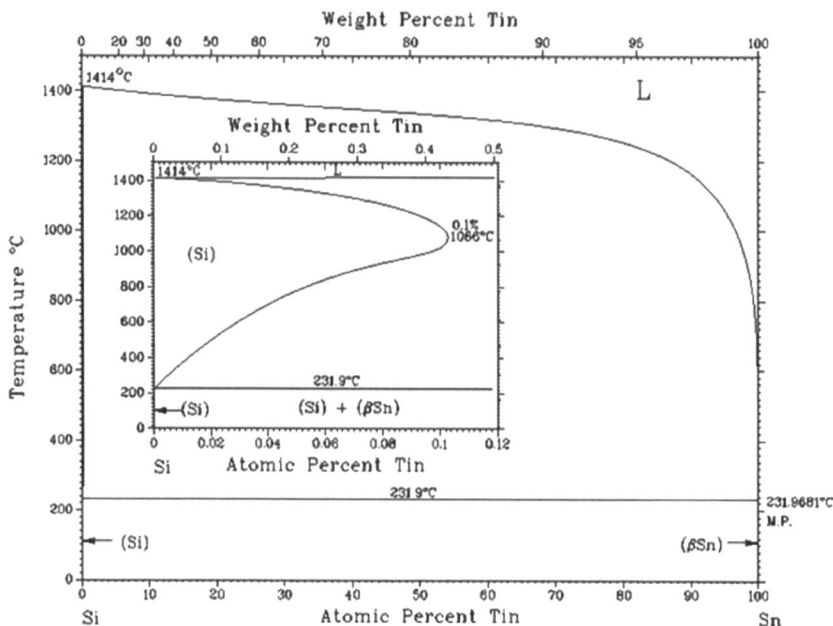

**Figure 5.26** Phase diagram of the system Si–Sn. Reproduced from ref. 87 with permission from Springer Nature, Copyright 1984.

The concomitant effect of growth rate and anisotropic stress on the crystallographic structure of silicon nanowires was, instead, experimentally demonstrated by Liu and Wang,[89] who carried out the growth of Si NWs using $SiH_4$ in a $SiH_4$–He mixture, and Au nanoparticles as seeds, at a process temperature of 475 °C.

They showed, see Figure 5.27d, that at very low growth rates[§§§] ($<20$ nm min$^{-1}$) the wires take only the cubic structure, while an increasing density of hexagonal and defective (flawed) wires is observed with the increase in the growth rate. An inspection of the HRTEM images of the Au/Si interface at increasing growth rates shows that epitaxial conditions occur at low growth rates (0.2 μm min$^{-1}$), see Figure 5.27a, in the absence of detectable amounts of volume defects, since stacking fault (red lines) arising from anisotropic stress can be spontaneously annihilated at the Au/Si interface, as is shown in Figure 5.27b. At higher growth rates, see Figure 5.27c, stacking faults are, instead, trapped at the Au/Si interface, and induce the formation

---

[§§§] Which was varied by increasing the precursor pressure.

**Table 5.4** Role of process conditions on the crystallographic structure of silicon nanowires grown using the CVD technique.

| Growth temperature (°C) | Catalyst | Process conditions | Precursor | NW diameter (nm) | Crystal structure | Ref. |
| --- | --- | --- | --- | --- | --- | --- |
| 400 | Sn | VLS | $SiH_4 - H_2$ | 2–4 | Hexagonal | 86 |
| 450–460 | Au nanoclusters | VLS | $SiH_4 - He$ | 6–31 | Cubic | 73 |
| 475 | Au nanoparticles (10–100 nm) | VLS | $SiH_4 - He$ | ~30 | Cubic at low rate, hex at higher rates | 89 |
| 470–550 | $Cu_3Si$ | VSS | $Si_2H_6 - Ar$ | n.d. | Cubic | 65 |
| 500–650 | Au | VLS | $SiH_4 + H_2$ | <20–100 | Cubic + hex | 75 |
| 600 | Au nanocolloids (3 nm) | VLS | $SiH_4 - H_2$ | 10 | Cubic + hex | Fabbri |
| 600–850 | Au film, 2–3 nm | VLS | $SiCl_4$ | 150–200 | Cubic | 69 |
| 1230 | $SiS_2$ | VLS | Silicon sulphides | 26 | Cubic | 74 |

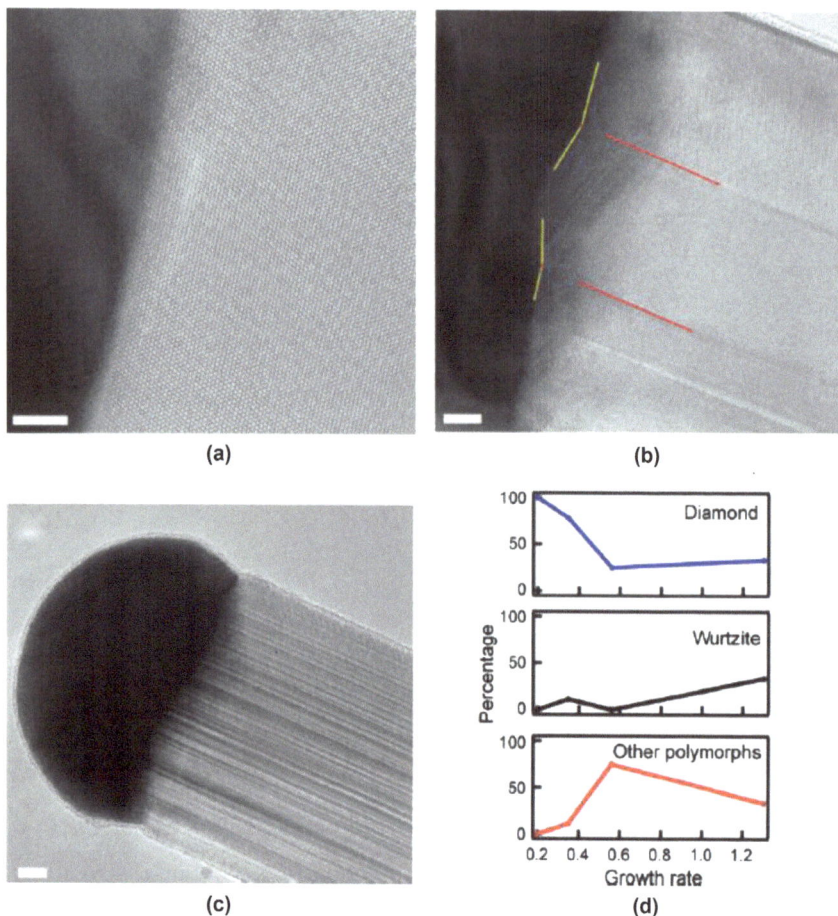

**Figure 5.27**  (a–c) HRTEM images of the Au/Si heterogeneous interface. (d) Effect of the growth rate (in $\mu m\ min^{-1}$) on the crystallographic structure of the wires grown at 475 °C. Reproduced from ref. 89 with permission from Springer Nature, Copyright 2009.

of the hexagonal wurtzite phase and of other polymorphs,[¶¶¶] that differ from the cubic phase by their stacking sequence.

It can be, therefore, conclusively assumed that growth rate, structural defects and surface strain dominate the conditions of formation of cubic or hexagonal silicon NWs. In fact, only at very low growth rates does the system run close to thermodynamic equilibrium conditions, with the formation of the equilibrium, diamond-cubic phase of silicon, in the absence of surface strain-activated extended defects that favour the phase transformation.

---

[¶¶¶] A detailed discussion on the stability of silicon polymorphs will be given in Appendix 5.1.

### 5.2.3.2 MACE Grown Nanocrystalline Silicon Wires

It is well known that electroless processes occur with the spontaneous oxidation and dissolution of a metal electrode dipped in a suitable electrolytic solution, as an aqueous solution of HCl

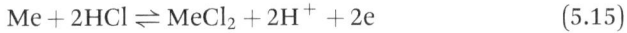

$$Me + 2HCl \rightleftharpoons MeCl_2 + 2H^+ + 2e \tag{5.15}$$

once it is connected to a noble metal electrode on which a coupled redox reaction occurs

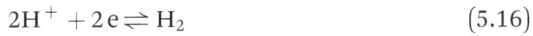

$$2H^+ + 2e \rightleftharpoons H_2 \tag{5.16}$$

with the evolution of hydrogen.

The dissolution process of the metal occurs spontaneously provided the Gibbs energy of the reaction

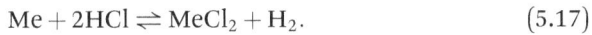

$$Me + 2HCl \rightleftharpoons MeCl_2 + H_2. \tag{5.17}$$

$\Delta G_{5.17}$, obtained by the electroneutral coupling of reactions (5.15) and (5.16), is less than zero.

In full analogy with the previous example, the metal assisted chemical etching (MACE) process, used to prepare arrays of silicon (and germanium) nanowires,[90-94] is an electroless, self-catalysed or metal-catalysed electrochemical process that provides the spontaneous dissolution of a silicon sample on top of which is deposited a noble metal electrode when the system is dipped in an aqueous solution of HF, see Figure 5.28, where electrons are supposed to mediate the charge transfer.

Under these conditions, the silicon oxidation reaction occurs at an n-type silicon electrode

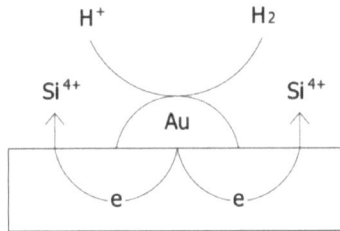

$$Si + 4HF \rightleftharpoons SiF_4 + 4H^+ + 4e \tag{5.18}$$

**Figure 5.28** Schematic illustration of the MACE process at an n-type silicon electrode, where the silicon oxidation occurs, associated with the simultaneous evolution of hydrogen at the gold electrode. The process is supposed to be mediated by an electron transfer.

to which pertains an electrochemical potential of $-1.27$ V *vs.* the standard hydrogen electrode (SHE), represented here by the noble metal electrode that works as a hydrogen electrode

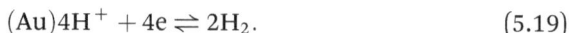

$$(Au)4H^+ + 4e \rightleftharpoons 2H_2. \tag{5.19}$$

Given that eqn (5.18) is coupled to the electroneutral reaction

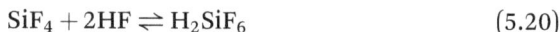

$$SiF_4 + 2HF \rightleftharpoons H_2SiF_6 \tag{5.20}$$

the overall cell process is given by the reaction

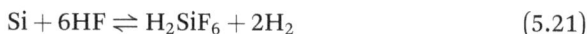

$$Si + 6HF \rightleftharpoons H_2SiF_6 + 2H_2 \tag{5.21}$$

for which a Gibbs energy of reaction $\Delta G_{5.21} = -490.22$ kJ mol$^{-1}$ at 298 K can be calculated if the electrode processes occur without overvoltage drops, from a theoretical cell EMF $= 1.27$ V, which drives the simultaneous anodic oxidation of silicon and the hydrogen evolution at the gold electrode.

Different from the case of two generic metal electrodes in short circuit, eqn (5.15) and (5.16), the gold electrode and the silicon electrode of Figure 5.28 are, however, not in short circuit because a Schottky junction forms at the Au–Si interface,[95,96] see Figure 5.29.

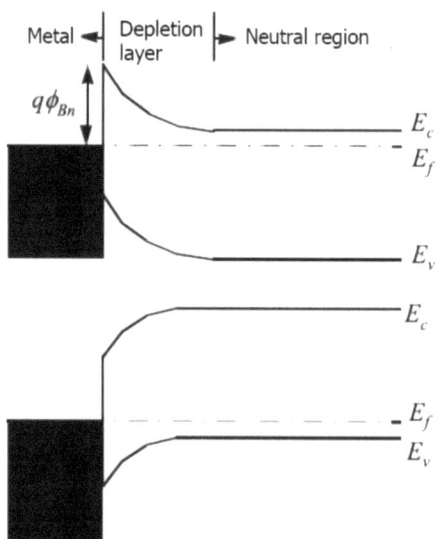

**Figure 5.29** A Schottky metal/semiconductor junction for an n-type semiconductor (top) and for a p-type semiconductor (bottom).

It can be further expected that the thermodynamically allowed Au silicide formation also occurs at the Au/Si interface, with even higher Schottky barrier values (up to 1.35 eV).[97]

In both cases, a barrier-limited current flows across the two electrodes under a bias voltage of 1.27 V, the EMF of the electro-chemical cell, with electrons flowing from the silicon electrode to the hydrogen electrode, across the Schottky junction, as is shown in Figure 5.28.

To identify the nature of the current carriers (electrons or holes) that are actually transferred due to the redox reaction occurring at the semiconductor electrode dipped in an electrolytic solution, it is necessary to know the equilibrium conditions, in corresponding energy scales (eV and absolute potentials), between a semiconductor and a hydrogen electrode in an electrolytic solution, as schematically displayed in Figure 5.30,[98] where $U_{F(FB)}$ represents the absolute potential difference between the Fermi level (flat band) of the semi-conductor and the standard hydrogen electrode (SHE).

The $U_{F(FB)}$ values cannot be obtained directly by means of EMF measurements but might be determined by capacitance or alternating current resistance measurements[99,100] or by photoelectrochemical

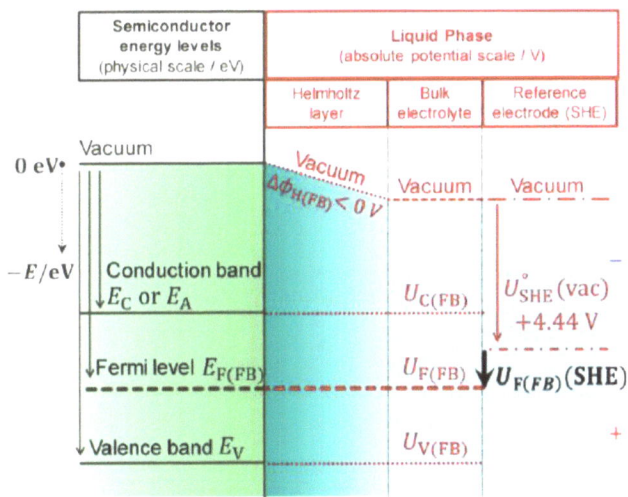

**Figure 5.30** Schematic representation of the energy level diagram of a semiconductor and of the corresponding potentials for an arbitrary semiconductor/electrolyte junction, *vs.* the standard hydrogen electrode (SHE). Reproduced from ref. 98, https://doi.org/10.1039/C9TA09569A, under the terms of the CC BY 3.0 license, https://creativecommons.org/licenses/by/3.0/.

methods,[101] with a large spread of measured values, especially for p-type semiconductors.

The values of $U_{F(FB)}$ *vs.* a standard calomel electrode (SCE) (+0.268 V *vs.* SHE) obtained with ac resistance measurements for silicon in a HF solution by Ottow *et al.*,[100] which is the case of MACE operation with silicon, are +0.14 V for p-type Si and −0.54 V for n-type Si, both of which satisfactory compare with other literature values.

Since the level of the Si/Si$^{4+}$ redox couple is 1.27 V below that of the SHE, thus deep in the valence band of silicon, see Figure 5.31, holes are injected in silicon to activate its oxidation

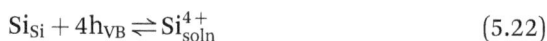

$$Si_{Si} + 4h_{VB} \rightleftharpoons Si^{4+}_{soln} \qquad (5.22)$$

in good agreement with the arguments of Chattopadhyay *et al.*,[90] Gösele *et al.*[92] and Irrera *et al.*[93,94]

In the MACE practice different chemistries are, however, credited in the literature, including that involving ammonium fluoride (NH$_4$F) instead of HF, which is also the selective wet etchant of silicon for its ability to dissolve its surface oxide, but is not environment-friendly,[102] or hydrogen peroxide (H$_2$O$_2$), used to provide the initial oxidation of silicon

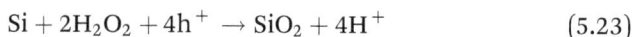

$$Si + 2H_2O_2 + 4h^+ \rightarrow SiO_2 + 4H^+ \qquad (5.23)$$

and its further dissolution by HF.

In its practical application, the MACE process proceeds with the oxidation and dissolution of silicon in correspondence with the

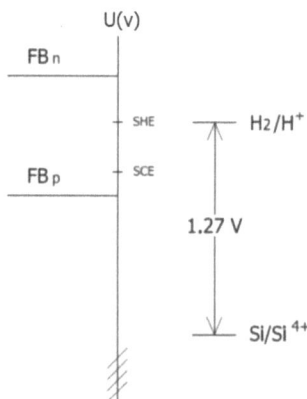

**Figure 5.31**  Qualitative energy scheme illustrating the level of the Si/Si$^{4+}$ redox couple *vs.* the SHE and the flat bands (FB) potentials of n-type and p-type silicon.

regions covered by the noble metal film, which sinks deep into the silicon, generating an array of pores or wires, whose actual morphology depends on the morphology of the patterned noble metal film at the sample surface and on the doping of the silicon sample.[92]

It is, however, not yet experimentally known whether the anodic reaction occurs in correspondence with the noble metal surface regions or HF diffuses through the nanometric thin gold layer, and the electrochemical oxidation of Si occurs at the Au/electrolyte interface, which sinks inside the silicon volume.

The morphology of the NWs arrays depends, in fact, on the surface texture of the noble metal catalyst,[103] as can be seen in Figure 5.32, on the nature of the metal working as cathode, and on the preliminary chemical treatments to which the silicon surface has been subjected, which affect its roughness and wettability.

Eventually, it depends on the kinetics of the electrode reaction at the silicon electrode,[92,104] which could occur with overvoltage contributions, depending on the nature of the metal, on the chemistries involved and on the temperature, as shown, as an example, by Gonchar *et al.*[102,105] who used a solution of ammonium fluoride $NH_4F$ and $H_2O_2$, and observed that the wire morphology strongly depends on the pH of the solution, see Figure 5.33.

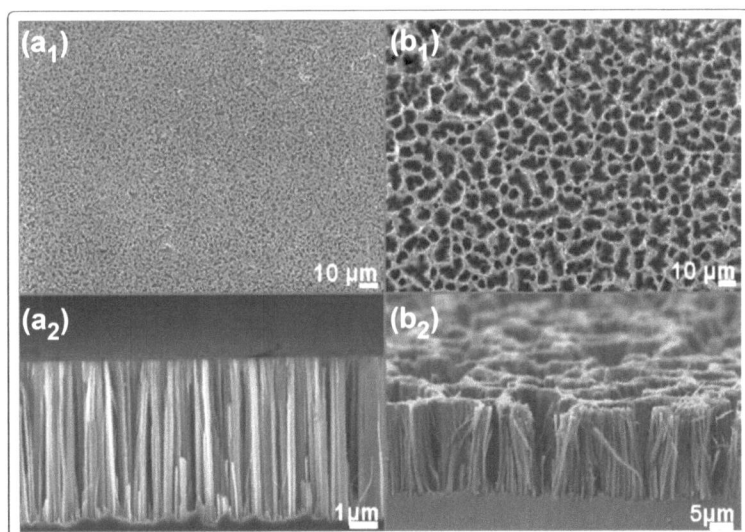

**Figure 5.32**  SEM images in plan view ($a_1$ and $b_1$) and in cross section ($a_2$ and $b_2$) of the surface morphology and silicon nanowire distribution. Reproduced from ref. 103, https://doi.org/10.1186/1556-276X-6-597, under the terms of the CC BY 2.0 license, https://creativecommons.org/licenses/by/2.0/.

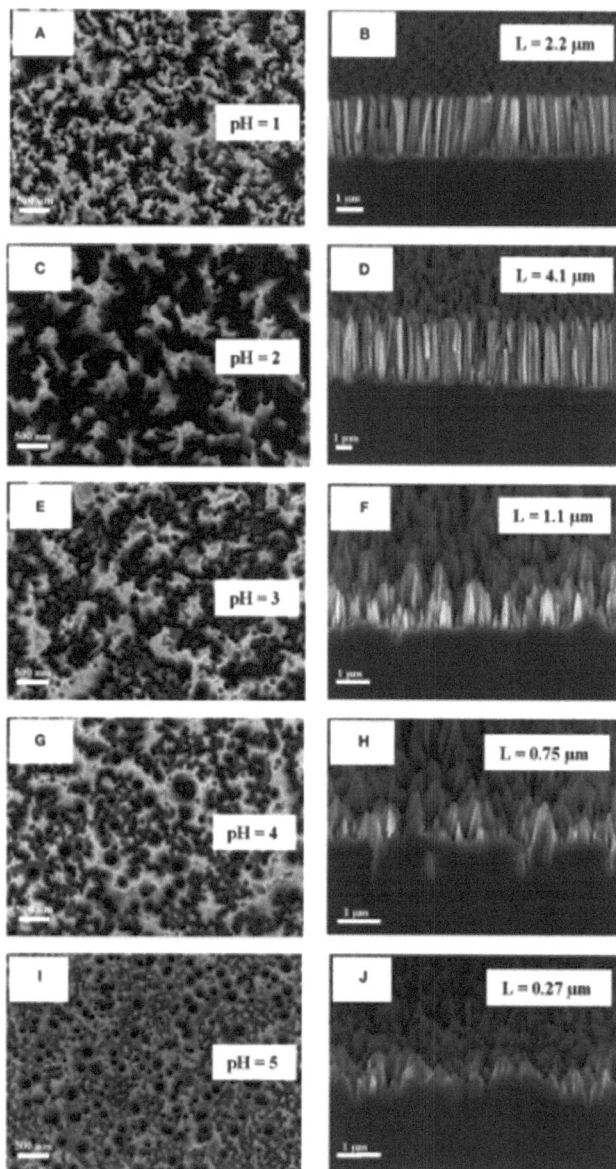

**Figure 5.33**   SEM micrographs of Si NWs prepared with the MACE process in a solution of ammonium fluoride and hydrogen peroxide. $L$ is the length of the wire. A, C, E, G, and I: Top views of the samples. B, D, F, H, and J: Lateral views. Reproduced from ref. 105, https://doi.org/10.3389/fchem.2018.00653, under the terms of the CC BY 4.0 license, https://creativecommons.org/licenses/by/4.0/.

The manufacturing process of Si NWs starts with the deposition at the surface of a clean, crystalline silicon sample a metallic (Ag or Au) film with an honeycomb-shaped array of holes, which defines the region of the crystal where the dissolution process does not operate, realized using a nano-imprinting-assisted metal patterning method,[106] a colloidal lithographic method,[107] or a lithographic method that implies the use of an array of polymeric nanospheres (NS) to mask a continuous metallic film in correspondence with the holes.[108]

This last process is illustrated in Figure 5.34, which shows that after the deposition of a thin layer of gold over the full surface of the sample, an array of polymeric NS is placed onto the gold surface to work as a template. Then, the NS diameter is reduced to the required dimension (A), and the metal is perforated (B) in correspondence with the NS obtained by reactive ion etching (RIE) or a plasma etching process. Eventually (C), the MACE process is operated, with the production of a vertical array of Si NWs.

A systematically good performance of the NW growth process with MACE procedures is not at all obvious, as we have seen while discussing the chemistry of the MACE process and considering that also the accurate control of the patterning process is critical, since it determines the diameter of the wires and their spacing.[108]

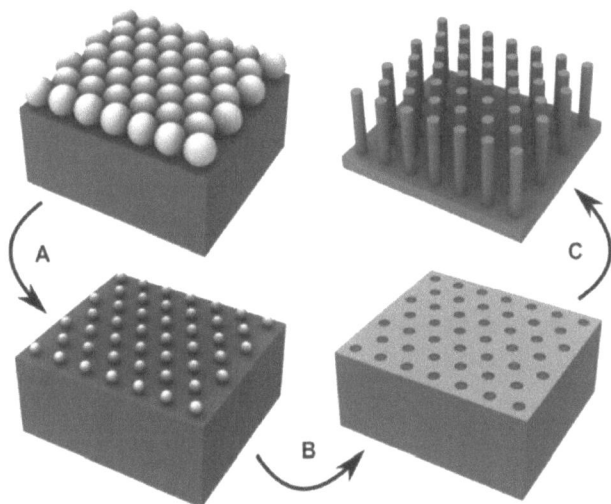

**Figure 5.34** Schematic flowsheet of the surface patterning and MACE Si NW fabrication process. Reproduced from ref. 108 with permission from John Wiley & Sons, Copyright © 2013 Wiley-VCH Verlag GmbH & Co. KGaA, Weinheim.

Considering that MACE-grown wires are prepared by etching diamond cubic silicon, one expects that the optoelectronic properties of these NWs should correspond to those of cubic silicon, possibly with quantum confinement effects arising at the nanometric size.

The experimental results available do not support this picture.

As a first example, the PL spectra measured at room temperature of the NWs prepared by Gonchar *et al.*,[105] with their diameters confined within 50–100 nm, far from the range at which one expects QC effects, present a broad emission peaked at 750 nm (1.65 eV), see Figure 5.35. To explain the strong blue-shift ($> 0.5$ eV) above the optical gap of silicon of the PL emission of these wires, the authors suggest that the optical activity of these wires is due to the QC emission of nanometric crystallites of undefined structure and size, located at the nanowire sidewalls.

This view is shared by Sivakov *et al.*[109] and Fakhri *et al.*,[110] who carried out room temperature PL measurements on MACE nanowires grown using as the noble metal catalysts Ag particles deposited from $AgNO_3$ solutions of different molarities to change the metal surface texture. The etching was carried out in a conventional $HF–H_2O_2$ solution. For both authors the PL spectra present a broad band peaking at 750 nm (1.65 eV), which is attributed, again, to a QC emission of silicon nanocrystals or of surface states of nanometric pores at the surface of the wires, as also suggested by Naffeti *et al.*[111] and by Choi *et al.*,[112] who also found PL emission at 750 nm (1.65 eV) in their mesoporous nanowires, see Figure 5.36.

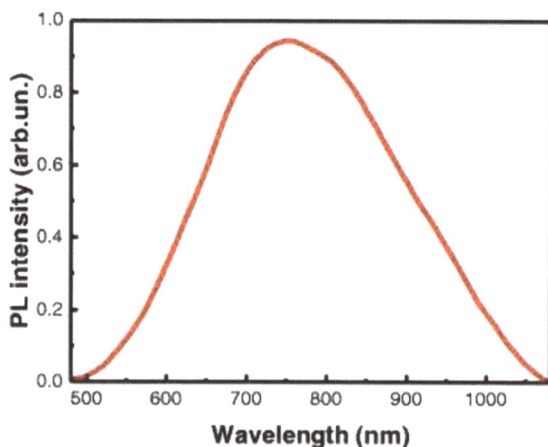

**Figure 5.35** Photoluminescence spectra of MACE grown Si NWs. Reproduced from ref. 102, https://doi.org/10.1186/s11671-016-1568-5, under the terms of the CC BY 4.0 license, https://creativecommons.org/licenses/by/4.0/.

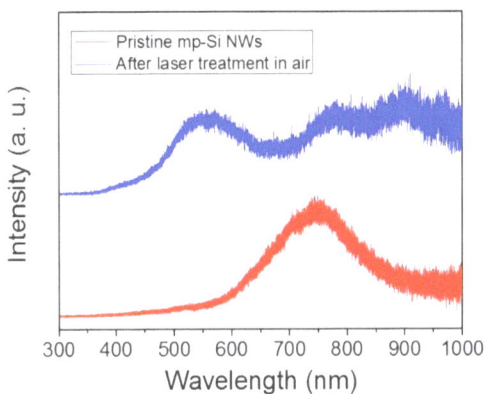

**Figure 5.36** PL emission of mesoporous silicon nanowires. Reproduced from ref. 112 with permission from Springer Nature, Copyright 2014.

The hypothesis that the PL emission arises from QC nanocrystals at the NW surfaces is supported by the experimental, and theoretically confirmed, QC emission of silicon nanocrystals (Si NCs),[45,46] which vary between 1.1 and 1.9 eV for NCs diameters ranging between 9.5 and 2.6 nm, as shown in Figure 5.11. However, an almost constant value of the PL emission at 1.65 eV (see Table 5.5), fitting with the optical gap of hexagonal 4H Si (lonsdaleite) observed in most of the works discussed above, suggests that a different explanation should be envisaged, such as the onset of a cubic–hexagonal phase transition stimulated by surface stress, as demonstrated by Lopez *et al.*,[113] and Dixit and Shukla.[84,114] The latter authors observed, in fact, that the prolonged HF-etching of 100 nm in diameter Si NWs, prepared by the MACE process with silver as the metal catalyst, leads to their reduction in diameter down to tens of nm, and to the phase transformation to hexagonal silicon induced by surface stress, demonstrated by the evidence of the Raman active modes of lonsdaleite $NH_4$ silicon, spectroscopically measured on the etched material.

The PL spectrum of a sample of etched NWs, displayed in Figure 5.37, shows that the PL emission consists of a broad band, which could be deconvolved in several peaks ranging within 2 and 2.2 eV, interpreted as due to stress-induced surface states of hexagonal silicon. If we consider that the sample consists of an array of wires of different size, the presence of several PL peaks could also be explained by the QC emission of hexagonal wires of different sizes.

This conclusion is also supported by the PL spectra of the MACE-grown silicon nanowires with diameters of 5, 7 and 9 nm prepared by

**Table 5.5**  Photoluminescence emission energies of CVD-grown hexagonal nanowires and of MACE grown silicon nanowires.

| Growth process | NW diameter (nm) | PL-emission energy (eV) | Ref. |
| --- | --- | --- | --- |
| CVD | 30 | 1.59 | 83 |
| MACE | 50–100 | 1.107 + 1.65 | 102 |
| MACE | 90–135 | 1.65 | 110 |
| MACE | n.d. | 1.65 | 109 |
| MACE | 9, 7, 5 | 1.55–1.907 | 93 |
| MACE | ~10 | 2.06, 2.11, 2.16 and 2.20 | 84 |
| MACE | Mesoporous | 1.65 | 112 |

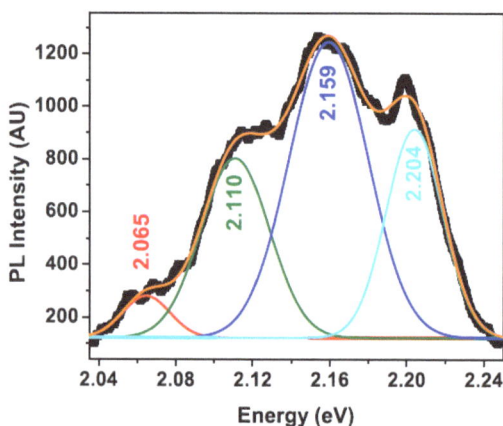

**Figure 5.37**  PL spectrum of MACE-grown Si NWs after 90 min of HF etch. Reproduced from ref. 114 with permission from AIP Publishing, Copyright 2018.

Irrera *et al.*,[93] see Figure 5.38, which consist of single PL bands peaked between 1.55 and 1.907 eV, and that could be interpreted, in contrast to the authors, as the QC emissions of hexagonal silicon wires, with a blue-shift that increases with the decrease of the wire diameter.

# 5.3  Nanocrystalline Silicon Films and Si/SiO$_2$ Superlattices

## 5.3.1  Nanocrystalline Silicon Films

Nanocrystalline silicon (NC-Si) films, deposited with variants of a low temperature (<600 °C) plasma enhanced chemical vapor deposition (PECVD) process on top of a glass or a low-cost substrate, using a

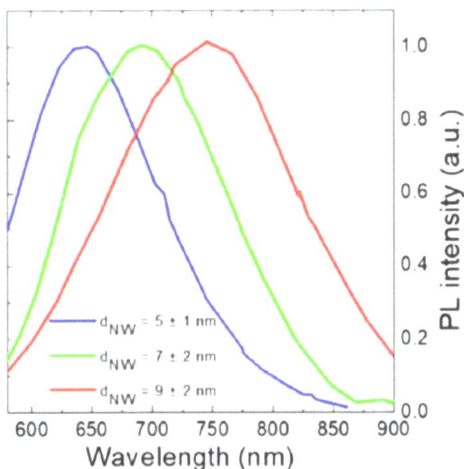

**Figure 5.38** Size dependence of the photoluminescence spectra of MACE grown Si NWs. Reproduced from ref. 93, https://doi.org/10.3390/nano11020383, under the terms of the CC BY 4.0 license, https://creativecommons.org/licenses/by/4.0/.

mixture of silane and hydrogen, gained substantial interest as low-cost solar cells materials at the beginning of this century, on the hypothesis that this kind of material would allow a better use of solar light for PV applications due a combination of cost and optoelectronic advantages. Furthermore, the possibility could be envisaged that its optoelectronic properties might be tuned with an appropriate reduction of the size of the silicon crystallites towards the nanometric size, leading to the set-up of quantum confinement effects, with a shift of the absorption spectrum toward higher energies and a better fit with the solar spectrum.[115–126]

Nanocrystalline silicon (NC-Si) films, also called microcrystalline silicon (µc-Si) films, present a typical morphology, characterized by the presence of arrays of columnar silicon nanocrystals embedded in an amorphous silicon tissue, as shown in Figure 5.39, whose diameters might range from a few nm to 100 nm, depending on the details of the growth process. The amorphous silicon tissue that embeds each nanocrystal prospectively plays a major role in the minimization of the grain boundary recombination losses of the light-induced minority carriers, with potential superior photovoltaic properties with respect to polycrystalline silicon, where recombination losses might be dramatic, depending on the morphology.[127]

The encouraging achievements obtained with NC-Si films grown with conventional PECVD processes, which succeeded in arriving at a convincing material quality and at a reasonable (∼10%) photovoltaic

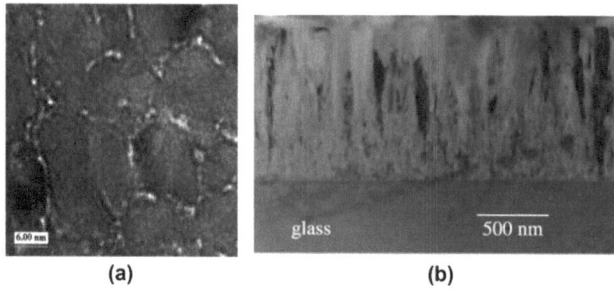

(a)                                          (b)

**Figure 5.39**  (a) HRTEM micrograph of a nanocrystalline silicon film, showing the silicon nanocrystallites embedded in amorphous silicon. Reproduced from ref. 119 with permission from Elsevier, Copyright 2006. (b) TEM cross section of a nanocrystalline silicon layer. Reproduced from ref. 115 with permission from Elsevier, Copyright 2003.

efficiency, still not comparable with that obtained with single crystal silicon, have stimulated interest in the development of dedicated, novel growth processes.

To this end, the macroscopic input parameters of these processes (growth temperature, $SiH_4/H_2$ ratio, $SiH_4$ flux, ion energy), which are empirically managed in the common practice in order to tune the NC-Si properties, should be optimized by theoretical studies addressed at the control of the composition of the plasma phase, and at the understanding of the role of the silane radicals ($SiH_3^\bullet$, $Si_2H_5^\bullet$, $Si_3H_8^\bullet$, ...) on the rate of the NC-Si deposition reaction, as a function of the temperature and of the concentration of the silane radicals.

These novel processes should also be capable of overcoming one of the critical problems arising with the use of conventional PECVD processes that occur under the assistance of high energy ions with large impact on their microstructural and physical properties.

It is known[128] that the deposition process, which occurs from neutral radicalic species behaving as precursors, is influenced by the bombardment of hydrogen-containing ions on the surface of the growing film, which induces the hydrogenation of the surface and the formation of surface and bulk defects when the energy released by the ion impact is of the order of magnitude of the formation energy of structural defects. For energies approaching a critical value $E_{i,c}$,[||||] ions penetrate below the surface and contribute to atomic displacement, sputtering and redeposition phenomena with the suppression of the columnar structure and material densification.

---

[||||] Which depends on the nature of the material.

For energies $E_i > E_{i,c}$ the ions penetrate deep under the surface, leading to gas entrapment and chemisorption at internal surfaces, and structural changes.

Two main solutions, hot wire CVD (HWCVD) and low energy plasma enhanced CVD (LEPECVD) were applied to limit the damage arising from the impact of high energy ions.

## 5.3.2 Growth of Nanocrystalline Silicon Films Using the HWCVD Process

The HWCVD process[128–135] studied by Schropp, Rath *et al.*[115,129–132] and by several other authors[133–136] works in the absence of a plasma phase, with a filament of a high melting metal (W or Ta), heated at $T \geq 1500\,^{\circ}\text{C}$, which catalytically cracks silane molecules to silicon and atomic hydrogen in the gas phase

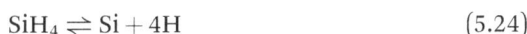

$$\text{SiH}_4 \rightleftharpoons \text{Si} + 4\text{H} \tag{5.24}$$

and provides the thermodynamic conditions for the formation of $\text{SiH}_3^{\bullet}$

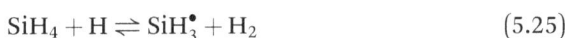

$$\text{SiH}_4 + \text{H} \rightleftharpoons \text{SiH}_3^{\bullet} + \text{H}_2 \tag{5.25}$$

the radicalic species that is the immediate precursor of NC-Si, and decomposes at the surface of the NC-Si deposits with the formation of silicon and hydrogen

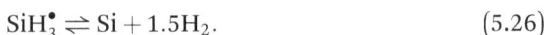

$$\text{SiH}_3^{\bullet} \rightleftharpoons \text{Si} + 1.5\text{H}_2. \tag{5.26}$$

The hydrogen plays an important role in the silicon nucleation.[115,132]

In turn, atomic silicon vapours react with silane with the formation of other radicalic species, as is the case of $\text{Si}_2\text{H}_4^{\bullet}$

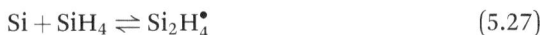

$$\text{Si} + \text{SiH}_4 \rightleftharpoons \text{Si}_2\text{H}_4^{\bullet} \tag{5.27}$$

which can be further reduced by hydrogen with the formation of another radicalic species

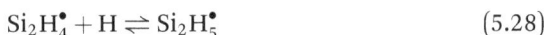

$$\text{Si}_2\text{H}_4^{\bullet} + \text{H} \rightleftharpoons \text{Si}_2\text{H}_5^{\bullet} \tag{5.28}$$

which is also a precursor of silicon.

Although the electrons emitted by the hot filament could induce the formation of ionic species, their effect on the properties of the Si-NC deposited is shown to be negligible.

The application of this process for the deposition of micro-crystalline**** silicon results in films presenting a typical columnar morphology, as is seen in Figure 5.39b.[115] The result of structural studies on NC-Si films grown within 200 and 400 °C, using Ta wires heated between 1600 and 1800 °C, characterized by a crystallinity fraction $\chi_c = \dfrac{Si_c}{Si_a + Si_c}$ in the range 60–80%, and grain size between 15 and 30 nm, supports the view of Brühne *et al.*[134] of a morpho-logical model of (110)-oriented silicon nanocrystals, clustered in micron-size columns, embedded in an a-Si coat. Depending on dif-ferent growth conditions, (111) and (220) oriented silicon crystallites are also experimentally found.

As a consequence of this morphology, the typical room temperature dark conductivities $\sigma_d$ of HWCVD grown NC-Si films are in the range of $\sigma_d = 10^{-7} \, \Omega^{-1} \, cm^{-1}$, insensitive to storage in air or to heat treat-ments, slightly less than the intrinsic conductivity of bulk silicon $(3.12 \times 10^{-6} \, \Omega^{-1} \, cm^{-1})$, indicating poor contact among crystallites.

The photoluminescence (PL) of HWCVD-grown NC-Si films shows a complex dependence on the silane concentration, on the growth $T_g$ and wire $T_w$ temperature, and on the crystallinity, which suggests the presence of delicate equilibria between defect centres, and be-tween the crystalline/amorphous silicon phase, most of which is still without a theoretical explanation.

As a first example, the room temperature PL spectra of NC-Si films, HWCVD-grown at 300 °C on a glass substrate, at different wire tem-peratures $T_w$, characterized by crystal sizes less than 5 nm with (111) preferential orientation, measured by Torchynska *et al.*,[136] are dis-played in Figure 5.40.

The main feature of these PL spectra is a broad band peaking be-tween 1.5 and 1.6 eV, which could be deconvolved into three bands peaking at 1.39, 1.5 and 1.75 eV, with the peak energy that decreases with the wire temperature $T_w$.

PL spectra with peak energies ranging between 1.55 and 1.78 eV, increasing with the increase in $T_w$ from 1850 to 1900 °C, were also found by Ferreira *et al.*[135] in nanocrystalline silicon films HWCVD-grown at 200 °C. These films present a highly porous morphology, consisting of agglomerates of silicon crystallites with sizes in the range 10–60 nm preferentially oriented in the (220) and (111) dir-ections and separated by voids.

---

**** Polycrystalline is used by the authors as a synonym of microcrystalline.

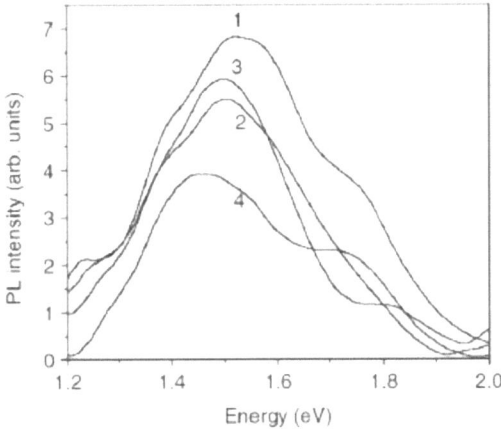

**Figure 5.40**  PL spectra at 300 K of a-Si/NC-Si films prepared at a substrate temperature of 300 °C and different hot wire temperatures (1, 1650 °C; 2, 1750 °C; 3, 1850 °C; 4, 1950 °C). Reproduced from ref. 136 with permission from Elsevier, Copyright 2006.

The porous morphology of these films accounts for their low dark conductivity[††††] ($\sigma_d \approx 10^{-8}\ \Omega^{-1}\ cm^{-1}$), and probably also accounts for their optical properties if we suggest that the porosities of the film are decorated by nanometric crystallites with sizes lower than few nm, since crystallites having sizes within 10–60 nm cannot account for the experimental PL energies.

The previous results could be compared with those found by Golubev et al.[137] for thin, porous, amorphous-nanocrystalline silicon films grown using a PECVD process, with average diameter ranging between 3 and 5 nm, see Figure 5.41, curve 1, who demonstrated that the dual peak features of the PL spectra of these films are the result of Fabry–Pérot interference phenomena, and that the true reconstructed, Pl spectrum is given in Figure 5.41 curve 2, whose intensity decreases with the increase of the crystallinity $\chi_c$ and becomes negligible for $\chi_c = 25\%$. The PL emission is associated with only one recombination centre, responsible for the emission at 1.5 eV, with a FWHM of 0.3 eV, almost independent of the NC size (3–5 nm) and crystallinity. It is, therefore, concluded that this emission is due to QC effects, given the size of the crystallites, and that the decrease in the PL intensity with the increase of the crystallinity is due to clustering of the nanocrystals.

The PL spectra measured by Merdzhanova[138] add further information on the role of structural defects on the PL features of NC-Si films

---

[††††] Measured in the absence of illumination.

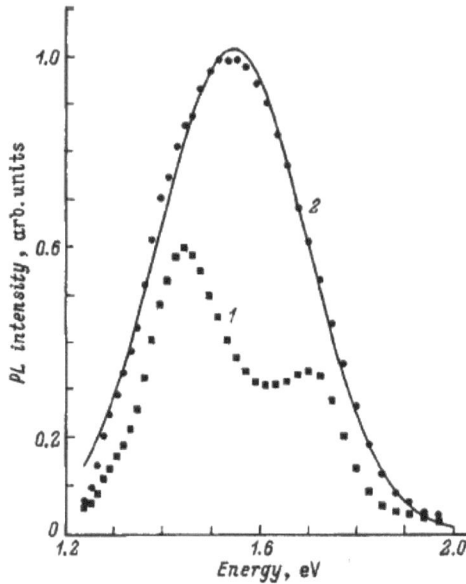

**Figure 5.41** Photoluminescence spectra ($T = 77$ K) of thin amorphous-nanocrystalline silicon films grown using a PECVD process. (1) Experimental spectrum. (2) Reconstructed spectrum. Reproduced from ref. 137 with permission from Springer Nature, Copyright 1999.

HWCVD-grown at a substrate temperature of 185–330 °C, with Ta wires held at 1650–1730 °C.

In fact, these PL spectra show the presence of a defect band at 0.7 eV (due to recombination at grain boundaries), a band at 1.01–1.03 eV typical of μc-Si and a band at 1.35 eV, attributed to a-Si, which is the dominant feature of PL spectra of a-Si films.

Eventually, the PL spectra measured by Dutt *et al.*[139] on NC-Si films, HWCVD-grown at a substrate temperature of 200 °C, consisting of a mixture of nanocrystals ($\leq 5$ nm in diameter) embedded in amorphous silicon oxide, show the influence of the dielectric matrix where the Si crystallites are distributed. The PL spectra of this material present a band in the blue peaking at 450–500 nm (2.77–2.48 eV), and a second band peaking at 850 nm (1.45 eV).

The preliminary conclusion arising from these results is the extreme variability of the optoelectronic properties of HWCVD grown NC-Si films arising from different morphologies, nanocrystal sizes and from the number of different phases (NC-Si *vs.* a-Si), which in turn depend on the large number of process variables that cannot be entirely controlled.

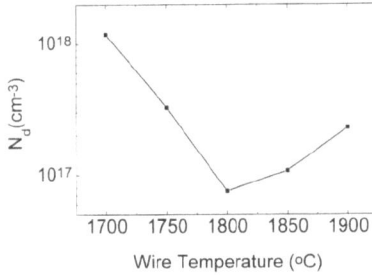

**Figure 5.42** Dangling bond defect density as a function of the wire temperature for a HWCVD growth process. Reproduced from ref. 130 with permission from Elsevier, Copyright 1998.

This is the primary condition that limits the application of HWCVD process to solar cell technologies, together with the presence of grain boundaries and thus of dangling bond (DB) defects in the HWCVD-grown material, which play a minor role in the transport of photo-generated carriers[130] but a major influence as recombination centres being, together with light induced defects, responsible for the limited efficiency of μc-Si solar cells fabricated with the HWCVD process.

The maximum efficiency of these solar cells was, in fact, measured to range between 7.1%[140] and 9.4%[141] while numerical simulations show that efficiencies up to 12% could be reached by reducing the defect density down to $10^{14}$ cm$^{-3}$ [142] almost three orders of magnitude lower than the average dangling bond (DB) defect density of HWCVD-grown NC-Si, measured by Rath,[131] amounting, see Figure 5.42, to a minimum value of $7.8 \times 10^{16}$ cm$^{-3}$ for $T_w = 1800$ °C.

Only amorphous silicon allows reaching defect densities of this order of magnitude and achieving, thanks also to its excellent opto-electronic properties, efficiencies comparable with those of c-Si, as demonstrated by the use of heterojunction amorphous/crystalline silicon solar cells[143] whose maximum efficiency ranges between 14.8 and 18.2%[144] but still much below the conversion efficiency of fully crystalline silicon solar cells.

### 5.3.3 Growth of Nanocrystalline Silicon Films Using the LEPECVD Process

The LEPECVD process, originally developed and successfully applied for the epitaxial deposition of Si-Ge heterostructures on Si,[145–147] is a low-voltage (10–12 V), high current (20–70 A) arc discharge, high density plasma CVD process, see Figure 5.43, that minimizes surface damage and reconstruction arising from high energy ions. The dense

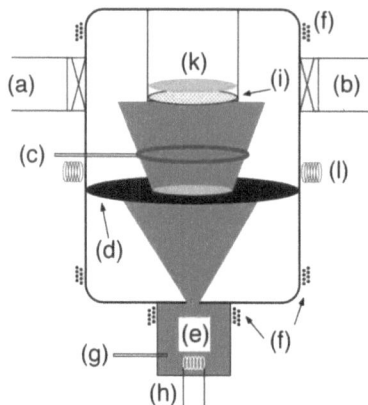

**Figure 5.43** Schematic drawing of the LEPECVD system. The shaded area represents the plasma columns with (a) turbo pump connection, (b) load lock, (c) gas inlet for silane, (d) auxiliary anode, (e) plasma source, (f) plasma focusing coils, (g) gas inlet for Ar, (h) heated filament, (i) substrate, (k) sample heater, (l) deflection magnets. The input voltage is applied between (e) and (i). Reproduced from ref. 147 with permission from Elsevier, Copyright 2002.

Ar plasma leads, instead, to a very efficient cracking of silane, and the intense flux of low energy ions (10–12 eV) leads to high deposition rates and low surface state density.

With this method, epitaxial silicon deposition on silicon substrates was carried out with excellent results, as well as the deposition of fully relaxed, several μm thick Si–Ge substrates, with Ge concentrations between 10 and 100% and with growth rates of 5–7 nm s$^{-1}$,[147] well above the growth rates typical of MBE and CVD methods that are of the order of 1 nm s$^{-1}$ and make these processes unacceptable for mass-production use.

The LEPECVD process was also used to grow nanocrystalline silicon films by Pizzini et al.,[116–122,148] in the frame of a multi-partner project, implemented by the theoretical simulation of the growth process and by measurement and calculation of its optoelectronic properties by Mattoni, Colombo et al.,[149–155] that allowed them to arrive at a complete structural and optoelectronic characterization of the material.

Nanocrystalline silicon films were deposited on thermally oxidized Si substrates and on clean HF-etched silicon substrates, at temperatures between 210 and 280 °C, using silane and hydrogen as reactive gas, with growth rates ranging between 0.5 and 1.3 nm s$^{-1}$ and silane dilution ratios $\dfrac{\phi(SiH_4)}{\phi(SiH_4) + \phi(H_2)}$ ranging between 1.0 and 10%. The

**Table 5.6** Growth conditions and structural properties of a set of NC-Si films prepared using LEPECVD on clean silicon substrates and on oxidized silicon substrates. Data from ref. 119.

| Substrate | Sample | $T_g$ (°C) | $d$ | $\chi_{cr}$ | $L_{(111)}$ (nm) | $L_{(220)}$ (nm) | $L_{(011)}$ (nm) |
|---|---|---|---|---|---|---|---|
| Si(001) | 7363 | 210 | 4.2 | 73 | 22.5 | 20.5 | 16.6 |
| Si(001) | 7502 | 210 | 1.4 | 74.5 | 22.5 | 10.8 | 16.5 |
| Si(001) | 7503 | 210 | 7 | 76 | | | |
| Si(001) | 6956 | 230 | 4.2 | 75 | 18.1 | 10.7 | 12 |
| Si(001) | 7956 | 230 | 4.2 | 75 | 18.1 | 10.7 | 12 |
| S(i001)/SiO$_2$ | 6731 | 250 | 4.2 | 75 | | 13 | 16 |
| S(i001)/SiO$_2$ | 56 173 | 280 | 1 | 70 | 28.1 | 19.4 | 18.9 |
| S(i001)/SiO$_2$ | 56 172 | 280 | 3 | 73 | 25.8 | 22.5 | 23.7 |
| S(i001)/SiO$_2$ | 56 171 | 280 | 6 | 72.5 | | | |
| S(i001)/SiO$_2$ | 56 170 | 280 | 10 | 70.5 | 23.9 | 32.6 | 19.6 |
| S(i001)/SiO$_2$ | 7575 | 280 | 10 | 72 | 17.3 | 12.8 | 17.9 |

silane dilution ratio$^{‡‡‡‡}$ was always held well below 25%, at which pure amorphous Si is instead deposited.[156]

The preparation conditions of a selection of NC-Si samples prepared using this method, presenting an average thickness of 3 µm and the corresponding structural properties that well represent the average properties of films prepared with the LEPECVD process, are displayed in Table 5.6, where $\chi_{cr} = \dfrac{A_{cr}}{A_{cr} + \sigma A_{am}}$ is the crystalline volume fraction of the sample, $A_{cr}$ and $A_{am}$ are the experimental peak areas of the Raman peaks for crystalline and amorphous silicon, taken as the measure of their volume amounts,[122] $\sigma$ is a constant,[157] and $L_{(111)}$, $L_{(220)}$, $L_{(011)}$ are the integrated XRD peak intensities that depict the crystallographic texture of the films as a function of the preparation conditions.

It is apparent that the crystallinity decreases with the increase in the temperature with little impact in this temperature range on the silane dilution factor.

Different from the case of HWCVD-grown NC-Si films, with their extreme variableness of morphological and structural properties, the LEPECVD process runs smoothly, almost fully preventing the problems of the HWCVD process. Moreover, the crystallographic structure of the NC-Si films is invariably diamond cubic.

As an example, Figure 5.44 (bottom)[118] displays the typical XRD spectrum of NC-Si samples grown on a thermally oxidized substrate

---

$^{‡‡‡‡}$ The silane dilution ratio is a very important process parameter since it determines the amount of amorphous silicon in the deposit.

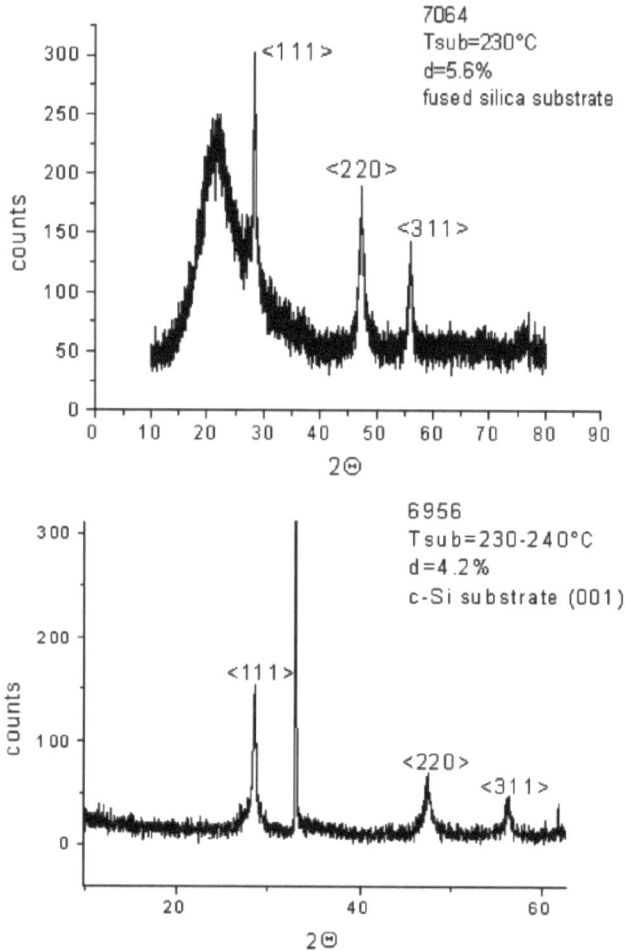

**Figure 5.44** XRD spectrum of a NC Si sample grown at 230 °C on a fused silica substrate (top) or on a thermally oxidized silicon substrate (bottom). Reproduced from ref. 118 with permission from Elsevier, Copyright 2005.

with the expected reflection patterns of diamond cubic silicon.[§§§§] This kind of spectrum is observed for all the NC samples grown on oxidized silicon substrates, independent of the growth temperature. If the growth is carried out on a glassy substrate, the $\langle 111 \rangle$ reflection is partially covered by the broad reflection of the glassy substrate, see Figure 5.44 (top). Therefore, diamond cubic silicon is the statistically ascertained crystallographic structure of NC-Si films, LE-PECVD grown with a growth rate of 0.5–1.3 nm s$^{-1}$, that apparently

---

[§§§§] The strongest reflection after the $\langle 111 \rangle$ is the $\langle 001 \rangle$ reflection of the CZ silicon substrate.

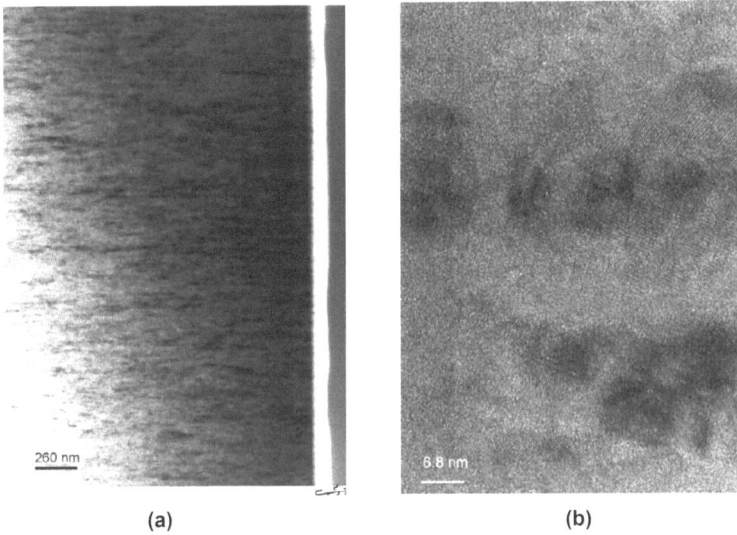

(a)                                    (b)

**Figure 5.45** (a) TEM image of a section of a film grown at $T = 250\ °C$ on an oxidized silicon sample. (b) HRTEM image of a section of the same sample. Reproduced from ref. 117 with permission from Elsevier, Copyright 2005.

allows the growth to occur close to thermodynamic equilibrium conditions.

The TEM images of a LEPECVD Si film grown at 250 °C given in Figure 5.45[117] display the typical morphology of NC-Si films. They present a columnar structure, see Figure 5.45a, with elongated columns of amorphous Si in which oriented silicon nanocrystals are distributed, as can be seen in Figure 5.45b. The loose contact among nanocrystallites explains the low, room temperature, dark electrical conductivities $\sigma_d$ $(1.8 \times 10^{-7}\ \Omega^{-1}\,\mathrm{cm}^{-1} \leq \sigma_d \geq 4 \times 10^{-6}\ \Omega^{-1}\,\mathrm{cm}^{-1})$ of films grown at 250 °C.

The HRTEM image of a section of a sample grown at 280 °C, displayed in Figure 5.46,[119] shows that the effect of a higher growth temperature is not only the slight increase in the crystallite size, but also the onset of a clustering process of individual silicon crystallites, still individually separated by a thin layer of a-Si.

This clustering phenomenon was well modelled using molecular dynamics by Mattoni *et al.*,[149,150] who succeeded in simulating the initial formation of cylindrical silicon crystallites embedded in amorphous silicon Figure 5.47 (top), followed by a clustering process, activated by heating the sample at 1200 K, that well reproduces the final morphology of our samples, as can be seen by comparing Figure 5.46 with Figure 5.47 (middle and bottom).

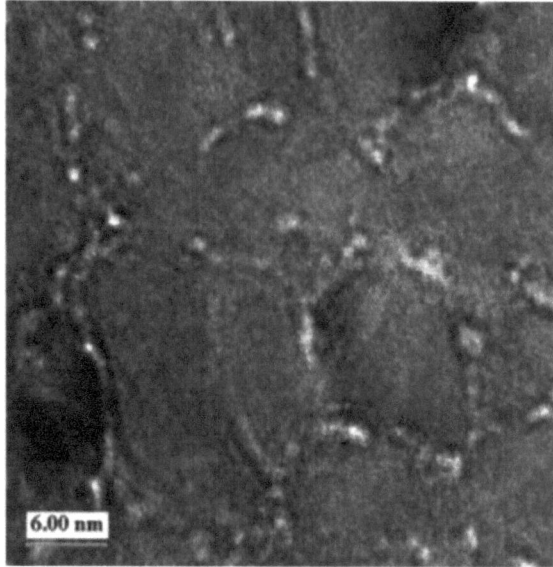

**Figure 5.46**  HRTEM image of a section of a sample grown at 280 °C, showing a cluster of crystallites with an average size of 3–6 nm, separated by an amorphous Si layer. Reproduced from ref. 119 with permission from Elsevier, Copyright 2006.

The photoluminescence spectra of NC-Si films well reflect the influence of the crystallographic texture on their optoelectronic properties, see Figure 5.48.[119] In fact, the sample grown at 210 °C, with a crystallinity $\chi_{cr} = 76\%$, exhibits two sub-band emissions at 0.7 and 0.9 eV, both typical of polycrystalline silicon,[158] of which the B4 one at 0.7 eV might be attributed to intra-grain and surface defect recombination and the B3 one (at 0.9 eV) due to excitonic transition. The sample grown at 280 °C, with a crystallinity $\chi_{cr} = 72\%$, *i.e.* with a slightly higher volume content of a-Si, exhibits instead a strong emission at 1.3 eV and another one, more feeble, at 1.85 eV, both strongly blue shifted above the energy gap of Si.

Considering that the TEM image of Figure 5.46 gives a mean size of 3–6 nm for the crystallites of this sample, lower than the Bohr radius of silicon (4.2 nm) at least for some crystallites, the occurrence of QC conditions could be suggested to explain the PL emissions above the Si band gap, of which that at 1.4 eV corresponds to the blue-shifted optical gap of c-Si and that at 1.85 eV is the emission of nanostructured a-Si.[159]

The onset of QC conditions for a set of NC-Si samples grown at 280 °C, was also experimentally demonstrated by Bagolini *et al.*,[153]

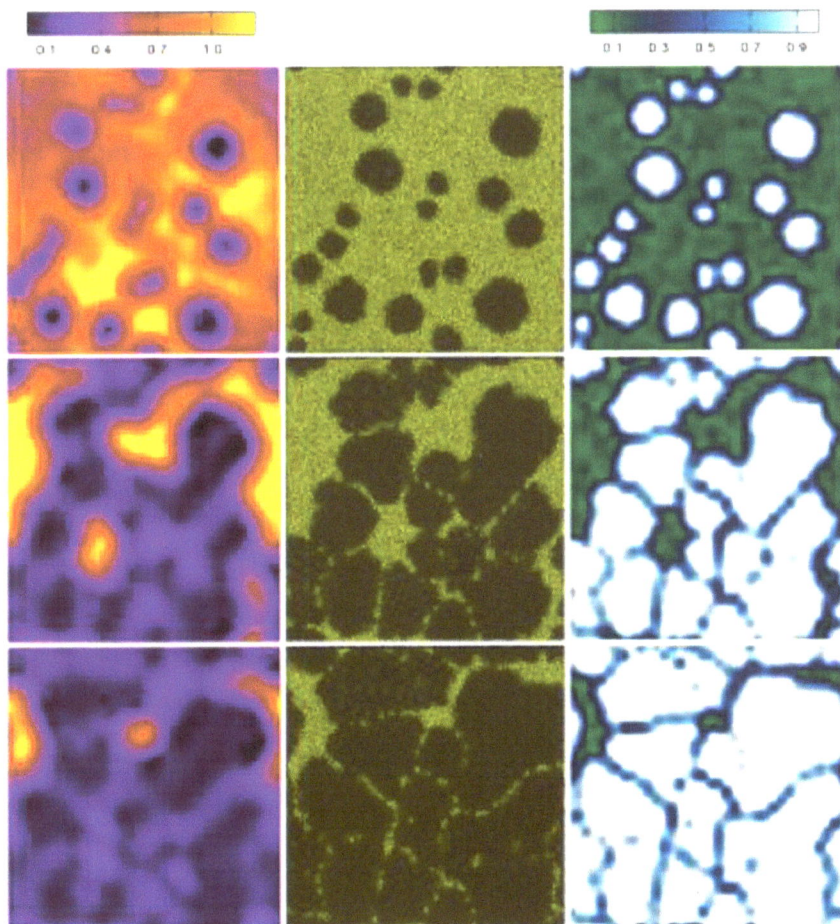

**Figure 5.47** Center: Snapshots of the nc silicon structure after different times during the annealing at 1200 K. Left and right: The corresponding colour maps. Reproduced from ref. 150 with permission from American Physical Society, Copyright 2009.

who showed that the peak emission energy in their PL spectra, see Figure 5.49, lies above the energy gap of silicon for average crystal sizes of 3.2 and 3.5 nm, measured by XRD measurements.

This broad configuration of the PL spectra is the outcome of emissions from the crystallites, the embedding of an a-Si matrix and from the matrix–crystallites interfaces. Since the peak PL emission increases from 1.1 to 1.3 eV with the decrease of the average crystallites size from 5.6 to 3.2 nm, but remains almost unchanged with the change of crystallinity, see details in Figure 5.49, this is a good sign that the emission is controlled only by the crystal size.

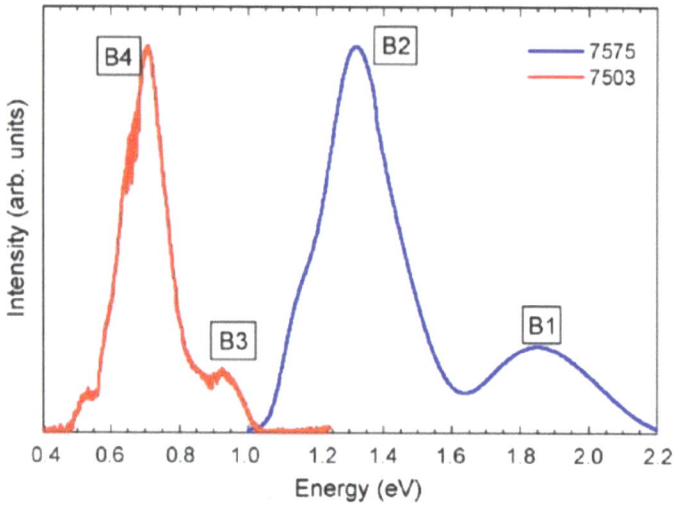

**Figure 5.48**   Photoluminescence spectra at 12 K of the sample 7503 grown at 210 °C with $\chi_{cr} = 76\%$ and of the samples 7575 grown at 280 °C with $\chi_{cr} = 72\%$. Reproduced from ref. 119 with permission from Elsevier, Copyright 2006.

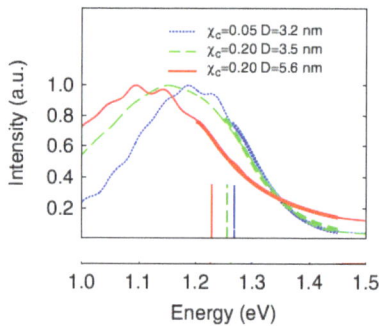

**Figure 5.49**   Effect of grain size $D$ and crystallinity $\chi_c$ on the PL emission at 14 K of silicon nanocrystals grown at 280 °C. Reproduced from ref. 153 with permission from American Physical Society, Copyright 2010.

## 5.4   Si/SiO$_2$ Superlattices

These films exhibit a superlattice (SL) morphology,[160–162] see Figure 5.50, where each layer of the SL is an ordered dispersion of quantum-confined nanocrystalline silicon dots in a dielectric matrix of SiO$_2$.

Each single layer of the superlattice is prepared by reactive sublimation of SiO powder in oxygen atmosphere or by PECVD,[162]

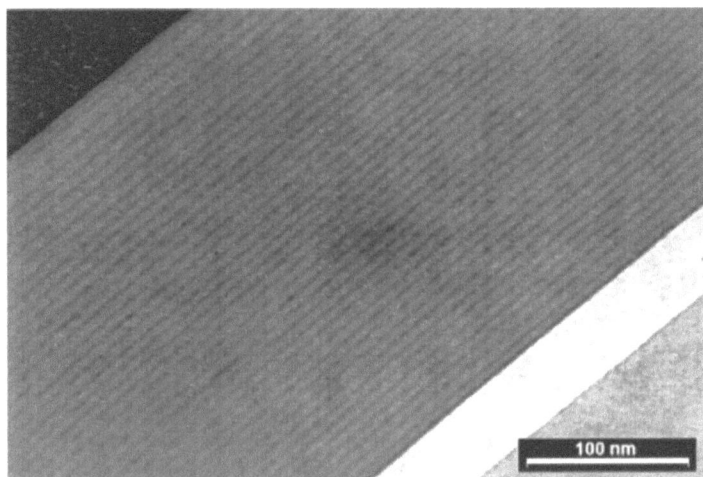

**Figure 5.50** TEM image of a Si/SiO$_2$ superlattice after crystallization. Reproduced from ref. 160 with permission from American Physical Society, Copyright 2000.

leading to the formation of a layer of non-stoichiometric silicon monoxide.

Once the complete superlattice, layer by layer, is deposited, it is annealed at 1100 °C in inert gas atmosphere to decompose the non-stoichiometric silicon monoxide, which results in the formation of crystalline Si dots embedded in SiO$_2$[161]

$$SiO_x \rightarrow \left(1 - \frac{x}{2}\right)Si + \frac{x}{2}SiO_2. \qquad (5.29)$$

The SL approach provides a promising route for the direct use of silicon in photonic applications (LEDs, lasers and PV cells), since the optical bandgap can be tuned between 1.3 and 1.7 eV as a function of the Si dots size, leading, as an example, to an intense, blue-shifted light emission at 1.6 eV for 2 nm nanocrystals.[161]

The electroluminescence (EL)[¶¶¶¶¶] efficiency of Si/SiO$_2$ SLs is, however, limited by charge transport problems and by charge recombination at surface defects.[162,163]

In fact, efficient charge transport is only achievable if SiNCs of comparable sizes are almost in intimate contact with each other and, thus, charge transport problems might only be managed by the improvement of the size and distance distribution of the silicon

---

[¶¶¶¶¶] Light emission of SLs for LEDs or laser applications must be electrically excited.

nanocrystals present in each layer of the superlattice. A satisfactory answer to this requirement is, however, not yet available.

It is, instead, already known that the EL efficiency depending on charge-recombination at defects might be improved by defect hydrogenation[164] and by low ion beam treatments,[165] although further improvements are still possible.

## 5.5  Compound Semiconductor Nanostructures, the Case of ZnO

Nanostructured compound semiconductors, SiC, GaAs, CdS, CdSe and CdTe, as a few examples, have received attention worldwide in the last 20 years for their particularly attractive potential in photonic applications,[4] and for their almost immediate utilization with traditional microelectronic technologies.

Different from fully covalent compound semiconductors, the family oxide semiconductors received minor attention, despite their already important side-application as wide band gap, optically transparent and electrically conducting materials, as is the case of indium tin oxide (ITO), or as chemical sensors (see Section 1.5.3), given that their room temperature conductivity might be modulated by reducing gases.

More interesting, is that oxide semiconductors are also considered possible candidates to replace the silicon-based CMOS technology, when the Moore law approaches definitely its limit.[166]

Among oxide semiconductors we intend to discuss here ZnO, with its wide and direct band-gap (3.37 eV at room temperature) and its hexagonal wurtzite structure, because it is considered a serious competitor to GaN, with several potential applications in short-wavelength optoelectronic devices, solid-state display devices, solar cells, sensors, electrochromic windows, and as a source of acoustic waves at microwave frequency.

Given its exciton binding energy (60 meV) larger than that of GaN (25 meV), ZnO is in fact considered a material of excellence for the fabrication of nanowires addressed at the manufacture of low-threshold excitonic lasers and light-emitting diodes (LEDs),[167,168] without forgetting its almost standard use for chemical nanosensors.[169]

ZnO nanostructures have been grown by several processes, including vapour deposition, pulsed laser deposition, molecular beam epitaxy, metal organic chemical vapour deposition (MOCVD), sputtering, electron beam evaporation, spray pyrolysis, sol–gel processing, chemical, and electrochemical deposition processes.[167–177]

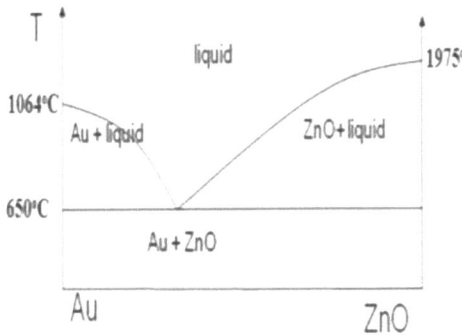

**Figure 5.51**  Phase diagram of the system Au–ZnO. Reproduced from ref. 178 with permission from Penerbit UTM Press, Copyright 2014.

A VLS process was used by Wang[170] to deposit ZnO nanorods and nanobelts, using ZnO as the sublimation source and Au droplets as the catalysts, taking advantage of the condition that ZnO is soluble in liquid Au above 650 °C, see Figure 5.51. One can therefore foresee that the growth process occurs with a strong analogy of the growth of Si NWs with Au as the catalyst.

The process is carried-out under a vacuum of 200–600 Torr in an alumina tube, and a flux of Ar is used[178] to address the ZnO vapours to the deposition substrate, which in general consists of polycrystalline α-$Al_2O_3$, with the same hexagonal structure of ZnO, which could work as a natural template given by the grain boundary surface texture.

Sublimation of ZnO is carried out at 1350 °C, and the growth process is carried out above 650 °C, the eutectic temperature of the Au–ZnO system.

Since ZnO partially decomposes to Zn vapours and oxygen

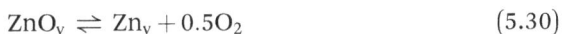

$$ZnO_v \rightleftharpoons Zn_v + 0.5O_2 \qquad (5.30)$$

at the sublimation temperature, with a Zn equilibrium pressure of $10^{-3}$ Torr, it is supposed that Zn is also transported to the Au catalyst, where it is dissolved together with ZnO, giving origin, under supersaturation conditions, to the growth of arrays of constant diameter ZnO nanorods, see Figure 5.52. It is apparent that the alumina substrate does not work as a template since the NWs are not vertically oriented and aligned.

A mixture of ZnO and graphite was instead used as a precursor by Rafique *et al.*[177] to grow ZnO nanostructures on a GaN layer deposited on a sapphire substrate, using Au droplets as the catalyst. The

**Figure 5.52**   SEM image of ZnO nanorods grown using Au as the catalyst (a) length scale of 500 nm (b) length scale of 200 nm. Reproduced from ref. 170 with permission from IOP Publishing, Copyright 2004.

presence of a GaN layer should favour the quasi-epitaxial deposition of ZnO. The process was carried out under a controlled flow of oxygen and argon between 880 and 950 °C, a temperature much lower than that used by Wang[170] to sublimate ZnO. The chemistry of the process is only qualitatively discussed by the authors, who suggest that ZnO is reduced by graphite to Zn vapours, which are transferred to the Au catalyst, where Zn dissolution occurs with the formation of a super-saturated Zn–Au solution. The Zn atoms that segregate at the surface of the catalysts are then oxidized in the oxygen-rich atmosphere, leading to the final segregation of ZnO NWs. Depending on the chemistry of the surface oxidation, the growth of ZnO NWs could be associated with the presence of nanowalls (two dimensional, vertically oriented platelets) that connect the NWs.

As expected, it was observed that the morphology of the nanostructures obtained with this process (density, orientation and nanowall formation) depends both on the oxygen content in the gas phase and on the temperature, which take their optimum values for 30% of $O_2$ and 880 °C, respectively, where the growth of vertically oriented NWs occurs.

The PL emission of these NWs is a narrow band emission at 380 nm (3.26 eV), in correspondence with the optical gap of ZnO, and a broad emission in the green at 520 nm (2.38 eV), supposed to be associated to oxygen vacancies in ZnO.

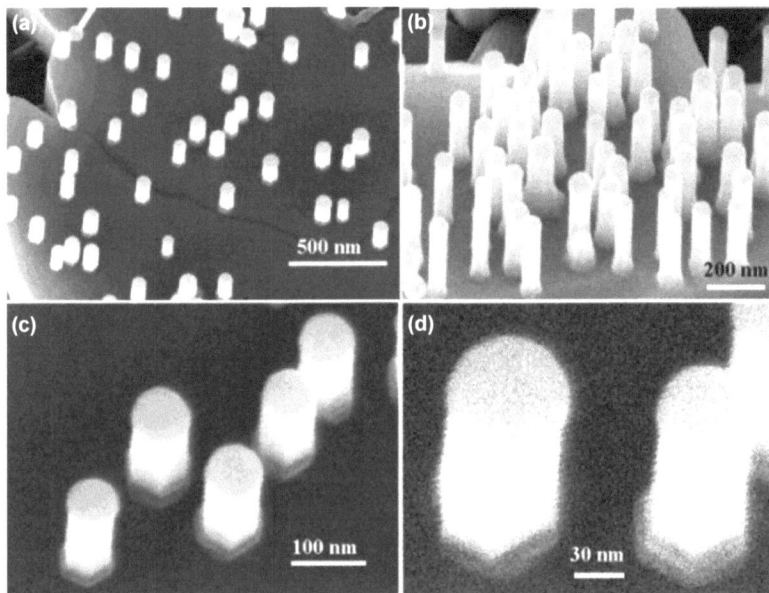

**Figure 5.53** A distribution of ZnO nanorods grown at 600 °C on an alumina substrate using Sn as the catalyst. The magnification of the images increases from (a) to (d). Reproduced from ref. 179 with permission from American Chemical Society, Copyright 2003.

As demonstrated by Gao *et al.*,[179] Sn also works well as a catalyst for the deposition of vertically oriented ZnO nanorods. The results are displayed in Figure 5.53 for a growth process carried out at 1100 °C using a polycrystalline alumina substrate. Here the substrate works well as a template and the reason could be the high growth temperature and the wetting of ZnO on the alumina substrate.

According to the authors, the diameter of the oriented nanorods is well confined by the diameter of the liquid Sn ball on the tip.

In contrast to VLS processes, the thermal decomposition at 300 °C of Zn acetate was considered by Chandraiahgari *et al.*[176] and Tyona *et al.*[180] who demonstrated that Zn acetate also works well as a pre-seeding agent.

The chemical deposition method of ZnO NWs in aqueous solutions[171] and a seedless electrochemical (EC) method proposed by Lupan *et al.*[167] have also been proven to be suitable for the growth of ZnO nanowires, despite their apparent simplicity and low cost, but only with the electrochemical method aligned arrays of NWs were obtained.

The EC process is based on the use of an electrically conductive substrate, consisting of fluorine-doped $SnO_2$ (FTO)-coated glass substrates or of indium doped tin oxide (ITO), which presents a lattice mismatch of only 3% between the oxygen–oxygen (O–O) distance on the $\langle 111 \rangle$ ITO planes and on the $\langle 0001 \rangle$ ZnO planes,[167] which would allow the best conditions for a quasi-epitaxial deposition.

The seedless deposition of ZnO NWs is carried out between 70 and 90 °C in a diluted (5 mM) aqueous $ZnCl_2$ solution, using as supporting electrolyte a 0.1 M solution of KCl, or in an aqueous solution of Zn perchlorate $Zn(ClO_4)_2$. A Zn wire works as the counter-electrode. The solution is saturated with pure oxygen, which is bubbled in it during the entire process.

Once a ZnO crystal is electrochemically nucleated at the surface of the conductive substrate, the growth process continues with the electroreduction of the molecular oxygen that saturates the solution and is chemisorbed at the surface of the ZnO crystal that works as an electrode[181]

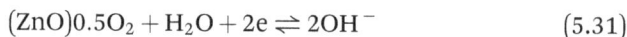

$$(ZnO)0.5O_2 + H_2O + 2e \rightleftharpoons 2OH^- \tag{5.31}$$

followed by the segregation of ZnO

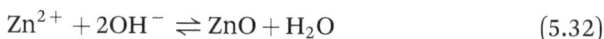

$$Zn^{2+} + 2OH^- \rightleftharpoons ZnO + H_2O \tag{5.32}$$

which allows the further vertical growth of a wire, occurring in the absence of any lateral growth, leading to wires presenting an average width of 120 nm.

After electrodeposition, the samples are subjected to a hydrothermal treatment at 150 °C, which was shown to improve the electronic properties of the as-grown ZnO, with water that behaves as a successful passivating agent for structural defects.

Figure 5.54 displays a SEM plane-view image (a) and a tilted image (b) of an array of vertically aligned ZnO nanowires grown using the EC method and an ITO substrate, which demonstrates the occurrence of a thermodynamically allowed self-aligning process.

While detailed information on the influence of the process temperature, of the solution composition and pH on the morphological, structural, and physical properties of the nanowires grown with the EC method is available,[173–175] a problem remains concerning the quasi-epitaxial wire nucleation process on ITO or FTO substrates, on which no detailed information is available.

We suggest, on the hypothesis that both ITO and FTO are polycrystalline, that nucleation occurs at the GB triple points (nodes), where three grains meet along each junction line.[182] GBs and GB

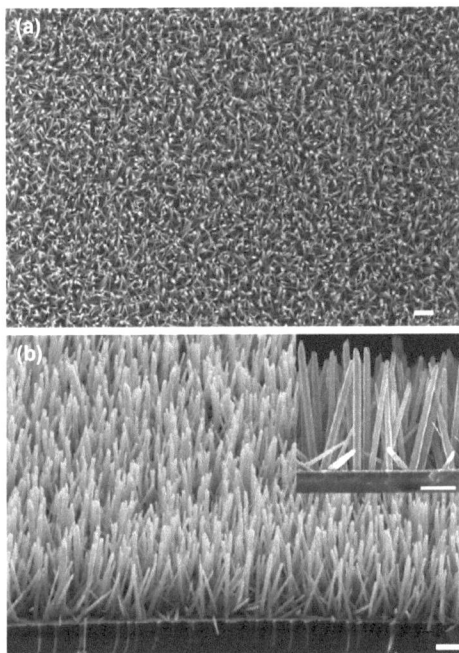

Figure 5.54  SEM images of an array of ZnO nanowires (scale bar 1 μm) (a) top view (b) lateral view. Reproduced from ref. 167 with permission from Elsevier, Copyright 2010.

nodes, in fact, are known to be preferred sites for the segregation of new phases, thanks to their excess of interface energy.

XRD and TEM measurements on the thermally annealed EC-grown ZnO nanorods show that the all the Bragg reflections are consistent with the space group $P6_3mc$ of the wurtzite ZnO lattice.

Both the texture and the photoluminescence of EC-grown ZnO NWs and nanorods are strongly affected by the preparation conditions. As is shown in Figure 5.55, as an example, at 70 °C with a 5 mM $ZnCl_2$ solution a dense surface film is deposited, with modest photo-luminescence properties, while nanorods grown from a diluted (0.2 mM) $ZnCl_2$ solution exhibit a PL intensity that is strongly enhanced when the growth is carried out at 85 °C.

As the average width of the nanorods ($>100$ nm) is largely above the Bohr radius of excitons in ZnO (0.9 nm), QC effects are not expected and the luminescence peak energy fits well with that of bulk ZnO.

It seems, therefore, to be conclusively shown that the EC method is a valuable competitor of the high temperature VLS processes,

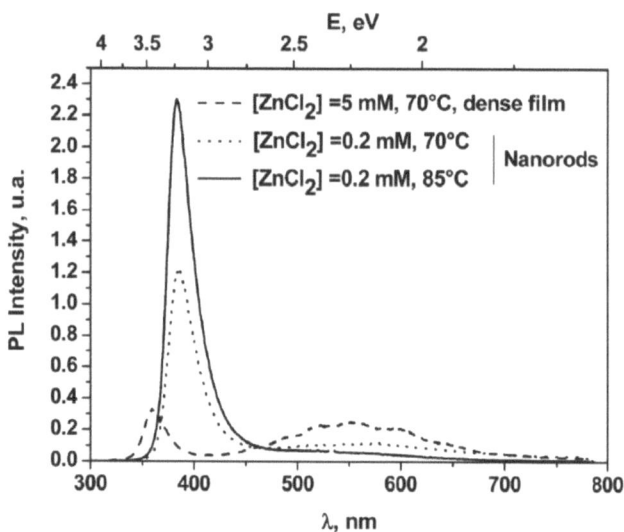

**Figure 5.55**  PL emission of EC-grown ZnO as a function of the temperature and ZnCl$_2$ concentration. Reproduced from ref. 173 with permission from American Chemical Society, Copyright 2009.

both opening the door to ZnO NWs to photonic applications, but p-type doping problems that exist also for bulk ZnO,[183] see Chapter 2, inhibit, at the time this book has been written, the preparation of n–p junctions and the use of ZnO nanowires for photonic applications.[184]

Like in the case of bulk ZnO, as-grown nanocrystalline ZnO is n-type self-doped, and substitutional Ag (Ag$_{Zn}$) is considered a promising p-type dopant both of ZnO nanoparticles[185] and of ZnO NWs grown using the electrochemical route.[186]

According to Thomas and Cui[186] effective electrochemical p-type doping can only be obtained using high Ag concentrations (>0.1 mol%) after a thermal annealing at 500 °C, because the as-grown, Ag-doped ZnO NWs are still n-type doped.

This behaviour could be understood if we recall, see again Chapter 2, that, according to Van de Walle[187,188] and Koßmann and Hättig,[189] interstitial hydrogen in various configurations could be considered the donor responsible for the n-type self-doping of ZnO. Since hydrogen is likely incorporated as interstitial hydrogen in ZnO in the EC process, and can be removed by thermal annealing, it is reasonable to suppose that interstitial hydrogen is responsible for the n-type conductivity of EC as-grown ZnO NWs.

# Appendix 5.1 High Pressure Silicon Allotropes and Metastable Allotropes at Ambient Pressure

Silicon is credited with more than 12 high pressure polymorphs that can be obtained by the application of hydrostatic pressure to diamond cubic (DC) silicon samples in special cells or by using micro-indentation techniques.[190–195]

Upon compression, DC silicon undergoes a series of first order phase transitions,[193] of which the first step is the conversion of the cubic structure to the body centred tetragonal structure of β-tin, with the contextual conversion of covalent bonds to metallic bonds,[194] see Table 5.7. At further compression, additional phase transitions occur, leading to a simple hexagonal, an orthorhombic, a close packed hexagonal and face-centered cubic structure, with a continuous increase in density, as is seen in Figure 5.56, where curves 1 and 2 are the calculated isotherms for diamond cubic Si, and curves 3 and 4 are the calculated isotherms for β-tin and orthorhombic silicon.

Figure 5.57 displays, instead, the phase diagram of silicon, calculated from first principles by Paul et al.,[196] where it is possible to see the regions of stability of the different polymorphs identified by their space groups or by their structures.

Many other phases of silicon have been predicted and recently produced, as is the case of the $Si_{24}$ allotrope grown by Guerrette et al.[197] by high vacuum annealing at 400 K of its sodium silicide $Na_4Si_{24}$ precursor, which is prepared with high temperature (1123 K) and pressure (9 GPa) synthesis. The $Na_4Si_{24}$ phase contains channels along its $a$ crystallographic axis, which facilitate the removal of sodium and favour its potential use as the anode in Na-ion batteries.[199]

**Table 5.7** Structural properties of high pressure, metastable and silicon polymorphs. Data from ref. 193.

| Symbol | Structure | Space group | Transition pressure (GPa) |
|---|---|---|---|
| DC Si; Si-I | Diamond cubic | $Fd\bar{3}m$ | |
| β-tin; Si-II | Body-centered tetragonal | $I4_1/amd$ | 11.7 |
| Si-XI | Orthorhombic | $Imma$ | 13.2 |
| SD Si; Si-V | Simple hexagonal | $P6/mmm$ | 15.3 |
| Si-VI | Orthorhombic | $Cmca$ | 38 |
| Si-VII | Hexagonal close packed | $P6_3/mmc$ | 42 |
| Si-X | Face centred cubic | $Fm\bar{3}m$ | 79 |

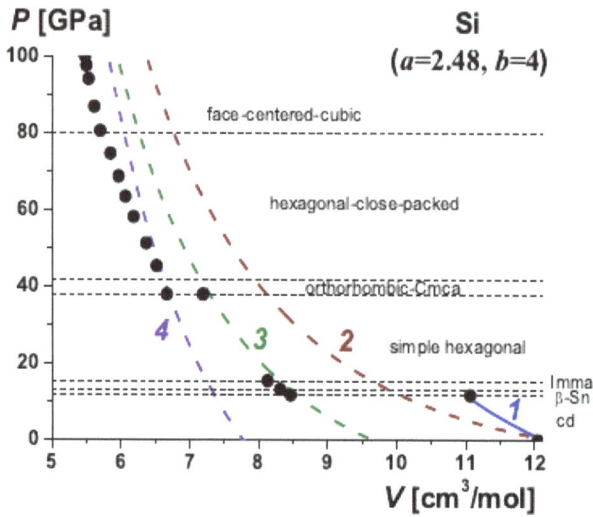

**Figure 5.56** Calculated PV diagram of Si at 300 K (black points display experimental results). Reproduced from ref. 194, http://dx.doi.org/10.1088/1742-6596/918/1/012031, under the terms of the CC BY 3.0 license, https://creativecommons.org/licenses/by/3.0/.

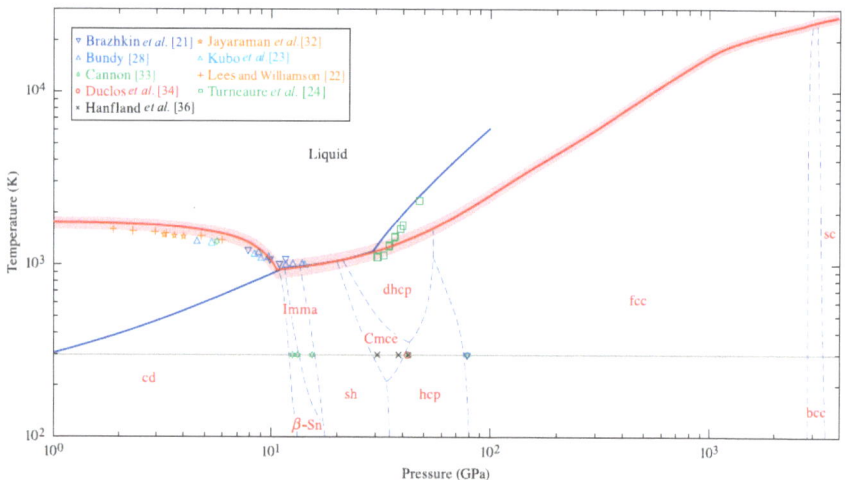

**Figure 5.57** Temperature–pressure phase diagram of silicon, calculated from first-principles. Reproduced from ref. 196 with permission from American Physical Society, Copyright 2019.

After high vacuum treatment, the effective removal of Na leaves a lightly Na-doped $Si_{24}$ phase with 99.9985 at% Si.

The equation of state of silicon at ambient temperature up to 105.2 GPa has been experimentally determined by synchrotron X-ray

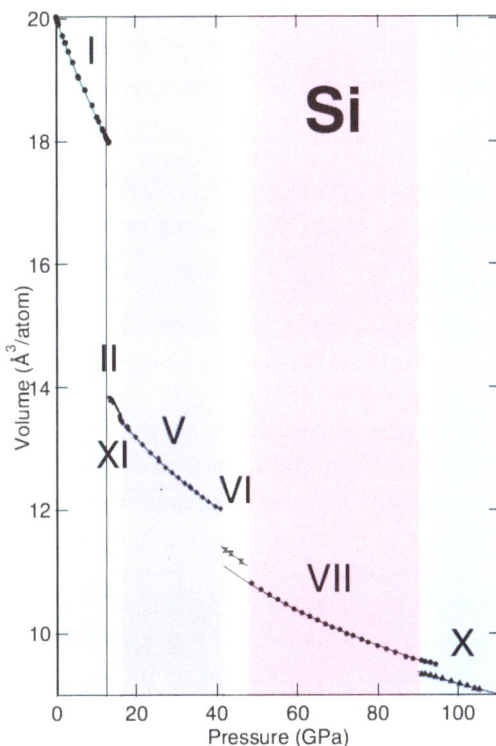

**Figure 5.58** Isothermal PV diagram of silicon, determined by synchrotron X-ray diffraction experiments. Reproduced from ref. 198, https://doi.org/10.1038/s41598-019-51931-1, under the terms of the CC BY 4.0 license, https://creativecommons.org/licenses/by/4.0/.

diffraction experiments by Anzellini *et al.*[198] Results are displayed in the PV diagram of Figure 5.58,[‖‖‖] which confirms with accuracy, see Table 5.7, the stability range of the silicon phases up to the FCC Si-X allotrope.

Typical of these phase transformations is that they do not revert by pressure release to the parent phases, but to a series of other metastable phases. This is the case of the β-tin phase, which forms by pressure release an intermediate rhombohedral R8 phase (Si-XII) almost simultaneously with the conversion to the BC8 phase, which eventually transforms by annealing at 473 K to the HD-Si, Si-IV phase with a hexagonal diamond structure, known as lonsdaleite, or wurtzite silicon. The Si-IV phase gained interest for photonic applications thanks to its calculated indirect gap of 0.95 eV and an optically

---

[‖‖‖] See Table 5.7 for the nature of the phases.

forbidden direct gap at 1.63 eV, which is transformed to optically allowed direct gap by high biaxial strain.[80,84]

The Si-IV (HD Si) hexagonal silicon allotrope, in turn, is stable under thermal treatments up to 700 °C,[199] and thus is at least potentially capable of being submitted without damage to the high temperature steps of a microelectronic process, although the problem has been for years the availability of bulk samples of sufficient good quality and size to allow reliable X-ray diffraction and optical measurements.[199]

Bulk samples of Si-IV of very poor quality were, in fact, produced, among a few others, by Wentorf and Kasper[200] in the 1960s by annealing between 200 and 600 °C BC8-Si samples recovered from high-pressure conditions. The Si-IV samples produced this way show broad, ambiguous XR diffraction patterns, that suggest the co-presence of disordered amorphous silicon and of defective nanocrystalline regions.

The solution came only recently, starting from the work of Hauge et al.,[78] who succeeded in obtaining Si-IV samples of good quality produced by epitaxial deposition at 900 °C of a hexagonal silicon layer on a template with the wurtzite structure consisting of a gallium phosphide (GaP) crystal. The X-ray diffraction spectra on this material allow the successful determination of its lattice constants ($a = 3.89$ Å; $c = 6.34$ Å).

Bulk samples of Si-IV of exceptional crystallinity and of sufficiently large size ($\sim 100$ μm × 100 μm × 10 μm), were produced and XRD characterized by Shiell et al.[77] who obtained Si-4H crystals by annealing at 300 °C the $Si_{24}$ allotrope of silicon, originally synthesised by Guerrette et al.,[197] which we have already discussed.

The advantage of $Si_{24}$ as the precursor of Si-4H[201] is its ortho-rhombic structure, with lattice constants ($a = 3.82$ Å, $c = 12.63$ Å) close to those of hexagonal allotrope 4H-Si ($a = 3.84$ Å, $c = 12.59$ Å)*****
that allow the onset of a relatively facile phase transition. The optical characterization of samples of this 4H Si allotrope confirm the onset of the optical band gap at 1.2 eV, which fits with previous calculations and optical characterizations.

# References

1. J. N. Tiwari, R. N. Tiwari and K. S. Kim, Zero-dimensional, one-dimensional, two-dimensional, and three-dimensional nanostructured materials for advanced electrochemical energy devices, *Prog. Mater. Sci.*, 2012, **57**, 724.
2. L. Pavesi, Will be silicon be the photonic material of the third millennium?, *J. Phys. Condens. Matter.*, 2003, **15**, R1169.

***** Double the lattice constant measured by Hauge.

3. U. Gösele, Nanocrystals, Shedding new light to silicon, *Nat. Nanotechnol.*, 2008, **3**, 134.
4. S. Pizzini, *Defect in nanocrystals*, CRC Press, 2020.
5. E. Comini, C. Baratto, G. Faglia, M. Ferroni, A. Vomiero and G. Sberveglieri, Quasi-one dimensional metal oxide semiconductors: Preparation, characterization and application as chemical sensors, *Prog. Mater. Sci.*, 2009, **54**, 1.
6. C. R. Chandraiahgari, G. De Bellis, P. Ballirano, S. K. Balijepalli, S. Kaciulis, L. Caneve, F. Sarto and M. S. Sarto, Synthesis and systematic characterization of highly crystalline ZnO nanorods in Nanoscale excitations in emergent materials, *RSC Adv.*, 2015, **5**, 49861.
7. N. Kamarulzaman, M. F. Kasim and R. Rusdi, Band Gap Narrowing and Widening of ZnO Nanostructures and Doped Materials, *Nanoscale Res. Lett.*, 2015, **10**, 346.
8. F. Fabbri, E. Rotunno, L. Lazzarini, N. Fukata and G. Salviati, Visible and infra-red light emission in boron-doped wurtzite silicon nanowires, *Sci. Rep.*, 2014, **4**, 3603.
9. L. S. Zhang, W. Sun and P. Chen, Spectroscopic and theoretical studies of quantum and electronic confinement effects in nanostructured materials, *Molecules*, 2003, **8**, 207.
10. G. D. Stucky and J. E. McDougall, Quantum confinement and host guest chemistry probing a new dimension, *Science*, 1990, **247**, 669.
11. K. Pedersen, *Quantum size effects in nanostructures*, Lecture notes to the course Organic and Inorganic Nanostructures Department of Physics and Nanotechnology, Aalborg University, 2006.
12. V. Kuncser and L. Miu, *Size Effects in Nanostructures: Basics and Applications*, Springer, 2014.
13. H. Jang, *et al.*, Quantum Confinement Effects in Transferrable Silicon Nanomembranes and Their Applications on Unusual Substrates, *Nano Lett.*, 2013, **13**, 5600.
14. E. G. Barbagiovanni, D. J. Lockwood, P. J. Simpson and L. V. Goncharova, Quantum confinement in Si and Ge nanostructures. Theory and Experiment, *Appl. Phys. Rev.*, 2014, **1**, 011302.
15. J. I. Pankove, *Optical processes in semiconductors*, Dover Publ. Inc., New York, 1971.
16. V. Jovanovic, C. Biasotto, L. K. Nanver, J. Moers, D. Grützmacher, J. Gerharz, G. Mussler, J. van der Cingel, J. J. Zhang, G. Bauer, O. G. Schmidt and L. Miglio, n-Channel MOSFETs Fabricated on SiGe Dots for Strain-Enhanced Mobility, *IEEE Electron Dev. Lett.*, 2010, **31**(10), 1083.
17. Zh. V. Smagina, V. A. Zinoviev, S. A. Rudin, E. E. Rodyakina, P. L. Novikov, A. V. Nenashev and A. V. Dvurechenskii, Self-Organization of Ge(Si) Nanoisland Groups on Pit-Patterned Si(100) Substrates, *Semiconductors*, 2020, **54**(14), 1866.
18. M. Grydlik, G. Langer, T. Fromherz, F. Shaffler and M. Brehm, Recipes for the fabrication of strictly ordered Ge island on pit-patterned Si(001) substrates, *Nanotechnology*, 2013, **24**, 105601.
19. S. Pizzini, M. Acciarri, S. Binetti, D. Cavalcoli, A. Cavallini, D. Chrastina, L. Colombo, E. Grilli, G. Isella, M. Lancin, A. Le Donne, A. Mattoni, K. Peter, B. Pichaud, E. Poliani, M. Rossi, S. Sanguinetti, M. Textier and H. Von Kanel, Nanocrystalline silicon films as multifunctional material for optoelectronic and photovoltaic applications, *Mater. Sci. Eng., B*, 2006, **134**, 118.
20. D. H. Wang, D. Xu, Y.-J. Hao, G.-Q. Jin, X.-Y. Guo and K. N. Tu, Periodically twinned SiC nanowires, *Nanotechnology*, 2008, **19**, 215602.
21. J. Sun and S. L. Simon, The melting behavior of aluminum nanoparticles, *Thermochim. Acta*, 2007, **463**, 32.
22. I. V. Talyzin, M. V. Samsonov, V. N. Samsonov, M. Yu. Pushkar and V. V. Dronnikov, Size dependence of the melting point of silicon nanoparticles: Molecular Dynamics and Thermodynamic simulation, *Semiconductors*, 2019, **53**(7), 947.

23. P. Chattopadhyay, G. R. Patel and T. C. Pandy, Effect of Size and Shape on Thermo-Elastic Properties of Nano-Germanium, *Int. J. Nanosci. Nanotechnol.*, 2021, **17**(3), 141.
24. T. L. Hill, A different approach to nanothermodynamics, *Nano Lett.*, 2001, **1**, 273.
25. S. Pizzini, Physics and thermodynamics of nanostructures, in *Defects in Nanocrystals: Structural and physicochemical Aspects*, CRC Press, 2020.
26. W. Thomson, On the equilibrium vapor at a curved surface of liquid, *Philos. Mag.*, 1871, **42**, 448.
27. F. Ercolessi, W. Andreoni and E. Tosatti, Melting of small particles of gold. Mechanism and size effects, *Phys. Rev. Lett.*, 1961, **66**, 911.
28. I. Coropceanu, E. M. Janke, J. Portner, D. Haubold, T. D. Nguyen, A. Das, P. N. Christian, C. P. N. Tanner, J. K. Utterback, S. W. Teitelbaum, M. H. Hudson, N. A. Sarma, A. M. Hinkle, C. J. Tassone, A. Eychmüller, D. T. Limmer, M. O. de la Cruz, N. S. Ginsberg and D. V. Talapin, Self-assembly of nanocrystals into strongly electronically coupled all-inorganic supercrystals, *Science*, 2022, **375**(6587), 1422.
29. S. Pizzini, *Defects in nanocrystals*, CRC Press, 2020.
30. S. Pizzini, M. Donghi, S. Binetti, I. Gelmi, A. Cavallini, B. Fraboni and G. Wagner, Influence of the host composition on the equilibrium structure of Er-centres in Silicon, *Solid State Phenom.*, 1997, **54**, 86.
31. M. Kittler, W. Seifert and V. Higgs, Recombination activity of misfit dislocations in silicon, *Phys. Status Solidi A*, 1993, **137**(2), 327.
32. E. Leoni, L. Martinelli, S. Binetti, G. Borionetti and S. Pizzini, The origin of the photoluminescence from oxygen precipitates nucleated at low temperature in semiconductor silicon, *J. Electrochem. Soc.*, 2004, **151**(12), G866.
33. A. Castaldini, D. Cavalcoli, A. Cavallini and S. Pizzini, Defect states in Czochralski p-type silicon; the role of oxygen and dislocations, *Phys. Status Solidi A*, 2005, **5**, 889.
34. V. Kveder, B. Badylevich, W. Schröter, M. Seibt, E. Steinman and A. Izotov, Silicon light-emitting diodes based on dislocation-related luminescence, *Phys. Status Solidi A*, 2005, **202**(5), 901.
35. J. Li, J.-Y. Yang and S.-G. Lu, A scalable synthesis of silicon nanoparticles as high performance material for Li-ion batteries, *Rare Met.*, 2019, **38**, 199.
36. M. J. Llansola Portoles, R. P. Diez, M. L. Dell'Arciprete, P. Caregnato, J. R. Romero, D. O. Martiré, O. Azzaroni, M. Ceolín and M. C. Gonzalez, Understanding the Parameters Affecting the Photoluminescence of Silicon Nanoparticles, *J. Phys. Chem. C*, 2012, **116**, 11315.
37. T. Lopez and L. Mangolini, On the nucleation and crystallization of nanoparticles in continuous-flow nonthermal plasma reactors, *J. Vac. Sci. Technol., B: Nanotechnol. Microelectron.: Mater., Process., Meas., Phenom.*, 2014, **32**, 061802.
38. L. Mangolini, E. Thimsen and U. Kortshagen, High-yield plasma synthesis of luminescent silicon nanocrystals, *Nano Lett.*, 2005, **5**, 655.
39. L. Mangolini, Synthesis, properties, and applications of silicon nanocrystals, *J. Vac. Sci. Technol., B: Nanotechnol. Microelectron.: Mater., Process., Meas., Phenom.*, 2013, **31**, 020801.
40. A. Fojtik and A. Henglein, Luminescent colloidal silicon particles, *Chem. Phys. Lett.*, 1994, **221**, 363.
41. A. Fojtik and A. Henglein, Surface chemistry of luminescent colloidal silicon nanoparticles, *J. Phys. Chem. B*, 2006, **110**, 1994.
42. Z. Zhou, L. Brus and R. Friesner, Electronic structure, and luminescence of 1.1 and 1.4 nm silicon nanocrystals: oxide shell *versus* hydrogen passivation, *Nano Lett.*, 2003, **3**, 163.

43. U. Hansen and P. Vogl, Hydrogen passivation of silicon surfaces: A classical molecular-dynamics study, *Phys. Rev. B: Condens. Matter Mater. Phys.*, 1988, **57**(20), 13295.

44. P. Garnier, Silicon Surface Passivation in HF Solutions for Improved Gate Oxide Reliability, *Solid State Phenom.*, 2016, **255**, 8.

45. M. C. Beard, K. P. Knutsen, P. Yu, J. M. Luther, Q. Song, W. K. Metzger, R. J. Ellingson and A. J. Nozik, Multiple Exciton Generation in Colloidal Silicon Nanocrystals, *Nano Lett.*, 2007, **7**, 2506.

46. D. C. Hannah, J. Yang, P. Podsiadlo, K. Y. Maria, M. K. Y. Chan, A. Demortière, D. J. Gosztola, V. B. Prakapenka, G. C. Schatz, U. Kortshagen and R. D. Schaller, On the Origin of Photoluminescence in Silicon Nanocrystals: Pressure-Dependent Structural and Optical Studies, *Nano Lett.*, 2012, **12**(8), 4200.

47. M. Shiratani, S. Uchida, H. Seo, D. Ichida, K. Koga, N. Itagaki and K. Kamataki, Nanostructure control of Si and Ge quantum dots based solar cells using plasma processes, *Mater. Sci. Forum*, 2014, **783–786**, 2022.

48. D. Carolan and H. Doyle, Size Controlled Synthesis of Germanium Nanocrystals: Effect of Ge Precursor and Hydride Reducing Agent, *J. Nanomater.*, 2015, 506056.

49. A. A. Leonardi, M. A. Lo Faro and A. Irrera, Silicon Nanowires Synthesis by Metal-Assisted Chem ical Etching: A Review, *Nanomaterials*, 2021, **11**, 383.

50. J.-C. Harmand, L. Liu, G. Patriarche, M. Tchernycheva, N. Akopian, U. Perinetti and V. Zwiller, Potential of semiconductor nanowires for single photon sources *in* Quantum Sensing and Nanophotonic Devices VI, Manijeh Razeghi, Rengarajan Sudharsanan, ed. G. J. Brown, Proc. of SPIE, 2009, vol. 7222, p. 722219.

51. T. Ishiyama, S. Nakagawa and T. Wakamatsu, Growth of epitaxial silicon nanowires on a-Si substrated by a metal-catalyst-free process, *Sci. Rep.*, 2016, **6**, 30608.

52. V. Schmidt, J. V. Wittemann, S. Senz and U. Gösele, Silicon Nanowires: A Review on Aspects of their Growth and their Electrical Properties, *Adv. Mater.*, 2009, **21**, 2681.

53. V. Schmidt, J. V. Wittemann and U. Gösele, Growth, thermodynamics, and electrical properties of silicon nanowires, *Chem. Rev.*, 2010, **110**, 361.

54. Y. Wang, V. Schmidt, S. Senz and U. Gösele, Epitaxial growth of silicon nanowires using an aluminium catalyst, *Nat. Nanotechnol.*, 2006, **1**, 186.

55. J. Westwater, D. P. Gosain, S. Tomiya and S. Usui, Growth of silicon nanowires *via* gold/silane vapor–liquid–solid reaction, *J. Vac. Sci. Technol., B: Nanotechnol. Microelectron.: Mater., Process., Meas., Phenom.*, 1997, **15**, 554.

56. D. Hourlier, P. Lefebre-Legry and P. Perrot, Preparation of silicon-based nanowires and the thermochemistry of the process, *JEEP*, 2009, 00002.

57. D. Hourlier and P. Perrot, Au-Si and Au-Ge phases diagrams for nanosystems, *Mater. Sci. Forum*, 2010, **653**, 77.

58. M. Hyvl, M. Müller, T.-H. Stuchlikova, J. Stuchlik, M. Silavik, J. Kocha, A. Feifar and J. Cervenka, Nucleation and growth of metal-catalyzed silicon nanowires under plasma, *Nanotechnology*, 2020, **31**, 225601.

59. R. A. Puglisi, C. Bongiorno, S. Caccamo, E. Fazio, G. Mannino, F. Neri, S. Scalese, D. Spucches and A. La Magna, Chemical Vapor Deposition Growth of Silicon Nanowires with Diameter Smaller Than 5 nm, *ACS Omega*, 2019, **4**(19), 17967.

60. R. Rurali, Colloquium: Structural, electronic, and transport properties of silicon nanowires, *Rev. Mod. Phys.*, 2010, **82**, 427.

61. H. Zhao, S. Zhou, Z. Hasanali and D. Wang, Influence of Pressure on Silicon Nanowire Growth Kinetics, *J. Phys. Chem. C.*, 2008, **112**(15), 5695.

62. J. Qin, X. Li, J. Wang and S. Pan, The self-diffusion coefficients of liquid binary M-Si (M=Al, Fe, Mg and Au) alloy systems by first principles molecular dynamics simulation, *AIP Adv.*, 2019, **9**, 035328.

63. H. Jeong, T. E. Park, H. K. Seong and H. J. Choi, Growth kinetics of silicon nanowires by platinum assisted vapour-liquid-solid mechanism, *Chem. Phys. Lett.*, 2009, **467**(4), 331.

64. A. M. Beers and J. Bloem, Temperature dependence of the growth rate of silicon prepared through chemical vapor deposition from silane, *Appl. Phys. Lett.*, 1982, **41**, 153.

65. C.-Y. Wen, M. C. Reuter, J. Tersoff, E. A. Stach and F. M. Ross, Structure, growth kinetics, and ledge flow during vapor-solid-solid growth of copper-catalysed silicon nanowires, *Nano Lett.*, 2010, **10**, 514.

66. M. F. Hainey and J. M. Redwing, Aluminum-catalyzed silicon nanowires: Growth methods, properties, and applications, *Appl. Phys. Rev.*, 2006, **3**, 040806.

67. K. Sato, A. Castaldini, N. Fukata and A. Cavallini, Electronic Level Scheme in Boron- and Phosphorus-Doped Silicon Nanowires, *Nano Lett.*, 2012, **12**(6), 3012.

68. S. V. Bulyariskiv and V. V. Svetukhin, Solubility of impurities in nanoparticles and nanoclusters, *Mater. Sci. Eng., B*, 2021, **272**, 115337.

69. T.-W. Ho and F. C.-N. Hong, A Novel Method to Grow Vertically Aligned Silicon Nanowires on Si(111) and Their Optical Absorption, *J. Nanomater.*, 2012, 274618.

70. B. Tian, P. Xie, T. J. Kempa, D. C. Bell and C. M. Lieber, Single crystalline kinked semiconductor nanowire superstructures, *Nat. Nanotechnol.*, 2009, **4**, 824.

71. B. Tian, X. Zheng, T. J. Kempa, Y. Fang, N. Yu, G. Yu, J. Huang and C. M. Lieber, Coaxial silicon nanowires as solar cells and nanoelectronic power sources, *Nature*, 2007, **449**, 885.

72. Y. Wu, Y. Cui, L. Huynh, C. J. Barrelet, D. C. Bell and C. M. Lieber, Controlled Growth and Structures of Molecular-Scale Silicon Nanowires, *Nano Lett.*, 2004, **4**(3), 433.

73. Y. Cui, L. J. Lauhon, M. S. Gudiksen, J. Wang and C. M. Lieber, Diameter-controlled synthesis of single-crystal silicon nanowires, *Appl. Phys. Lett.*, 2001, **78**, 2214.

74. T. Ishiyama, S. Nakagawa and T. Wakamatsu, Growth of epitaxial silicon nanowires an a Si substrate by a metal-catalyst-free process, *Sci. Rep.*, 2016, **6**, 30608.

75. I. Foncuberta, A. Morral, J. Arbiol, J. D. Prades, A. Cirera and J. R. Morante, Synthesis of Silicon Nanowires with Wurtzite Crystalline Structure by Using Standard Chemical Vapor Deposition, *Adv. Mater.*, 2007, **19**, 1347.

76. J. M. Besson, E. H. Mokhtari, J. Gonzales and G. Weill, Electrical properties of semimetallic silicon III and semiconductive silicon IV at ambient pressure, *Phys. Rev. Lett.*, 1987, **59**, 473.

77. T. B. Shiell, L. Zhu, B. A. Cook, J. E. Bradby, D. G. McCulloch and T. A. Strobel, Bulk Crystalline 4H-Silicon through a Metastable Allotropic Transition, *Phys. Rev. Lett.*, 2021, **126**(21), 215701.

78. H. I. T. Hauge, M. A. Verheijen, S. Conesa-Boj, T. Etzelstorfer, M. Watzinger, D. Kriegner, I. Zardo, C. Fasolato, F. Capitani, P. Postorino, S. Kölling, A. Li, S. Assali, J. Stangl and E. P. A. M. Bakkers, Hexagonal Silicon Realized, *Nano Lett.*, 2015, **15**(9), 5855.

79. B. Haberl, T. A. Strobel and J. E. Bradby, Pathways to exotic metastable silicon allotropes, *Appl. Phys. Rev.*, 2016, **3**, 040808.

80. C. Rödl, T. Sander, F. Bechstedt, J. Vidal, P. Olsson and S. Laribi, Wurtzite silicon as a potential absorber in photovoltaics: Tailoring the optical absorption by applying strain, *Phys. Rev. B: Condens. Matter Mater. Phys.*, 2015, **92**, 045207.

81. NIST Thermophysical Properties, National Institute of Standards and Technology, 2018.

82. F. Fabbri, E. Rotunno, L. Lazzarini, D. Cavalcoli, A. Castaldini, N. Fukata, K. Sato, G. Salviati and A. Cavallini, Preparing the way for doping wurtzite silicon nanowires while retaining the phase, *Nano Lett.*, 2013, **13**, 5900.

83. Y. Guo, Q. Wang and Y. Kawazoe, *et al.*, A New Silicon Phase with Direct Band Gap and Novel Optoelectronic Properties, *Sci. Rep.*, 2015, **5**, 14342.
84. S. Dixit and A. K. Shukla, Optical properties of lonsdaleite silicon nanowires: A promising material for optoelectronic applications, *J. Appl. Phys.*, 2018, **123**, 224301.
85. J. Tang, J.-L. Maurice, I. Florea, F. Fossard, P. R. Cabarrocas, E. V. Johnson and M. Foldyna, Natural occurrence of the diamond hexagonal structure in silicon nanowires grown by a plasma-assisted vapour–liquid–solid method, *Nanoscale*, 2017, **9**, 8113.
86. J. Tang, J. L. Maurice, W. H. Chen, S. Misra, M. Foldyna, E. V. Johnson and P. R. Cabarrocas, Natural occurrence of the diamond hexagonal structure in silicon nanowires grown by a plasma-assisted vapour–liquid–solid method, *Nanoscale Res. Lett.*, 2016, **11**, 455.
87. R. W. Olesinski and G. J. Abbaschian, The Si-Sn (Silicon-Tin) System, *Bull. Alloy Phase Diagrams*, 1984, **5**(3), 273.
88. G. Weill, J. L. Mansot, G. Sagon, C. Carlone and J. M. Besson, Characterization of Si III and Si IV metastable forms of silicon at ambient pressure, *Semicond. Sci. Technol.*, 1989, **4**, 280.
89. X. Liu and D. Wang, Kinetically-Induced Hexagonality in Chemically Grown Silicon Nanowires, *Nano Res.*, 2009, **2**, 575.
90. S. Chattopadhyay, X. L. Li and P. W. Bohn, In-plane control of morphology and tunable photoluminescence in porous silicon produced by metal-assisted electroless chemical etching, *J. Appl. Phys.*, 2002, **91**, 6134.
91. M. L. Zhang, K. Q. Peng, X. Fan, J. S. Jie, R. Q. Zhang and S. T. Lee, Preparation of large-area uniform silicon nanowires arrays through metal-assisted chemical etching, *J. Phys. Chem. C*, 2008, **112**, 4444.
92. Z. Huang, N. Geyer, P. Werner, J. de Boor and U. Gösele, Metal-assisted chemical etching of silicon: A Review, *Adv. Mater.*, 2011, **23**, 285.
93. A. A. Leonardi, M. J. Lo Faro and A. Irrera, Silicon Nanowires Synthesis by Metal-Assisted Chemical Etching: A Review, *Nanomaterials*, 2021, **11**, 383.
94. A. A. Leonardi, M. J. Lo Faro, D. Morganti, B. Fazio, P. Musumeci, M. Miritello, G. Franzò, F. Nastasi, F. Puntoriero, C. Di Pietro, F. Priolo and A. Irrera, Silicon nanowires: a building block for future technologies, *Proceedings of SPIE, Volume 11800, Low-Dimensional Materials and Devices 2021*, 2021, p. 118000S.
95. R. Williams, Electrochemical reactions of semiconductors, *J. Vacuum Sci. Technol.*, 1976, **13**, 12.
96. J. Singh, *Semiconductor Devices: Basic principles*, J. Wiley & Sons, Singapore, 2001.
97. V. G. Zavodinsky and I. A. Kuyanov, Schottky barrier formation in a Au/Si nanoscale system: A local density approximation study, *J. Appl. Phys.*, 1997, **81**, 2715.
98. A. Hankin, F. E. Bedoya-Lora, J. C. Alexander, A. Regoutz and G. H. Kelsall, Flat band potential determination: avoiding the pittfalls, *J. Mater. Chem. A.*, 2019, **7**, 26162.
99. W. P. Gomes and F. Cardon, Electron energy levels in semiconductor electrochemistry, *Prog. Surf. Sci.*, 1982, **12**, 155.
100. S. Ottow, G. S. Popkirov and H. Föll, Determination of the flat-band potential of silicon electrodes in HF by means of ac resistance measurements, *J. Electroanal. Chem.*, 1998, **455**, 29.
101. J. O. M. Bockris and K. Uosaki, The rate of photoelectrical generation of hydrogen at p-type semiconductors, *J. Electrochem. Soc.*, 1997, **124**, 1348.
102. K. A. Gonchar, A. A. Zubairova, A. Schleusener, L. A. Osminkina and V. Sivakov, Optical Properties of Silicon Nanowires Fabricated by Environment-Friendly Chemistry, *Nanoscale Res. Lett.*, 2016, **11**, 357.
103. A. G. Nassiopoulou, V. Gianneta and C. Katsogridakis, Si nanowires by a single-step metal assisted chemical etching process on lithographically defined areas. Formation kinetics, *Nanoscale Res. Lett.*, 2011, **6**, 1.

104. L. T. Canham, Silicon quantum wire array fabrication by electrochemical and chemical dissolution of wafers, *Appl. Phys. Lett.*, 1990, **5**, 1046.
105. K. A. Gonchar, V. Y. Kitaeva, G. A. Zharik, A. A. Eliseev and L. A. Osminkina, Structural and Optical Properties of Silicon Nanowire Arrays Fabricated by Metal Assisted Chemical Etching with Ammonium Fluoride, *Front. Chem.*, 2018, **6**, 653.
106. J. Ju, X. Huang, S.-M. Kim and J. Yeom, Fabrication of Highly Ordered Silicon Nanowires by Metal Assisted Chemical Etching Combined with a Nanoimprinting Process, *J. Nanosci. Nanotechnol.*, 2017, **17**(10), 7771.
107. M. Rey, F. J. Wendisch, E. S. A. Goerlitzer, J. S. J. Tang, R. S. Bader, G. R. Bourret and N. Vogel, Anisotropic silicon nanowire arrays fabricated by colloidal lithography, *Nanoscale Adv.*, 2021, **3**(12), 3634.
108. J. Yeom, D. Ratchford, C. R. Field, T. H. Brintlinger and P. E. Pehrsson, Decupling Diameter and Pitch in Silicon nanowire Arrays by Metal assisted Chemical Etching, *Adv. Funct. Mater.*, 2014, **24**, 106.
109. V. Sivakov, F. Voigt, B. Hoffmann, V. Gerliz and S. Christiansen, Chemically Etched Silicon Nanowire Architectures: Formation and Properties, in *Nanowires - Fundamental Research*, ed. A. Hashim, 2011, ISBN: 978-953-307-327-9.
110. E. Fakhri, M. T. Sultan, A. Manolescu, S. Ingvarsson, N. Plugaru, R. Plugaru and H. G. Svavarsson, Synthesis, and photoluminescence study of silicon nanowires obtained by metal assisted chemical etching, *IEEE International Semiconductor Conference (CAS)*, 18 November 2021, DOI: 10.1109/CAS52836.2021.9604178.
111. M. Naffeti, P. A. Postigo, R. Chtourou and M. A. Zaibi, Elucidating the effect of etching time key parameter toward optically and electrically active silicon nanowires, *Nanomaterials*, 2020, **10**(3), 404.
112. Y. R. Choi, M. Zheng and F. Bai, *et al.*, Laser-induced Greenish-Blue Photoluminescence of Mesoporous Silicon Nanowires, *Sci. Rep.*, 2014, **4**, 4940.
113. F. J. Lopez, U. Givan, J. G. Connell and L. J. Lauhon, Silicon nanowire polytypes: Identification by Raman spectroscopy, generation mechanism, and misfit strain in homostructures, *ACS Nano*, 2011, **5**, 8958.
114. S. Dixit and A. K. Shukla, Optical properties of lonsdaleite silicon nanowires: A promising material for optoelectronic applications, *J. Appl. Phys.*, 2018, **123**, 224301.
115. J. K. Rath, Low temperature polycrystalline silicon: a review on deposition, physical properties, and solar cell application, *Sol. Energy Mater. Sol. Cells*, 2003, **76**, 431.
116. A. Gordijn, J. K. Rath and R. E. I. Schropp, Role of growth temperature and the presence of dopants in layer-by-layer plasma deposition of thin microcrystalline silicon doped layers, *J. Appl. Phys.*, 2004, **95**(12), 8290.
117. S. Binetti, M. Acciarri, M. Bollani, L. Fumagalli, H. von Känel and S. Pizzini, Nanocrystalline silicon films grown by Low Energy Plasma Enhanced Chemical vapor deposition for optoelectronic applications, *Thin Solid Films*, 2005, **487**, 19.
118. M. Acciarri, S. Binetti, M. Bollani, A. Comotti, L. Fumagalli, S. Pizzini and H. Von Känel, Nanocrystalline silicon films grown by LEPECVD for photovoltaic applications, *Sol. Energy Mater. Sol. Cells*, 2005, **87**, 11.
119. S. Pizzini, M. Acciarri, S. Binetti, D. Cavalcoli, A. Cavallini, D. Chrastina, L. Colombo, E. Grilli, G. Isella, M. Lancin, A. Le Donne, A. Mattoni, K. Peter, B. Pichaud, E. Poliani, M. Rossi, S. Sanguinetti, M. Textier and H. Von Känel, Nanocrystalline silicon films as multifunctional material for optoelectronic and photovoltaic applications, *Mater. Sci. Eng., B*, 2006, **134**, 118.
120. M. Textier, M. Acciarri, S. Binetti, D. Cavalcoli, A. Cavallini, D. Chrastina, G. Isella, M. Lancin, A. Le Donne, A. Tomasi, B. Pichaud, S. Pizzini and M. Rossi, Structural Characterization of Nanocrystalline Silicon Layers Grown by LEPECVD for Optoelectronic Applications, *Microscopy of Semiconducting Materials*,

Part of the Springer Proceedings, in Physics Book Series (SPPHY), 2007, vol. 120, p. 305.

121. A. Cavallini, D. Cavalcoli, M. Rossi, A. Tomasi, S. Pizzini, D. Chrastina and G. Isella, Defect analysis of hydrogenated nanocrystalline Si thin films, *Phys. B*, 2007, **401–402**, 519.

122. A. Le Donne, S. Binetti, G. Isella, B. Pichaud, M. Texier, M. Acciarri and S. Pizzini, Advances in structural characterization of thin film nanocrystalline silicon for photovoltaic applications, *Solid State Phenom.*, 2008, **131–133**, 33.

123. J. K. Rath, Nanocrystalline silicon solar cells, *Appl. Phys. A: Mater. Sci. Process.*, 2009, **96**, 145.

124. C. Garozzo, R. A. Puglisi and S. Lombardo, Structural and chemical characterization of nanocrystalline and amorphous hydrogenated Si films, *Phys. Status Solidi C*, 2012, **9**(10–11), 1892.

125. C.-H. Lee, M. Shin, M.-H. Lim, J.-Y. Seo, J.-E. Lee, B.-J. Kim and D. Choi, Material properties of microcrystalline silicon for solar cell application, *Sol. Energy Mater. Sol. Cells*, 2011, **95**, 207.

126. V. S. Waman, M. M. Kamble, S. S. Ghosh, A. H. Mayabadi, B. B. Gabhale, S. R. Rondiya, A. V. Rokade, S. S. Khadtare, V. G. Sathe, H. M. Pathan, S. W. Gosavi and S. R. Jadkar, Evolution of microstructure and opto-electrical properties in boron doped nc-Si:H films deposited by HW-CVD method, *J. Alloys Compd.*, 2014, **585**, 523.

127. D. P. Josh and D. P. Bhatt, Theory of grain boundary recombination and carrier transport in polycrystalline silicon under optical illumination, *IEEE Trans. Electron Devices*, 1990, **37**(1), 237.

128. L. Martinu, J. E. Klemberg-Sapieha, O. M. Küttel, A. Raveh and M. R. Wertheimer, Critical ion energy and ion flux in the growth of films by plasma-enhanced chemical vapor deposition, *J. Vac. Sci. Technol.*, 1994, **12**(4), 1360.

129. R. E. I. Schropp, K. F. Feenstra, E. C. Molenbroek, H. Meiling and J. K. Rath, Hot-wire deposited amorphous silicon thin-film transistors, *Philos. Mag. B*, 1997, **76**, 309.

130. J. K. Rath, A. Barbon and R. E. I. Schropp, Limited influence of grain boundary defects in hot-wire CVD polysilicon films on solar cell performance, *J. Non-Cryst. Solids*, 1998, **227**, 1277.

131. J. K. Rath, A. Barbon and R. E. I. Schropp, Clustered defects in hot wire chemical vapor deposited poly-silicon films, *J. Non-Cryst. Solids*, 2000, **266–269**, 548.

132. J. K. Rath, R. E. I. Schropp and W. Beyer, Hydrogen at compact sites in hot-wire chemical vapour deposited polycrystalline silicon films, *J. Non-Cryst. Solids*, 2000, **266**, 190.

133. I. M. M. Ferreira, R. F. P. Martins, A. M. F. Cabrita, E. M. C. Fortunato and P. Vilarinho, Nanocrystalline Undoped Silicon Films Produced by Hot Wire Plasma Assisted Technique, *Mater. Res. Soc. Symp. Proc.*, 2000, **609**, A22.4.1.

134. K. Brühne, M. B. Schubert, C. Köhler and J. H. Werner, Nanocrystalline silicon from hot-wire deposition—a photovoltaic material?, *Thin Solid Films*, 2001, **395**(1–2), 163.

135. I. M. M. Ferreira, A. M. F. Cabrita, F. B. Fernandes, E. M. C. Fortunato and R. F. P. Martins, Morphology and structure of nanocrystalline p-doped films produced by hot wire technique, *Vacuum*, 2000, **64**, 237.

136. T. V. Torchynska, A. Vivas Hernandez, Y. Matsumoto, L. Khomenkova and L. Shcherbina, Photoluminescence and structure investigations of Si nanocrystals in amorphous silicon matrix, *J. Non-Cryst. Solids*, 2006, **352**, 1188.

137. V. G. Golubev, A. V. Medvedev, A. B. Pevtsov, A. V. Sel'kin and N. A. Feoktistov, Photoluminescence of thin amorphous-nanocrystalline silicon films, *Phys. Solid State*, 1999, **41**(1), 1372006.

138. T. Merdzhanova, Microcrystalline Silicon Films and Solar Cells Investigated by Photoluminescence Spectroscopy, Schriften des Forschungszentrums Jülich Reihe Energietechnik/Energy Technology Band, 2005, vol. 41, ISSN 1433-5522, ISBN 3-89336-401-3.
139. A. Dutt, S. Godavarthi, Y. Matsumoto, G. Santana-Rodriguez, A. Avila, V. Sanchez and G. Raina, HW-CVD Deposited Nanocrystalline Silicon Thin Films at Low Substrate Temperature with White-Blue Luminescence, *Curr. Nanosci.*, 2015, **11**(5), 621.
140. J. K. Rath, B. Stannowski, P. A. T. T. van Veenendaal, M. K. van Veen and R. E. I. Schropp, Application of hot-wire chemical vapor-deposited Si:H films in thin film transistors and solar cells, *Thin Solid Films*, 2001, **395**, 320.
141. S. Klein, F. Finger, R. Carius, B. Rech, L. Houben, M. Luysberg and M. Stutzmann, Defects in Microcrystalline silicon prepard with Hot Wire CVD, *Mater. Res. Soc. Symp. Proc.*, 2002, **715**, A26.2.
142. M. Fonrodona, D. Soler, F. Villar, J. Escarré, J. M. Asensi, J. Bertomeu and J. Andreu, Progress in single junction microcrystalline silicon solar cells deposited by Hot-Wire CVD, *Thin Solid Films*, 2006, **501**(1–2), 247.
143. T. H. Wang, E. Iwaniczko, M. R. Page, D. Levi, H. M. Branz and Q. Wang, Effect of emitter deposition temperature on surface passivation in hot-wire chemical vapor deposited silicon heterojunction solar cells, *Thin Solid Films*, 2006, **501**(1–2), 284.
144. T. H. Wang, M. R. Page, E. Iwaniczko, Y. Xu, Y. Yan, L. Roybal, D. Levi, R. Bauer, H. M. Branz and Q. Wang, High-Efficiency p-Type Silicon Heterojunction Solar Cells, *2006 IEEE 4th World Conference on Photovoltaic Energy Conference*, 2006, vol. 2, p. 1439.
145. C. Rosenblad, H. R. Deller, A. Dommann, T. Meyer, P. Schroeter and H. von Känel, Silicon epitaxy by low-energy plasma enhanced chemical vapor deposition, *J. Vac. Sci. Technol., A*, 1998, **16**, 2785.
146. C. Rosenblad, H. von Känel, M. Kummer and A. Dommann, A plasma process for ultrafast deposition of SiGe graded buffer layers, *Appl. Phys. Lett.*, 2000, **76**, 427.
147. M. Kummer, C. Rosenblad, A. Dommann, T. Hackbarth, G. Höck, M. Zeuner, E. Müller and H. von Känel, Low energy plasma enhanced chemical vapor deposition, *Mater. Sci. Eng., B*, 2002, **89**, 288.
148. M. Bollani, S. Binetti, M. Acciarri, L. Fumagalli, A. Arcari, S. Pizzini and H. Von Känel, Characterization of Nanocrystalline Silicon Film grown by LEPECVD for Photovoltaic Applications. MRS Online Proceedings Library (OPL), Symposium A – Amorphous and Nanocrystalline Silicon-Based Films, 2003, vol. 762, p. A5.3.
149. S. Meloni, L. Ferraro, A. Federico, M. Rosati, A. Mattoni and L. Colombo, Computational Materials Science application programming interface (CMSapi): a tool for developing applications for atomistic simulations, *Comput. Phys. Commun.*, 2005, **169**, 462.
150. A. Mattoni, L. Ferraro and L. Colombo, Calculation of the local electronic properties of nanostructured silicon, *Phys. Rev. B: Condens. Matter Mater. Phys.*, 2009, **79**, 245302.
151. L. Bagolini, A. Mattoni and L. Colombo, Electronic localization, and optical absorption in embedded silicon nanonograins, *Appl. Phys. Lett.*, 2009, **94**, 053115.
152. A. Mattoni and L. Ferraro, Colombo l. Calculation of the local optoelectronic properties of nanostructured silicon, *Phys. Rev. B: Condens. Matter Mater. Phys.*, 2009, **79**, 245302.
153. L. Bagolini, A. Mattoni, G. Fugallo, L. Colombo, E. Poliani, S. Saguinetti and E. Grilli, Quantum Confinement by an order-disorder boundary in Nanocrystalline silicon, *Phys. Rev. Lett.*, 2010, **104**, 176803.

154. L. Bagolini, A. Mattoni, R. T. Collins and M. T. Lusk, Carrier Localization in Nanocrystalline Silicon, *J. Phys. Chem. C*, 2014, **118**, 13417.
155. L. Bagolini, A. Mattoni and M. T. Lusk, Confinement of vibrational modes within crystalline lattices using thin amorphous layers, *J. Phys. Condens. Matter.*, 2017, **29**, 145302.
156. G. Yue, J. D. Lorentzien, J. Lin, D. Han and Q. Wang, Photoluminescence and Raman Studies in Thin-Film Materials: Transition from Amorphous to Micro-crystalline silicon, *Appl. Phys. Lett.*, 1999, **75**, 49.
157. C. Smit, R. A. C. M. M. van Swaaij, H. Donker, A. Petit, W. Kessels and M. van der Sanden, Determining the material structure of microcrystalline silicon from Raman spectra, *J. Appl. Phys.*, 2003, **94**, 3582.
158. Y. Koshka, S. Ostapenko, I. Tarasov, S. McHugo and J. P. Kaleys, Scanning room temperature photoluminescence in polycrystalline silicon, *Appl. Phys. Lett.*, 1999, **74**(11), 1555.
159. Y. Kanemitsu, Efficient light emission from crystalline and amorphous silicon nanostructures, *J. Lumin.*, 2002, **100**, 209.
160. M. Zacharias and P. Streitenberger, Crystallization of amorphous superlattices in the limit of ultrathin films with oxide interfaces, *Phys. Rev.*, 2000, **62**, 8391.
161. M. Zacharias, J. Heitmann, R. Scholz, U. Kahler, M. Schmidt and J. Blasing, Size-controlled highly luminescent silicon nanocrystals: A $SiO/SiO_2$ superlattice approach, *J. Appl. Phys. Lett.*, 2002, **80**, 661.
162. S. Gutsch, J. Laube, A. M. Hartel, D. Hiller, N. Zakharov, P. Werner and M. Zacharias, Charge transport in Si nanocrystal/$SiO_2$ superlattices, *J. Appl. Phys.*, 2013, **113**, 133703.
163. J. Lopez-Vidrier, Y. Berencen, S. Hernandez, O. Blazquez, S. Gutsch, J. Laube, D. Hiller, P. Loper, M. Schnabel, S. Janz, M. Zacharias and B. Garrido, Charge transport and electroluminescence of silicon nanocrystals/$SiO_2$ superlattices, *J. Appl. Phys.*, 2013, **114**, 163701.
164. S.-W. Fu, H.-J. Chen, H.-T. Wu, S.-P. Chen and C.-F. Shih, Enhancing the electroluminescence efficiency of Si $NC/SiO_2$ superlattice-based light-emitting diodes through hydrogen ion beam treatment, *Nanoscale*, 2016, **8**, 7155.
165. C. F. Shih, C. Y. Hsiao and K. W. Su, Enhanced white photoluminescence in silicon-rich oxide/$SiO_2$ superlattices by low-energy ion-beam treatment, *Opt. Express*, 2013, **21**(13), 15888.
166. M. Coll, *et al.*, Towards oxide electronics: a roadmap, *Appl. Surf. Sci.*, 2019, **482**, 1.
167. O. Lupan, V. M. Guérin, I. M. Tiginyanu, V. V. Ursaki, L. Chow, H. Heinrich and T. Pauporté, Well-aligned arrays of vertically oriented ZnO nanowires electrodeposited on ITO-coated glass and their integration in dye sensitized solar cells, *J. Photochem. Photobiol., A*, 2010, **211**, 65.
168. T. Pauporté, D. Lincot, B. Viana and F. Pellé, Toward laser emission of epitaxial nanorod arrays of ZnO grown by electrodeposition, *Appl. Phys. Lett.*, 2006, **89**, 233112.
169. O. Lupan, V. V. Ursaki, G. Chai, L. Chow, G. A. Emelchenko, I.-M. Tiginyanu, A. N. Gruzintsev and A. N. Redkin, Selective hydrogen gas nanosensor using individual ZnO nanowire with fast response at room temperature, *Sens. Actuators, B*, 2010, **144**, 56.
170. Z. L. Wang, Zinc oxide nanostructures: growth, properties, and applications, *J. Phys. Condens. Matter.*, 2004, **16**, R829.
171. O. Lupan, L. Chow, G. Chai, R. Roldan, A. Naitabdi, A. Schulte and H. Heinrich, Nanofabrication and characterization of ZnO nanorod arrays and branched microrods by aqueous solution route and rapid thermal processing, *Mater. Sci. Eng. B*, 2007, **145**(1–3), 57.

172. N. Wang, Y. Cai and R. Q. Zhang, Growth of nanowires, *Mater. Sci. Eng., R*, 2008, **60**(1–6), 1.
173. T. Pauporté, E. Jouanno, F. Pellé, B. Viana and B. P. Ashehoung, Key growth parameters for the electrodeposition of ZnO films with an intense UV-light emission at room temperature, *J. Phys. Chem. C*, 2009, **113**, 10422.
174. T. Pauporté, G. Bataille, L. Joulaud and F. J. Vermersch, Well-aligned ZnO nanowire arrays prepared by seed-layer-free electrodeposition and their Cassie–Wenzel transition after hydrophobization, *J. Phys. Chem. C*, 2010, **114**(1), 194.
175. Y. Leprince-Wang, A. Yacoubi-Ouslim and G. Y. Wang, Structure study of electrodeposited ZnO nanowires, *Microelectron. J.*, 2005, **36**, 625.
176. C. R. Chandraiahgari, G. De Bellis, P. Ballirano, S. K. Balijepalli, S. Kaciulis, L. Caneve, F. Sarto and M. S. Sarto, Synthesis and systematic characterization of highly crystalline ZnO nanorods in Nanoscale excitations in emergent materials, *RSC Adv.*, 2015, **5**, 49861.
177. S. Rafique, L. Han and H. Zhao, Density Controlled Growth of ZnO Nanowall–Nanowire 3D Networks, *J. Phys. Chem. C*, 2015, **119**, 12023.
178. M. A. Khan, S. Sakrani, S. Suhaima, Y. Wahab and R. Muhammad, Synthesis of $Cu_2O$ and ZnO Nanowires and their Heterojunction Nanowires by Thermal Evaporation: A Short Review, *J. Teknol. (Sci. Eng.)*, 2014, **71**(5), 83.
179. P. X. Gao, Y. Ding and Z. L. Wang, Crystallographic Orientation-Aligned ZnO Nanorods Grown by a Tin Catalyst, *Nano Lett.*, 2003, **3**, 1315.
180. M. D. Tyona, R. U. Osuji and F. I. Ezema, A review of zinc oxide photoanode films for dye-sensitized solar cells based on zinc oxide nanostructures, *Adv. Nano Res.*, 2013, **1**(1), 43.
181. A. Goux, T. Pauporte and D. Lincot, Oxygen reduction reaction on electro-deposited zinc oxide electrodes in KCl solution at 70 °C, *Electrochem. Acta*, 2006, **51**, 3168.
182. J. P. Hirth and J. Lothe, *Theory of dislocations*, Mc Graw-Hill, 1968.
183. K. Ellmer and A. Bikowski, Intrinsic and extrinsic doping of ZnO and ZnO alloys, *J. Phys. D: Appl. Phys.*, 2016, **49**, 413002.
184. J. Cui, Zinc oxide nanowires, *Mater. Charact.*, 2012, **64**, 43.
185. T. C. Bharat, Shubham, S. Mondal, H. S. Gupta, P. K. Singh and A. K. Das, Synthesis of Doped Zinc Oxide Nanoparticles: A Review, *Mater. Today: Proc.*, 2019, **11**, 767.
186. M. A. Thomas and J. B. Cui, Electrochemical Route to p-Type Doping of ZnO Nanowires, *J. Phys. Chem. Lett.*, 2010, **1**(7), 1090.
187. C. G. Van de Walle, Hydrogen as a Cause of Doping in Zinc Oxide, *Phys. Rev. Lett.*, 2000, **85**(5), 1012.
188. C. G. Van de Walle, Defect analysis and engineering in ZnO, *Phys. B*, 2001, **308–310**, 899.
189. J. Koßmann and C. Hättig, Investigation of interstitial hydrogen and related defects in ZnO, *Phys. Chem. Chem. Phys.*, 2012, **14**, 16392.
190. F. P. Bundy, Phase diagram of silicon and germanium to 200k Bar at 1000°C, *J. Chem. Phys.*, 1964, **41**, 3809.
191. J. Z. Hu, L. D. Merkle, C. S. Menoni and I. L. Spain, Crystal data for high-pressure phases of silicon, *Phys. Rev. B: Condens. Matter Mater. Phys.*, 1986, **34**, 4679.
192. T. C. Pandya, N. A. Thakar and A. D. Bhatt, Analysis of equations of state and temperature dependence of thermal expansivity and bulk modulus for silicon, *J. Phys.: Conf. Ser.*, 2012, **377**, 012097.
193. E. B. Jones and V. Stevanovic, Polymorphism in elemental silicon: Probabilistic interpretation of the realizability of elemental structures, *Phys. Rev. B*, 2017, **96**, 184101.
194. M. N. Magomedov, On the polymorphism of silicon and germanium, *J. Phys.: Conf. Ser.*, 2017, **918**, 012031.

195. L. Fan, D. Yang and D. A. Li, A review on metastable silicon allotropes, *Materials*, 2021, **14**, 3964.
196. R. Paul, S. X. Hu and V. V. Karesiev, Crystalline phase transitions and vibrational spectra of silicon up to multiterapascal pressures, *Phys. Rev. B*, 2019, **100**, 144101.
197. M. Guerrette, M. D. Ward, L. Zhu and T. A. Strobel, Single crystal synthesis of the open-framework allotrope Si 24, *J. Phys.: Condens. Matter*, 2020, **32**, 194001.
198. S. W. Anzellini, M. T. F. Miozzi, A. Kleppe, D. Daisenberger and H. Wilhelm, Quasi-hydrostatic equation of state of silicon up to 1 MBar at ambient temperature, *Sci. Rep.*, 2019, **9**, 15537.
199. L. Fan, D. Yang and D. Li, A Review on Metastable Silicon Allotropes, *Materials*, 2021, **14**, 3964.
200. R. H. Wentorf and J. S. Kasper, Two new forms of silicon, *Science*, 1963, **139**, 338.
201. X. Yang, C. He, X. Shi, J. Li, C. Zhang, C. Tang and J. Zhong, First-principles prediction of two hexagonal silicon crystals as potential absorbing layer materials for solar-cell application, *J. Appl. Phys.*, 2018, **124**, 163107.

# Subject Index

Figures in **bold**

www.ingramcontent.com/pod-product-compliance
Lightning Source LLC
Chambersburg PA
CBHW040243230326
41458CB00103B/6467